Dienerian (Early Triassic) ammonoids from the Northern Indian Margin

by David Ware, Hugo Bucher, Thomas Brühwiler,
Elke Schneebeli-Hermann, Peter A. Hochuli[†], Leopold Krystyn,
Ghazala Roohi, Khalil Ur-Rehman and Amir Yaseen

Acknowledgement
Financial support for the publication of this issue of
Fossils and Strata was provided by the Lethaia Foundation

Contents

Foreword

Dienerian (Early Triassic) ammonoids and the Early Triassic biotic recovery: a review

DAVID WARE AND HUGO BUCHER

FOSSILS AND STRATA

THE LETHAIA FOUNDATION

It has been estimated that about 90% of all marine species disappeared during the end-Permian mass extinction (Raup & Sepkoski 1982). It is the biggest known biodiversity crisis in the history of Phanerozoic life, and it led to the replacement of typical Palaeozoic faunas by typical modern communities (Sepkoski 1984). The recovery which followed in the Early Triassic is an intensively studied topic. This recovery is traditionally considered as delayed in comparison with other mass extinctions (Erwin 1998, 2006) as several major marine clades such as corals (Stanley 2003), foraminifers (Tong & Shi 2000) or radiolarians (Racki 1999) recovered only in the late Spathian (Early Triassic) or in the Anisian (Middle Triassic), ca. 5 My after the Permian–Triassic boundary. This delay is interpreted as the consequence of persisting anoxic conditions (Wignall & Twitchett 2002) and unstable environmental conditions during the entire Early Triassic (Payne et al. 2004). However, several recent studies suggest a more complex scenario, with pulses of recovery interrupted by periods of additional extinctions. For example, conodonts (Orchard 2007; Goudemand et al. 2008) first underwent an important turnover at the Griesbachian–Dienerian boundary, followed by an explosive radiation in the early–middle Smithian, a dramatic extinction in the late Smithian, and another radiation during the early Spathian. Ammonoids also recovered very fast compared to other groups, reaching pre-extinction levels of diversity already during the Smithian (Fig. 1; Brayard et al. 2009). Hofmann et al. (2014) showed that benthic ecosystems started to recover already in the

Griesbachian, but this recovery has been interrupted by a return to harsh environmental conditions (e.g. anoxia, warm temperatures) during the Dienerian. Recovery of the benthos resumed during the Smithian. Based on palynological and carbon isotopes analysis, Hermann et al. (2011a,b, 2012a,b) and Schneebeli-Hermann et al. (2012, 2015) contradicted the idea of persistent widespread anoxia and showed that this anoxia was restricted to the middle–late Dienerian and late Smithian. Late Permian and Early Triassic ecological crises of terrestrial plants also immediately predate extinction crises of marine organisms, and the Dienerian diversity low is no exception as documented by Hochuli et al. (2016).

Many studies addressing the recovery are based on insufficiently resolved age controls. The construction of a detailed time-scale for the Early Triassic is the cornerstone on which any study addressing this biotic recovery must be based. Ovtcharova et al. (2006) and Galfetti et al. (2007) established a duration of ca. 4.5 Myr for the Early Triassic and showed that the four Early Triassic ages were of very uneven duration, the Spathian representing more than half of this interval (Fig. 2). Galfetti et al. (2007) obtained a maximal duration of ca. 1.4 ± 0.4 Myr for the Griesbachian–Dienerian time interval. No duration of the Dienerian alone is available, but it can be reasonably assumed that it is <1 Myr. A new generation of high-resolution U-Pb ages for the Permian–Triassic boundary (Burgess et al. 2014) and for the Early–Middle Triassic boundary (Ovtcharova et al. 2015) indicate a duration of 4.83 ± 0.19 Myr

DOI 10.1111/let.12275 © 2018 Lethaia Foundation. Published by John Wiley & Sons Ltd

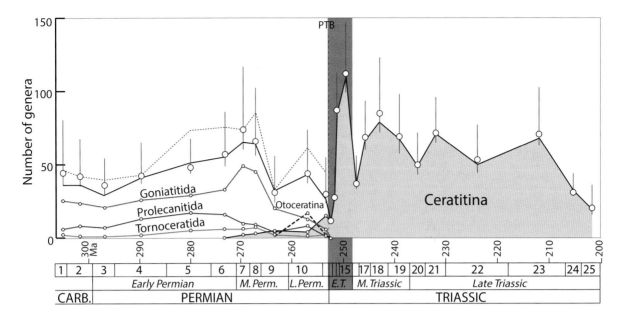

Fig. 1. Total generic richness (black bold line, all ammonoids; grey lines, major ammonoid groups) and mean Chao2 estimate of the overall generic richness with its 95% confidence interval (large circles with vertical bars). The Early Triassic is highlighted in dark grey. PTB, Permian–Triassic boundary; 1, Kasimovian; 2, Gzhelian; 3, Asselian; 4, Sakmarian; 5, Artinskian; 6, Kungurian; 7, Roadian; 8, Wordian; 9, Capitanian; 10, Wuchiapingian; unlabelled successive intervals, Changhsingian, Griesbachian, Dienerian, Smithian; 15, Spathian; 16, Early Anisian; 17, Middle Anisian; 18, Late Anisian; 19, Ladinian; 20, Early Carnian; 21, Late Carnian; 22, Early Norian; 23, Middle Norian; 24, Late Norian; 25, Rhaetian. Modified after Brayard *et al.* (2009).

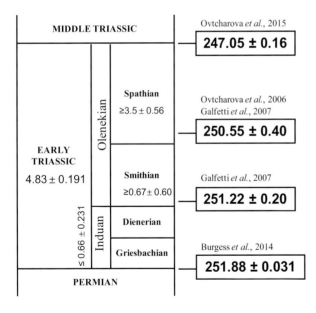

Fig. 2. Lower Triassic stage and substage subdivision (Ogg 2012) calibrated with recently published radiometric ages from South China.

for the Early Triassic. However, the respective duration of each of the four Lower Triassic substages may not be significantly changed because these new U-Pb ages are consistently younger than those of the previous generation.

Brayard & Bucher (2008) proposed a new detailed biostratigraphical scheme based on ammonoids for the Smithian of South China. Brühwiler *et al.* (2010a) constructed the most highly resolved biostratigraphical scheme for the Smithian based on ammonoids from the Northern Indian Margin. This work showed that ammonoids underwent an explosive radiation in the early Smithian, with constant high diversity associated with extremely high turnover rates throughout the middle Smithian and a major extinction in the late Smithian. This extinction could not be detected by Brayard *et al.* (2009) due to the coarser time-scale of this study. Conodonts also suffered from a drastic extinction in the late Smithian.

The stage subdivisions of the Lower Triassic are a subject of debate (Fig. 2). In the latest version of The Geologic Time Scale (Ogg 2012), the twofold subdivision of the Lower Triassic introduced by Kiparisova & Popov (1956), with the Induan and Olenekian, is endorsed. The four stages defined by Tozer (1965) are then considered as substages, the Induan being subdivided into Griesbachian and Dienerian, and the Olenekian into Smithian and Spathian. This twofold scheme is however strongly criticized (e.g. Shevyrev 2006), mainly as it does not reflect the end Smithian crisis, the most important extinction event known for both ammonoids and conodonts within the entire Triassic in every locality where this question has been addressed (e.g. Stanley

2009; for ammonoids: Brayard *et al.* 2006; Brühwiler *et al.* 2010a; Brayard & Bucher 2015; for conodonts: Orchard 2007; Chen *et al.* 2013; Komatsu *et al.* 2016). Moreover, having the Induan defined in the Tethyan realm and the Olenekian in the Boreal Realm makes the correlation of this stage boundary across such a broad palaeolatitudinal range an arduous task. Tozer (1965) provided a broad definition of the Dienerian–Smithian boundary (corresponding to the Induan–Olenekian boundary), explaining that it was probable that in Canada, the oldest known Smithian fauna may be younger than typical Smithian faunas from other areas. Krystyn *et al.* (2007a,b) proposed the Mud section (Spiti valley, India) as a GSSP candidate for this boundary. They based their definition of the boundary on the first occurrence of the conodont *Novispathodus waageni* (Sweet 1970) *sensu lato*. However, Brühwiler *et al.* (2010b) demonstrated the presence in the same section of ammonoid genera typical of the Smithian below the boundary as defined by Krystyn *et al.* (2007a,b) and thus proposed to use the first occurrence of *Flemingites bhargavai* Brühwiler *et al.* (2010b) as the index fossil for this boundary.

Tozer (1965, 1994) originally subdivided the Dienerian of Canada into two parts (lower and upper), each composed of one zone (the *Proptychites candidus* Zone and the *Vavilovites sverdrupi* Zone), and with the second one being further subdivided into three subzones. However, this zonation is based on scattered occurrences of the faunas, often without superpositional information (a fact which can be checked in the list of localities provided in Tozer 1994). Other zonations have been proposed for Northern Siberia (Dagys & Ermakova 1996) and Primorye, Russia (Shigeta & Zakharov 2009), but uncertainties in correlating these persist. Based on material from Nepal, a biozonation has been proposed by Waterhouse (1994, 1996) for the northern Gondwana margin. Unfortunately, this Nepalese material is poorly preserved and does not allow constructing a robust taxonomy (see Gaetani *et al.* 1995). Jenks *et al.* (2015) presented a review of the biostratigraphy of Triassic ammonoids and mentioned that Dienerian ammonoid faunas were still poorly known and in need of an extensive revision.

The Northern Indian Margin has long been recognized as a key area for the study of Early Triassic ammonoids and the establishment of the Early Triassic time-scale. During the Early Triassic, it was situated in southern Tethys, at a palaeolatitude of ca. 40°S (Fig. 3). The very first Dienerian ammonoids were discovered in the Salt Range by Andrew Fleming in the mid-19th century, and this material was described by de Koninck (1863). Waagen (1895)

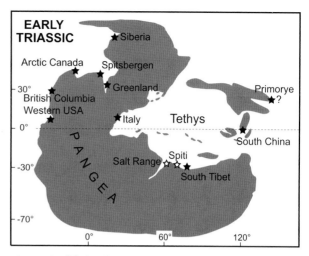

Fig. 3. Simplified palaeogeographical map of the Early Triassic with the palaeopositions of the studied localities (white stars) and of other localities mentioned in the text (black stars). Modified after Brayard *et al.* (2006).

conducted the most impressive and exhaustive study on Early Triassic ammonoids from the Salt Range. Diener (1897) and von Krafft & Diener (1909) published two monographs describing ammonoids from the Early Triassic of the Indian Himalayas, many of them from the Spiti Valley. The very first ammonoid biozonation of the Lower Triassic has been published by Mojsisovics *et al.* (1895) based on these two regions. They recognized only two zones in what we consider here as Dienerian, and this biozonation did not change in the absence of any subsequent detailed work on Dienerian ammonoids of the Salt Range and Spiti. For the Salt Range, Noetling (1905) and Spath (1934) added a few species and proposed slightly different classifications. Griesbachian ammonoids were first discovered by Schindewolf (1954) in the Salt Range. Kummel provided a detailed history of the stratigraphical and palaeontological investigations on the Permian and Triassic of the Salt Range (Kummel 1966; Kummel & Teichert 1966, 1970). He also mentioned that he collected numerous Dienerian ammonoids and that he intended to publish this material later on, a task he unfortunately never completed. Since Kummel's work, only two contributions on Early Triassic ammonoids from the Salt Range were published. The works by Guex (1978) and by the Pakistani-Japanese Research Group (PJRG 1985) only include scarce material of Griesbachian and Dienerian ages. Concerning ammonoids from the Indian Himalayas, only Bando (1981) described a few ammonoids from Kashmir, and Krystyn & Orchard (1996) and Krystyn *et al.* (2004, 2007a,b) gave some details concerning ammonoid biostratigraphy of Spiti, but without any description of ammonoids. A few poorly preserved ammonoids from South Tibet were also described by Wang & He (1976), and Dienerian ammonoids from Nepal were described by Waterhouse (1996). Outside the Northern Indian Margin, well-preserved Dienerian ammonoid faunas have been studied from British Columbia and Arctic Canada (e.g. Tozer 1994), the Verkhoyansk basin (Siberia, e.g. Dagys & Ermakova 1996), Primorye (Russia, e.g. Shigeta & Zakharov 2009), South China (e.g. Brühwiler *et al.* 2008) and Nevada (Ware *et al.* 2011).

Present work

The two articles included in this volume re-investigate the taxonomy of Dienerian ammonoids from the Salt Range (Pakistan) and Spiti (India). The data presented in the two contributions form the basis of the biochronology and diversity study published by Ware *et al.* (2015).

From 2007 to 2010, the research group at the University of Zürich carried out intensive fieldwork in the Salt Range and in Spiti. Bedrock-controlled high-resolution sampling (i.e. sampling ammonoids bed-by-bed, with the drawing and measuring of each section) of several sections in these regions was performed to revise the Dienerian ammonoid taxonomy and build a new, highly resolved biostratigraphical scheme. Whenever permitted by the sample size, great care was taken to integrate ontogenetic changes and intraspecific variation within the definition of species, in contrast to the traditional typological approach. Hence, the number of resulting valid taxa is more conservative and more robust than that of previous work.

The first article of this volume mainly addresses Dienerian ammonoids from the Salt Range. As the Salt Range is the type locality of most Dienerian ammonoid taxa, it was the region where the most intensive fieldwork was done, and where the most abundant material was found. Hence, this article includes an in-depth revision of Dienerian ammonoid taxonomy, including emended diagnoses of families, genera and species whenever possible. It also includes the description of the few rare Griesbachian ammonoids found in this region. The second article addresses Dienerian ammonoids from Spiti (Himachal Pradesh, India) and their comparison with the Salt Range.

Definitions of stages and substages

In the present work, the stage and substage subdivision of the Lower Triassic follows the recommendations of Ogg (2012), lowering the rank of the four Lower Triassic stages of Tozer (1965) to that of substage. The Induan stage is then subdivided into the Griesbachian and Dienerian substages, and the Olenekian into the Smithian and Spathian. There is however presently no consensus regarding the definitions of the substages boundaries relevant for the present work (the Griesbachian–Dienerian and Dienerian–Smithian boundaries).

The problem of the Dienerian–Smithian boundary was already briefly discussed. Here, the definition proposed by Brühwiler *et al.* (2010b) is adopted, the base of the Smithian being then defined by the first occurrence of *F. bhargavai* and its co-occurring species characterizing UA-Zone SM-1 (Brühwiler *et al.* 2010a). This zone has been recognized in every section described herein.

The Griesbachian–Dienerian boundary is problematic and less well documented. It was originally defined by Tozer (1965) with the first occurrence of 'Meekoceratidae' (i.e. Gyronitidae). In Spiti, Krystyn & Orchard (1996) placed this boundary within the Lower Limestone Member, considering the fauna from the upper part of this interval (their *Pleurogyronites planidorsatus* Zone) as typically Dienerian, based on the presence of abundant Gyronitidae, the absence of the typically Griesbachian *Ophiceras* and the occurrence of the conodont *Sweetospathodus kummeli* (Sweet 1970). In Spiti, this definition of the boundary coincides generally with a minor facies change, with the appearance of thin shale intervals and less massive, finer grained limestones than in the lower part of the Lower Limestone Member. This definition of the Griesbachian–Dienerian boundary was subsequently questioned by Krystyn *et al.* (2004), who considered these beds as late Griesbachian but proposed to use instead the Gangetian substage for the whole Lower Limestone Member. Krystyn *et al.* (2004) also reported some rare Gyronitidae (listed under the genus name '*Pleurogyronites*') associated with typical Griesbachian *Ophiceras* in the bed just below the Griesbachian–Dienerian boundary as defined by Krystyn & Orchard (1996). The same ammonoid association is here documented in the Salt Range (Ware *et al.* 2018) where it is interpreted as condensation. However, both Krystyn & Orchard (1996) and Krystyn *et al.* (2004) reported exclusively Griesbachian conodonts in this bed, thus questioning the condensed nature of this association. The definition of the Griesbachian–Dienerian boundary is thus still problematic, and more expanded sections spanning this boundary are necessary to solve this question. Here, we decided to place the Griesbachian–Dienerian boundary in agreement with the definition of Krystyn & Orchard (1996). This definition can easily be applied both in the Salt Range and in Spiti, the first Dienerian faunal association corresponding in both cases to the *Gyronites dubius* Regional Zone (equivalent of UA-Zone DI-1 of Ware *et al.* 2015).

Biochronology: the Unitary Association Method and terminology

Following the recommendations of Monnet *et al.* (2015), the construction of the biochronological scheme is here based on the Unitary Association Method of Guex (1991) and Guex *et al.* (2016). As already mentioned, the two papers presented in this volume constitute the base of the biozonation established by Ware *et al.* (2015). As the present work is only the first step in building this biozonation, we here present the construction of 'Regional Zones', which are then used to construct 'Unitary Association Zones' of Ware *et al.* (2015). Regional Zones correspond to 'Unitary Association Zones' built for only one basin, without addressing the lateral reproducibility of these zones outside of the studied basin. Regional Zones are customarily termed 'beds' by previous authors (e.g. Brayard & Bucher 2008; Brühwiler *et al.* 2012). Here, the term 'Regional Zone' is preferred to avoid any confusion with the term 'beds' as it was used in Spiti by previous authors to designate small lithological subdivisions of the different units within the Mikin Formation (e.g. the '*Otoceras*' and '*Meekoceras*' beds of von Krafft & Diener 1909; '*Gyronites*' beds of Krystyn & Orchard 1996). The term 'Local Maximal Horizon' designates maximal associations of species as directly observed within each section.

In the two papers presented here, the list of characteristic species and pairs of species is given for each regional zone. A species is said to be characteristic if and only if its range is equal to the zone. A pair of species is said to be characteristic if and only if the overlapping part of their ranges is equal to the zone. Additionally, the number of specimens of each species within each zone is indicated in brackets. This latter information gives an idea of the robustness of the different species ranges. When a species is restricted to one regional zone but represented only by a few rare specimens, its relevance for correlations must be *a priori* taken with caution as its stratigraphical range is likely to be longer. On the other hand, a species represented by many specimens and restricted to one regional zone provides *a priori* more robust information.

Each regional zone presented here is named after the most abundant of the species whose range is strictly restricted to the corresponding zone. Formal 'Unitary Association Zones' for the Dienerian of the Northern Indian Margin were constructed by Ware *et al.* (2015), and their correlation with each regional zones described herein is systematically provided here.

References

Bando, Y. 1981: Lower Triassic ammonoids from Guryul Ravine and the spur three kilometres north of Barus. *In* Nakazawa, K. & Kapoor, H.M. (eds): *The Upper Permian and Lower Triassic Faunas of Kashmir. Palaeontologia Indica*, Vol. *46*, 135–178. Geological Survey of India, Calcutta (India).

Brayard, A. & Bucher, H. 2008: Smithian (Early Triassic) ammonoid faunas from northwestern Guangxi (South China): taxonomy and biochronology. *Fossils and Strata 55*, 179 pp.

Brayard, A. & Bucher, H. 2015: Permian–Triassic extinctions and rediversifications. *In* Klug, C., Korn, D., De Baets, K., Kruta, I. & Mapes, R.H. (eds): *Ammonoid Paleobiology: From Macroevolution to Paleogeography*, 465–473, 628 pp. Topics in Geobiology 44, Springer Verlag, Berlin and Heidelberg.

Brayard, A., Bucher, H., Escarguel, G., Fluteau, F., Bourquin, S. & Galfetti, T. 2006: The Early Triassic ammonoid recovery: paleoclimatic significance of diversity gradients. *Palaeogeography, Palaeoclimatology, Palaeoecology 239*, 374–395.

Brayard, A., Escarguel, G., Bucher, H., Monnet, C., Brühwiler, T., Goudemand, N., Galfetti, T. & Guex, J. 2009: Good genes and good luck: ammonoid diversity and the end-permian mass extinction. *Science 325*, 1118–1121.

Brühwiler, T., Brayard, A., Bucher, H. & Guodun, K. 2008: Griesbachian and Dienerian (Early Triassic) ammonoid faunas from Northwestern Guangxi and Southern Guizhou (South China). *Palaeontology 51*, 1151–1180.

Brühwiler, T., Bucher, H., Brayard, A. & Goudemand, N. 2010a: High-resolution biochronology and diversity dynamics of the Early Triassic ammonoid recovery: the Smithian faunas of the Northern Indian Margin. *Palaeogeography, Palaeoclimatology, Palaeoecology 297*, 491–501.

Brühwiler, T., Ware, D., Bucher, H., Krystyn, L. & Goudemand, N. 2010b: New Early Triassic ammonoid faunas from the Dienerian/Smithian boundary beds at the Induan/Olenekian GSSP candidate at Mud (Spiti, Northern India). *Journal of Asian Earth Sciences 39*, 724–739.

Brühwiler, T., Bucher, H., Ware, D., Schneebeli-Hermann, E., Hochuli, P.A., Roohi, G., Rehman, K. & Yaseen, A. 2012: Smithian (Early Triassic) ammonoids from the Salt Range, Pakistan. *Special Papers in Palaeontology 88*, 1–114.

Burgess, S.D., Bowring, S. & Shen, S.-Z. 2014: High-precision timeline for Earth's most severe extinction. *Proceedings of the National Academy of Sciences of the United States of America 111*, 3316–3321.

Chen, Y., Twitchett, R.J., Jiang, H., Richoz, S., Lai, X., Yan, C., Sun, Y., Liu, X. & Wang, L. 2013: Size variation of conodonts during the Smithian–Spathian (Early Triassic) global warming event. *Geology 41*, 823–826.

Dagys, A.S. & Ermakova, S. 1996: Induan (Triassic) ammonoids from North-Eastern Asia. *Revue de Paléobiologie 15*, 401–447.

de Koninck, L.G. 1863: Description of some fossils from India, discovered by Dr. A. Fleming, of Edinburgh. *The Quarterly Journal of the Geological Society of London 19*, 1–19.

Diener, C. 1897: Part I: the Cephalopoda of the lower trias. *Palaeontologia Indica, Series 15. Himalayan fossils 2 2*, 1–181.

Erwin, D.H. 1998: The end and the beginning: recoveries from mass extinctions. *Trends in Ecology & Evolution 13*, 344–349.

Erwin, D.H. 2006: *Extinction: How Life on Earth Nearly Ended 250 Million Years Ago*, 296 pp. Princeton University Press, Princeton (USA).

Gaetani, M., Balini, M., Garzanti, E., Nicora, A., Tintori, A., Angiolini, L. & Sciunnach, D. 1995: Comments on Waterhouse, J.B., 1994. The Early and Middle Triassic ammonid succession of the Himalayas in western and central Nepal. Part 1. Stratigraphy, classification and Early Scythian ammonoid systematics. Palaeontographica, Abt. A, 232(1-3): 1–83. *Albertiana 15*, 3–9.

Galfetti, T., Bucher, H., Ovtcharova, M., Schaltegger, U., Brayard, A., Brühwiler, T., Goudemand, N., Weissert, H., Hochuli, P., Cordey, F. & Guodon, K. 2007: Timing of the Early Triassic carbon cycle perturbations inferred from new U-Pb ages and ammonoid biochronozones. *Earth and Planetary Science Letters 258*, 593–604.

Goudemand, N., Orchard, M., Bucher, H., Brayard, A., Brühwiler, T., Galfetti, T., Hochuli, P.A., Hermann, E. & Ware, D. 2008: Smithian-Spathian boundary: The biggest crisis in Triassic conodont history. *Abstracts with Program, Geological Society of America 40*, 505.

Guex, J. 1978: Le Trias inférieur des Salt Ranges (Pakistan): problèmes biochronologiques. *Eclogae Geologia Helvetica 71*, 105–141.

Guex, J. 1991: *Biochronological Correlations*, 252 pp. Springer, Berlin.

Guex, J., Galster, F. & Hammer, Ø. 2016: *Discrete Biochronological Time Scales*, 160 pp. Springer, Cham, Heidelberg, New York, Dordrecht, London.

Hermann, E., Hochuli, P.A., Bucher, H., Brühwiler, T., Hautmann, M., Ware, D. & Roohi, G. 2011a: Terrestrial ecosystems on North Gondwana following the end-Permian mass extinction. *Gondwana Research 20*, 630–637.

Hermann, E., Hochuli, P.A., Méhay, S., Bucher, H., Brühwiler, T., Ware, D., Hautmann, M., Roohi, G., ur-Rehman, K. & Yaseen, A. 2011b: Organic matter and palaeoenvironmental signals during the Early Triassic biotic recovery: the Salt Range and Surghar Range records. *Sedimentary Geology 234*, 19–41.

Hermann, E., Hochuli, P.A., Bucher, H., Brühwiler, T., Hautmann, M., Ware, D., Weissert, H., Roohi, G., Yaseen, A. & ur-Rehman, K. 2012a: Climatic oscillations at the onset of the Mesozoic inferred from palynological records from the North Indian Margin. *Journal of the Geological Society, London 169*, 227–237.

Hermann, E., Hochuli, P.A., Bucher, H. & Roohi, G. 2012b: Uppermost Permian to Middle Triassic palynology of the Salt Range and Surghar Range, Pakistan. *Review of Palaeobotany and Palynology 169*, 61–95.

Hochuli, P.A., Sanson-Barbera, A., Schneebeli-Hermann, E. & Bucher, H. 2016: Severest crisis overlooked – worst disruption of terrestrial environments postdates the Permian-Triassic mass extinction. *Scientific Reports 6*, 28372. https://doi.org/10.1038/srep28372.

Hofmann, R., Hautmann, M., Brayard, A., Nützel, A., Bylund, K.G., Jenks, J., Vennin, E., Olivier, N. & Bucher, H. 2014: Recovery of benthic marine communities from the end-Permian mass extinction at the low-latitudes of Eastern Panthalassa. *Palaeontology 57*, 547–589.

Jenks, J.F., Monnet, C., Balini, M., Brayard, A. & Meier, M. 2015: Biostratigraphy of Triassic ammonoids. In Klug, C., Korn, D., De Baets, K., Kruta, I. & Mapes, R.H. (eds): *Ammonoid Paleobiology: From Macroevolution to Paleogeography*, 277–298, 628 pp. Topics in Geobiology 44, Springer Verlag, Berlin and Heidelberg.

Kiparisova, L.D. & Popov, Y.N. 1956: Subdivision of the lower series of the Triassic system into stages. *Doklady Academy Sciences U.S.S.R. 109*, 842–845 [In Russian].

Komatsu, T., Takashima, R., Shigeta, Y., Maekawa, T., Tran, H.D., Cong, T.D., Sakata, S., Dinh, H.D. & Takahashi, O. 2016: Carbon isotopic excursions and detailed ammonoid and conodont biostratigraphies around Smithian–Spathian boundary in the Bac Thuy Formation, Vietnam. *Palaeogeography, Palaeoclimatology, Palaeoecology 454*, 65–74.

Krafft, A. von & Diener, C. 1909: Lower Triassic Cephalopoda from Spiti, Malla, Johar, and Byans. *Palaeontologia Indica 6*, 1–186.

Krystyn, L. & Orchard, M.J. 1996: Lowermost Triassic ammonoid and conodont biostratigraphy of Spiti, India. *Albertiana 17*, 10–21.

Krystyn, L., Balini, M. & Nicora, A. 2004: Lower and Middle Triassic stage and substage boundaries in Spiti. *Albertiana 30*, 40–53.

Krystyn, L., Bhargava, O.N. & Richoz, S. 2007a: A candidate GSSP for the base of the Olenekian Stage: Mud at Pin Valley; district Lahul & Spiti, Himachal Pradesh (Western Himalaya), India. *Albertiana 35*, 5–29.

Krystyn, L., Richoz, S. & Bhargava, O.N. 2007b: The Induan-Olenekian Boundary (IOB) in Mud – an update of the candidate GSSP section M04. *Albertiana 36*, 33–45.

Kummel, B. 1966: The Lower Triassic formations of the Salt Range and Trans-Indus Ranges, West Pakistan. *Bulletin of the Museum of Comparative Zoology 134*, 361–429.

Kummel, B. & Teichert, C. 1966: Relations between the Permian and Triassic formations in the Salt Range and Trans-Indus ranges, West Pakistan. *Neues Jahrbuch für Geologie Paläontologie. Abhandlungen 125*, 297–333.

Kummel, B. & Teichert, C. 1970: Stratigraphy and paleontology of the Permian-Triassic boundary beds, Salt Range and Trans-Indus Ranges, West Pakistan. *In* Kummel, B. & Teichert, C. (eds): *Stratigraphic Boundary Problems: Permian and Triassic of West Pakistan*, 1–110. Special Publication of the Department of Geology, Vol. 4, University of Kansas, Lawrence.

Mojsisovics, E.V., Waagen, W. & Diener, C. 1895: Entwurf einer Gliederung der pelagischen Sedimente des Trias-Systems. *Sitzungberichte der Akademie der Wissenschaften in Wien (I) 104*, 1271–1302.

Monnet, C., Brayard, A. & Bucher, H. 2015: Ammonoids and quantitative biochronology – a unitary association perspective. *In* Klug, C., Korn, D., De Baets, K., Kruta, I. & Mapes, R.H. (eds): *Ammonoid Paleobiology: From Macroevolution to Paleogeography*, 277–298, 628 pp. Topics in Geobiology 44. Springer Verlag, Berlin and Heidelberg.

Noetling, F. 1905: Die asiatische Trias. *In* Frech, F. (ed.): *Lethaea Geognostica, Das Mesozoicum 107–221*, 623 pp. Verlag der E. Schweizerbart'schen Verlagsbuchhandlung (E. Nägele), Stuttgart, Germany.

Ogg, J.G. 2012: Triassic. *In* Gradstein, F.M., Ogg, J.G., Schmitz, M.D. & Ogg, G.M. (eds): *The Geologic Time Scale 2012*, 681–730, 1144 pp. Elsevier, Amsterdam.

Orchard, M.J. 2007: Conodont diversity and evolution through the latest Permian and Early Triassic upheavals. *Palaeogeography, Palaeoclimatology, Palaeoecology 252*, 93–117.

Ovtcharova, M., Bucher, H., Schaltegger, U., Galfetti, T., Brayard, A. & Guex, J. 2006: New Early to Middle Triassic U-Pb ages from South China: calibration with ammonoid biochronozones and implications for the timing of the Triassic biotic recovery. *Earth and Planetary Science Letters 243*, 463–475.

Ovtcharova, M., Goudemand, N., Hammer, Ø., Guodun, K., Cordey, F., Galfetti, T., Schaltegger, U. & Bucher, H. 2015: Developing a strategy for accurate definition of a geological boundary through radio-isotopic and biochronological dating: the Early-Middle Triassic boundary (South China). *Earth Science Reviews 146*, 65–76.

Pakistani-Japanese Research Group 1985: Permian and Triassic systems in the Salt Range and Surghar Range, Pakistan. In Nakazawa, K. & Dickins, J.M. (eds): *The Tethys, Her Paleogeography and Paleobiogeography from Paleozoic to Mesozoic*, 221–312, 317 pp. Tokai University Press, Tokyo.

Payne, J.L., Lehrmann, D.J., Wei, J.Y., Orchard, M.J., Schrag, D.P. & Knoll, A.H. 2004: Large perturbations of the carbon cycle during recovery from the end-Permian extinction. *Science 305*, 506–509.

Racki, G. 1999: Silica-secreting biota and mass extinctions: survival patterns and processes. *Palaeogeography, Palaeoclimatology, Palaeoecology 154*, 107–132.

Raup, D.M. & Sepkoski, J.J. 1982: Mass extinctions in the marine fossil record. *Science 215*, 1501–1503.

Schindewolf, O.H. 1954: Über die Faunenwende vom Paläozoikum zum Mesozoikum. *Zeitschrift der Deutschen Geologischen Gesellschaft 105*, 153–182.

Schneebeli-Hermann, E., Kürschner, W.M., Hochuli, P.A., Bucher, H., Ware, D., Goudemand, N. & Roohi, G. 2012: Palynofacies analysis of the Permian–Triassic transition in the Amb section (Salt Range, Pakistan): implications for the anoxia on the South Tethyan Margin. *Journal of Asian Earth Sciences 60*, 225–234.

Schneebeli-Hermann, E., Kürschner, W.M., Bomfleur, B., Hochuli, P.A., Ware, D., Roohi, G. & Bucher, H. 2015: Vegetation history across the Permian-Triassic boundary in Pakistan (Amb section, Salt Range). *Gondwana Research 27*, 911–924.

Sepkoski, J.J. 1984: A kinetic-model of Phanerozoic Taxonomic Diversity. 3. Post-Paleozoic families and mass extinctions. *Paleobiology 10*, 246–267.

Shevyrev, A.A. 2006: Triassic biochronology: state of the art and main problems. *Stratigraphy and Geological Correlation 14*, 629–641.

Shigeta, S. & Zakharov, Y.D. 2009: Cephalopods. *In* Shigeta, Y., Zakharov, Y.D., Maeda, H. & Popov, A.M. (eds): *The Lower Triassic System in the Abrek Bay Area, South Primorye, Russia*, 44–140, 218 pp. National Museum of Nature and Science Monographs 38, Tokyo.

Spath, L.F. 1934: *Catalogue of the Fossil Cephalopoda in the British Museum (Natural History), Part IV: The Ammonoidea of the Trias*, 521 pp. The Trustees of the British Museum, London.

Stanley, G.D. 2003: The evolution of modern corals and their early history. *Earth-Science Reviews 60*, 195–225.

Stanley, S.M. 2009: Evidence from ammonoids and conodonts for multiple Early Triassic mass extinctions. *Proceedings of the National Academy of Sciences of the United States of America 106*, 15264–15267.

Sweet, W.C. 1970: Uppermost Permian and Lower Triassic Conodonts of the Salt Range and Trans-Indus Ranges, West Pakistan. *In* Kummel, B. & Teichert, C. (eds): *Stratigraphic Boundary Problems: Permian and Triassic of West Pakistan*, 207–275. Special Publication of the Department of Geology, Vol. 4, University of Kansas, Lawrence.

Tong, J. & Shi, G.R. 2000: Evolution of the Permian and Triassic foraminifera in south China. *In* Yin, H., Dickins, J.M., Shi, G.R. & Tong, J. (eds): *Permian-Triassic Evolution of Tethys and Western Circum-Pacific. Developments in Palaeontology and Stratigraphy*, 291–307, 392 pp. Elsevier, Amsterdam.

Tozer, E.T. 1965: Lower Triassic stages and Ammonoid zones of Arctic Canada. *Paper of the Geological Survey of Canada 65-12*, 14 pp.

Tozer, E.T. 1994: Canadian Triassic ammonoid faunas. *Bulletin of the Geological Survey of Canada 467*, 1–663.

Waagen, W. 1895: Salt Ranges fossils. vol. 2: fossils from the Ceratites formation – Part I – Pisces, Ammonoidea. *Palaeontologia Indica 13*, 1–323.

Wang, Y.G. & He, G.X. 1976: Triassic ammonoids from the Mount Jolmo Lungma region. *In* Xizang scientific expedition team of Chinese Academy of Science (ed.): A Report of Scientific Expedition in the Mount Jolmo Lungma region (1966–1968), 223–502. Palaeontology, fascicule 3. Science Press, Beijing [In Chinese].

Ware, D., Jenks, J.F., Hautmann, M. & Bucher, H. 2011: Dienerian (Early Triassic) ammonoids from the Candelaria Hills (Nevada, USA) and their significance for palaeobiogeography and palaeoceanography. *Swiss Journal of Geoscience 104*, 161–181.

Ware, D., Bucher, H., Brayard, A., Schneebeli-Hermann, E. & Brühwiler, T. 2015: High-resolution biochronology and diversity dynamics of the Early Triassic ammonoid recovery: the Dienerian faunas of the Northern Indian Margin. *Palaeogeography, Palaeoclimatology, Palaeoecology 440*, 363–373.

Ware, D., Bucher, H., Brühwiler, T., Schneebeli-Hermann, E., Hochuli, P.A., Roohi, G., Rehman, K. & Yaseen, A. 2018: Griesbachian and Dienerian (Early Triassic) ammonoids from the Salt Range, Pakistan. *Fossil and Strata 63*. 13–175.

Waterhouse, J.B. 1994: The Early and Middle Triassic ammonoid succession of the Himalayas in western and central Nepal. Part 1. Stratigraphy, classification and Early Scythian ammonoid systematics. *Palaeontographica, Abteilung A 232*, 1–83.

Waterhouse, J.B. 1996: The Early and Middle Triassic ammonoid succession of the Himalayas in western and central Nepal. Part 2. Systematic studies of the Early Middle Scythian. *Palaeontographica Abteilung A 241*, 27–100.

Wignall, P.B. & Twitchett, R.J. 2002: Extent, duration, and nature of the Permian-Triassic superanoxic event. *Geological Society of America, Special Paper 356*, 395–413.

Griesbachian and Dienerian (Early Triassic) ammonoids from the Salt Range, Pakistan

by

David Ware, Hugo Bucher, Thomas Brühwiler,
Elke Schneebeli-Hermann, Peter A. Hochuli[†], Ghazala Roohi,
Khalil Ur-Rehman *and* Amir Yaseen

Acknowledgements
Financial support for the publication of this issue of
Fossils and Strata was provided by the Lethaia Foundation

Contents

Griesbachian and Dienerian (Early Triassic) ammonoids from the Salt Range, Pakistan

DAVID WARE, HUGO BUCHER, THOMAS BRÜHWILER, ELKE SCHNEEBELI-HERMANN, PETER A. HOCHULI,[†] GHAZALA ROOHI, KHALIL UR-REHMAN AND AMIR YASEEN

FOSSILS AND STRATA

THE LETHAIA FOUNDATION

Ware, D., Bucher, H., Brühwiler, T., Schneebeli-Hermann, E., Hochuli, P.A., Roohi, G., Ur-Rehman, K. & Yaseen, A. 2018: Griesbachian and Dienerian (Early Triassic) ammonoids from the Salt Range, Pakistan. Fossils and Strata, No 63, pp. 13–175. doi: 10.1111/let.12273

Intensive and bedrock controlled sampling of four areas (Nammal Nala, Chiddru, Amb and Wargal) in the Salt Range yielded abundant well-preserved Griesbachian and Dienerian (Early Triassic) ammonoids. This material allows establishing a new, high-resolution biostratigraphical frame and an extensive revision of the taxonomy. The Griesbachian is represented by (in ascending order) the *Hypophiceras* cf. *H. gracile* Regional Zone, the *Ophiceras connectens* Regional Zone and the *Ophiceras sakuntala* Regional Zone. The Dienerian comprises 12 distinct regional zones leading to a threefold subdivision into lower, middle and upper Dienerian. The lower Dienerian, based on the occurrence of the genus *Gyronites*, can be divided into the *Gyronites dubius* Regional Zone, the *Gyronites plicosus* Regional Zone and the *Gyronites frequens* Regional Zone, in ascending order. The middle Dienerian, based on the occurrence of the genus *Ambites*, can be divided into five zones: the *Ambites atavus* Regional Zone, the *Ambites radiatus* Regional Zone, the *Ambites discus* Regional Zone, the *Ambites superior* Regional Zone and the *Ambites lilangensis* Regional Zone. The upper Dienerian, whose base is defined by the earliest representatives of Paranoritidae, can be divided into four zones: the *Vavilovites* cf. *V. sverdrupi* Regional Zone, the *Kingites davidsonianus* Regional Zone, the *Koninckites vetustus* Regional Zone and the *Awanites awani* Regional Zone. Correlations with basins outside the Northern Indian Margin are difficult because of the scarcity of such highly resolved studies on Dienerian ammonoids. Emended diagnoses and detailed synonymy lists are provided for most previously known taxa. In addition, five new genera (*Kyoktites, Ghazalaites, Pashtunites, Awanites* and *Subacerites*) and 18 new species (*Kyoktites hebeiseni, Ghazalaites roohii, Gyronites schwanderi, Ambites tenuis, Ambites bojeseni, Ambites subradiatus, Ambites bjerageri, Awanites awani, Koiloceras sahibi, Bukkenites sakesarensis, Proptychites wargalensis, Mullericeras shigetai, Mullericeras indusense, Mullericeras niazii, Ussuridiscus ventriosus, Ussuridiscus ornatus, Pseudosageceras simplelobatum* and *Subacerites friski*) are described. □ *Ammonoidea, biostratigraphy, Dienerian, Early Triassic, Salt Range, Pakistan.*

David Ware [david.ware@mfn-berlin.de], Museum für Naturkunde, Leibniz Institute for Evolution and Biodiversity Science, Invalidenstrasse 43 10115 Berlin, Germany; Hugo Bucher [hugo.fr.bucher@pim.uzh.ch], Thomas Brühwiler [bruehwiler@pim.uzh.ch], Elke Schneebeli-Hermann [elke.schneebeli@pim.uzh.ch] and Peter A. Hochuli [peter.hochuli@pim.uzh.ch], Paläontologisches Institut und Museum der Universität Zürich, Karl Schmid-Strasse 4 CH-8006 Zürich, Switzerland; Ghazala Roohi [roohighazala@yahoo.com], Khalil Ur-Rehman [reman_geol@yahoo.com] and Amir Yaseen [geologistgeologist@yahoo.com], Earth Science Division, Pakistan Museum of Natural History, Garden Avenue Shakarparian, Islamabad 44000, Pakistan; manuscript received on 27/10/2015; manuscript accepted on 11/08/2017.

Introduction

The biotic recovery following the end-Permian mass extinction is an intensively studied topic, for which high accuracy and high precision time control is of paramount importance. Nekto-pelagic clades such as ammonoids and conodonts recovered very quickly compared to other marine clades (e.g. Brayard *et al.* 2006, 2009; Orchard 2007) and play the leading roles in dating of Lower Triassic marine sedimentary rocks. However, many studies addressing the recovery are based on insufficiently resolved palaeontological age controls. This is particularly the case for the Dienerian, where ammonoids and biochronology are still poorly understood (Jenks *et al.* 2015). A review of the current knowledge of Griesbachian and Dienerian ammonoids from the Salt Range is given in the foreword of this volume (Ware & Bucher 2018) to which the reader is referred. From 2007 to 2010, our research group carried out intensive field work in the Salt Range and the Surghar Range. Palynological and carbon isotope records have been recently published by Hermann *et al.* (2011a,b, 2012a,b) and Schneebeli-Hermann *et al.* (2012), oxygen isotopes from biogenic phosphates by Romano *et al.* (2013), Smithian ammonoids by Brühwiler *et al.* (2012) and bivalves of Smithian and Spathian ages by Wasmer *et al.* (2012).

The present work focuses on Griesbachian and Dienerian ammonoids from four different areas in the Salt Range. It is based on abundant and well-preserved material sampled bed by bed. This new material provides the basis for a comprehensive revision of the taxonomy and biostratigraphy of Griesbachian and Dienerian ammonoids in the Salt Range, where all relevant sections are found. Because of the incomparable quality of the Dienerian ammonoid record of the Salt Range, this taxonomic and biostratigraphical re-investigation is an essential contribution to the Lower Triassic ammonoid zonation of the Northern Indian Margin and to the understanding of the Early Triassic biotic recovery.

Geological framework

The Salt Range constitutes a long and narrow mountain range, approximately 150 km SSW of Islamabad, Pakistan (Fig. 1B,C). The southern limit of the Salt Range defines the Himalayan main frontal thrust, which exposes a northern Gondwanan rift margin succession ranging from the Cambrian to the Cenozoic. It typically consists of a stack of tectonic slices with a south vergence, thus repeatedly exposing Triassic rocks of the Mianwali, Tredian and Kingriali formations (Gee 1980–1981). During the Early Triassic, the Salt Range was situated in the southern Tethys on the northern Gondwana margin, at a palaeolatitude of ca. 30 °S (Fig. 1A).

Stratigraphy

In the Salt Range, Lower Triassic sedimentary rocks are referred to the Mianwali Formation (Kummel & Teichert 1966). This 120 m thick formation is composed of limestone and siliciclastic sedimentary rocks. It unconformably rests on the Changhsingian (Upper Permian) Chiddru Formation. In the eastern part of the Salt Range, the Mianwali Formation is truncated by post-Cretaceous erosion and directly capped by Paleocene marine sedimentary rocks. In the western part, it is overlain by the Middle Triassic Tredian Formation.

The Griesbachian and Dienerian are represented by three units of the Mianwali Fm: the Kathwai Member (which is further subdivided into a dolomitic unit and a limestone unit), the Lower Ceratite Limestone and the lower part of the Ceratite Marls. The thickness of each of these units is highly variable throughout the Salt Range. Their boundaries are

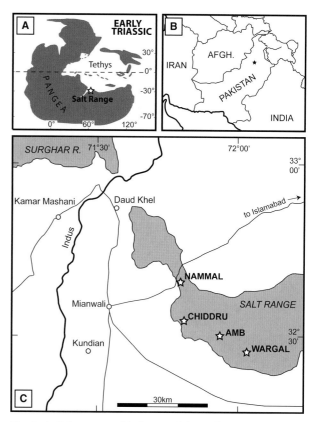

Fig. 1. **A.** Palaeogeographical map of the Early Triassic with the palaeoposition of the Salt Range (modified from Brayard *et al.* 2006). **B.** Map of Pakistan with position of the studied area (black star). **C.** Location map of sampled sections in the Salt Range (modified after Brühwiler *et al.* 2012).

here demonstrated to be diachronous across the different tectonic slices, thus suggesting that these thrusts might correspond to inverted Triassic normal faults. As a typical example among many others, Brühwiler *et al.* (2012) documented that the Dienerian–Smithian boundary coincides with the Lower Ceratite Limestone–Ceratite Marls boundary in Chiddru, whereas it is found within the lower third (ca. 7 m above the base) of the Ceratite Marls in Nammal. Most previous works (e.g. Mojsisovics *et al.* 1895; Spath 1934; Guex 1978; PJRG 1985) did not recognize this diachronism, and assumed that lithological boundaries are synchronous throughout the Salt Range and the Surghar Range.

Lithology and ammonoid preservation

Despite differences in thicknesses and ages in the different sections, the three units studied here show remarkably uniform facies throughout the Salt Range. Detailed lithological descriptions of these units were already published in Kummel & Teichert (1970) and in Hermann *et al.* (2011b). Therefore, only a summary of the lithological succession is given here, along with additional observations pertaining to the taphonomy and preservation of ammonoids.

Kathwai Member, dolomitic unit. – The dolomitic unit of the Kathwai Member (2–3.5 m thick) consists of a few massive beds of sandy dolomite. Fossils are very rare in this unit and are usually only represented by broken and unidentifiable shells. Only one ammonite (*Hypophiceras* aff. *H. gracile*) was found in this unit, in Nammal Nala. It occurred in a small lens rich in bivalves within a massive dolomitic bed.

Kathwai Member, limestone unit. – The limestone unit of the Kathwai Member (0.5–5 m thick) consists mostly of calcareous glauconitic sandstone beds alternating with thin beds of shale. Although fossils are not rare, they are generally very poorly preserved. Some beds contain accumulations of rhynchonellid brachiopods and echinid spines. Ammonites are rare, and usually represented only by extremely poorly preserved specimens. Identification, even at the genus level, is impossible. Therefore, they have not been included in the present study. The only exception is in Chiddru, where few, better preserved specimens assigned to *Ophiceras connectens*, were found at the base of this unit.

Lower Ceratite Limestone. – The Lower Ceratite Limestone (1–3 m thick) consists of thin, hard, coarse-grained coquinoid limestone beds. Glauconite and iron oxides are locally very abundant. Although very frequent, the fossils are mostly fragmented, and generally very difficult to prepare mechanically. The coarse grained sparitic matrix often crosses the shell boundaries. Ammonoids are very unevenly distributed, often imbricated and accumulated in lenses within the different layers of the Lower Ceratite Limestone. The body chambers of small specimens are generally broken, while large specimens (Proptychitidae) are represented by incomplete phragmocones, the upper side of outer whorls being corroded or eroded (see Pl. 21, figs 38, 39 for a good example). Phragmocones are often completely recrystallized, hence suture lines are only occasionally preserved. Considering the abundance of glauconite and the facies, the Lower Ceratite Limestone may be affected by condensation. However, in the absence of similar studies in sections where the Lower Ceratite Limestone is expanded, palaeontological condensation cannot be demonstrated. Only the second bed of the Lower Ceratite Limestone in Nammal is recognized as condensed. It contains both *Ophiceras sakuntala* and *Gyronites dubius*, an association of genera which has not been documented in any other section. Moreover, the species *Gyronites dubius* has also been

found in Amb, without any co-occurring representative of *Ophiceras*. In Nammal Nala, the two uppermost beds of the Lower Ceratite Limestone are different, being composed of fine grained limestone. The penultimate bed is only 1–2 cm thick, locally absent, and contains numerous and nearly complete ammonoids, some bivalve fragments and abundant fish scales and teeth. Ammonoids are accumulated at the top of the bed, often encrusted by worm tubes and only partially covered by a very thin limestone layer (Pl. 3, figs 1, 2). The last bed is ca. 7 cm thick and composed of fine-grained limestone with locally abundant, nearly complete but strongly recrystallized ammonoids. Its surface is encrusted by centimetric iron oxide concretions, indicating a stop in sedimentation.

Ceratite Marls. – The Ceratite Marls are composed of a ca. 30 m thick succession of marls with intercalated limestone and sandstone beds. Limestone beds are abundant at the base, while sandstone beds become gradually more abundant in the upper half of the Ceratite Marls. In the lower third, ammonoids are very abundant and well preserved in numerous limestone beds and lenses. The thickness of these beds is variable. Many of these beds show imbrication and size sorting of shells, indicative of bottom currents. A typical example of this facies is shown in Figure 2, with small imbricated, both complete and broken ammonoids accumulated at the base of the bed. The body chambers are generally partially broken. Large specimens (all belonging to Proptychitidae) often have their venter abraded. For example, the specimen illustrated on Plate 25, figures 10–14 is a complete phragmocone, the body chamber of which is missing, and whose venter has been abraded on almost the entire last preserved volution. In the Dienerian part of the Ceratite Marls, involute shells often have their narrow umbilicus encrusted by bivalves on both sides, similar to the ones already observed by Ware *et al.* (2011) in Dienerian ammonoid assemblages from Nevada. These bivalves may occasionally induce an irregular coiling of the umbilicus, thus indicating *in vivo* encrusting of epizoans. These bivalves are sometimes also present on the flanks of the whorl and, where overlapped by the body chamber, they are lined by a dorsal shell layer. Usually, only their cemented valve is preserved. The bivalves visible on the ammonoid figured on Plate 25, Figure 10, were described in detail by Hautmann *et al.* (2017), together with similar specimens from the Dienerian of Spiti (India) and the Griesbachian of Greenland. They identified them as *Liostrea* sp. ind., belonging to the sub-family Grypheinae and representing the oldest known representative of

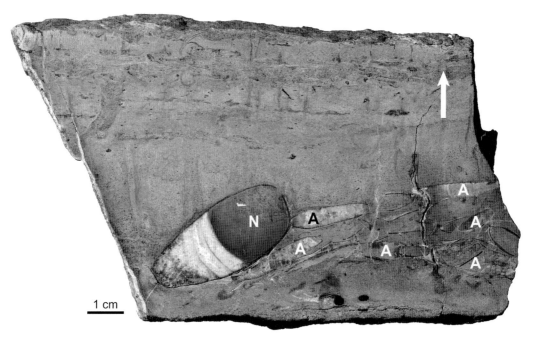

Fig. 2. Polished section of a bed belonging to the *Ambites lilangensis* Regional Zone (Nam100 and equivalents) from Nammal Nala, showing the typical facies of limestone beds at the base of the Ceratite Marls, with an accumulation of imbricated shells at its base with a nautiloid (N), several imbricated complete and broken ammonoids (A), and bioturbation increasing towards the top of the bed. Natural size, arrow indicates top of bed. Specimen PIMUZ30235.

Ostreidae. Besides this specimen, they are not well enough preserved to be formerly identified, but since no other coeval bivalve taxa are known to encrust ammonoids *in vivo*, they can reasonably be assumed to belong to the same taxa. It cannot be excluded that some of the intraspecific variability of the umbilical width observed on these involute shells is a consequence of the presence of these epizoans, but as no fauna without these encrusting bivalves could be found for comparison, this hypothesis cannot be tested. The presence or absence of bivalves in the umbilicus has therefore not been taken into account in the taxonomical analyses, except when they induced an obvious pathological coiling of the ammonoid shell. Such *in vivo* encrusting bivalves have so far not been recorded in older or younger ammonoid faunas in the Salt Range, even in early Smithian faunas which are found in similar facies. It is possible that this unusual high frequency of epizoans was caused by the coeval oxygen-poor water-sediment interface, but alternative hypotheses are also conceivable (see Hautmann *et al.* 2017, for details).

Hermann *et al.* (2011a,b, 2012a,b) and Schneebeli-Hermann *et al.* (2012) proposed a detailed palaeoenvironmental reconstruction based on several sections (including the ones studied here) throughout the Salt Range and Surghar Range. These studies documented that in Nammal, the middle to late Dienerian record a local peak of oxygen depletion, unlike previous (Griesbachian to early Dienerian) and subsequent (early and middle Smithian) time intervals. Romano *et al.* (2013) showed that temperature also peaked during the middle and late Dienerian (phase Ib of Romano *et al.* 2013).

Present work

The ammonoids presented in this study were collected in four different areas (Fig. 1C): Nammal Nala, Chiddru, Amb and Wargal, during four field work seasons from 2007 to 2010. Several sections were measured and sampled within a couple of square kilometres in each of these areas. Following recommendations of Guex (1991), a composite section for each area with the position of the samples was constructed based on lithological correlations. Therefore, several samples (each with a distinct number) may have been obtained from the same layer.

Nammal Nala

Nammal Nala is a narrow canyon situated about 5 km east of the village of Musa Khel (ca. 25 km ENE of Mianwali). This area was previously studied by Kummel (1966), Kummel & Teichert (1966, 1970), Guex (1978) and the Pakistani-Japanese Research Group (1985). The Lower Triassic

Fig. 3. Section near the entrance of Nammal Nala (N32°39′27.6″, E71°47′29.2″). KM.d: Kathwai Member, dolomitic unit; KM.l: Kathwai Member, limestone unit; LCL: Lower Ceratite Limestone; CM: Ceratite Marls. Note the presence of small normal faults (black lines).

Mianwali Formation is beautifully exposed and repeated by faulting. All the exposures reported here are from the northern side of the canyon, which is in part illustrated in Figure 3. A composite section with the ammonoid distribution is given in Figure 4. The most complete Griesbachian and Dienerian sequence of the Salt Range was found in Nammal Nala, especially for the base of the Ceratite Marls, where ammonoids are very abundant and well preserved. This canyon is also very easily accessible, thus allowing intensive sampling. The vast majority of the ammonoids (about 1200 specimens) described here come from Nammal Nala.

Chiddru

The sections studied in Chiddru are situated in a valley about 2.5 km east of the village (ca. 25 km ESE of Mianwali). Several sections were sampled in 2008 and 2010, but the best ones for the Dienerian were already described by Kummel (1966) and Kummel & Teichert (1966, 1970), especially the one on the west side of the valley (Fig. 5). A composite section showing ammonoid distribution is given in Figure 6. In this area, the Kathwai Member and Lower Ceratite Limestone are much

thicker than in other areas, and Dienerian faunas are restricted to the Lower Ceratite Limestone. The base of the Ceratite Marls is already early Smithian in age (Brühwiler *et al.* 2012). In the Lower Ceratite Limestone, ammonoids are abundant but generally broken and poorly preserved. Specimens well enough preserved for identification are rather rare. Despite intensive sampling, only ca. 50 specimens could be included in this study. However, this area is very important for the Dienerian, since it is the type locality of *Koninckites vetustus*, the type species of *Koninckites*. This area has been known since the beginning of the geological investigations in the Salt Range, and has been included in every study dealing with the Early Triassic of this region.

Amb

Amb is a small village situated ca. 35 km east of Mianwali. Three different outcrops were sampled during one field trip (2010) in the valley just south of the village, the best section being situated about 1 km south-east of the village (Fig. 7). A composite section showing the ammonoid distribution is given in Figure 8. The Kathwai Member and Lower Ceratite Limestone have thicknesses similar to those of Nammal, but the Dienerian part of the Ceratite Marls is thinner, with fewer limestone beds. Although often broken, ammonoids from the Lower Ceratite Limestone are very abundant and better preserved than in Nammal. In the Ceratite Marls ammonoids are rather rare, usually strongly recrystallized and difficult to prepare. A total of about 120 specimens from Amb could be included in this study. This locality is the type locality of the type species of the genus *Ambites* (*Ambites discus*). Since the pioneer work of Waagen (1895), the Early Triassic ammonoid fauna of Amb have not been studied.

Wargal

The sections near Wargal are situated in a syncline about 2.5 km west of the village, along Munta Nala. Although the Kathwai Member and Lower Ceratite Limestone are thicker than in Amb, and the Ceratite Marls thinner, the two sections are very similar. We spent only 3 days in this locality, mostly to collect specimens of the genus *Prionolobus*, as it is the type locality of its type species *Prionolobus atavus* and of *Prionolobus rotundatus*. About 100 specimens were collected, mostly in the Lower Ceratite Limestone and at the very base of the Ceratite Marls, easily accessible in the section previously described by

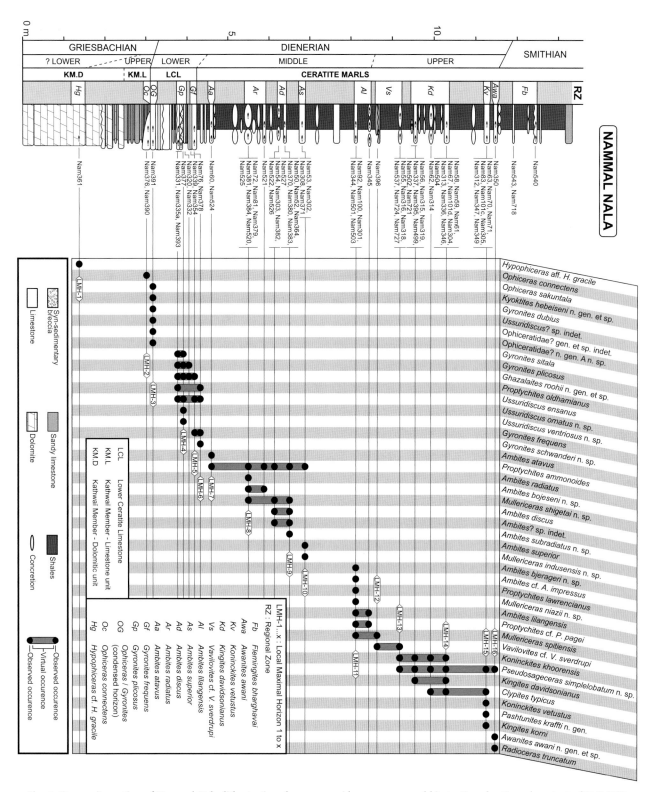

Fig. 4. Composite section of Nammal Nala: lithostratigraphy, ammonoid occurrences and biostratigraphy. Faunal content of NAM543, NAM718 and NAM540: see Brühwiler *et al.* (2012).

Kummel (1966) and Kummel & Teichert (1966, 1970), and illustrated here in Figure 9. Only the upper part of the lower third of the Ceratite Marls was sampled in another nearby tributary, where a

Koninckites vetustus fauna (sample War104) was found. A composite section showing the ammonoid distribution is given in Figure 10. As in Amb, ammonoids from the Lower Ceratite Limestone are

Fig. 5. Section above the village of Chiddru (N32°32′59.7″, E71°47′55.9″) on the West side of the gorge, previously described by Kummel & Teichert (1966, 1970) and Kummel (1966). Hammer highlighted for scale. Chiddru Fm., Chiddru formation; KM, Kathwai Member; LCL, Lower Ceratite Limestone; CM, Ceratite Marls.

abundant and rather well preserved, but they are also abundant and well preserved in the Ceratite Marls. This area has been investigated first by Waagen (1895) and later by Kummel (1966) and Kummel & Teichert (1966, 1970).

Biostratigraphy

The extensive bedrock controlled sampling on which the present work is based allows the recognition of a total of 15 regional zones: three from Griesbachian and 12 from Dienerian (Figs 11, 12). The resulting zonation is new with improved resolution. Previously, only three zones were recognized from the Salt Range (Guex 1978; Pakistani-Japanese Research Group 1985), and two zones and four subzones were established in the Dienerian succession of Canada (Tozer 1965, 1994). A preliminary version of this zonation was published in Romano *et al.* (2013). This regional zonation served as base for the construction of the formal biozonation for the Dienerian of the Northern Indian margin palaeoprovince of Ware *et al.* (2015).

The different regional zones are described herein and their correlation with ammonoid zonations from other basins is discussed. Synthetic range charts for Griesbachian and Dienerian ammonoid species and genera from the Salt Range are given in Figures 13 and 14, respectively. Technically, most of these regional biozones correspond directly to local maximal horizons, except for the *Gyronites plicosus* Regional Zone, the *Gyronites frequens* Regional Zone and the *Vavilovites* cf. *V. sverdrupi* Regional Zone

(see below). It should be noted that correlation of this new biozonation with previous ones (e.g. Mojsisovics *et al.* 1895; Guex 1978; Pakistani-Japanese Research Group 1985) cannot be provided because the undetected diachronicity of the different lithostratigraphical units led to mix ammonoid faunas of different (late Griesbachian to early Smithian) ages for the Lower Ceratite Limestone. The age distribution of the Lower Ceratite Limestone varies geographically, this unit being upper Griesbachian to lower Dienerian in Nammal Nala, lower Dienerian to lower middle Dienerian in Amb and Wargal, and Dienerian to lowermost Smithian in Chiddru. These results contrast with the previous age assignment of the Lower Ceratite Limestone to one single zone, namely the *Gyronites frequens* Zone.

For each regional zone, the list of co-occurring species is given, and the characteristic species and pairs of species are indicated in braces. Additionally, the number of specimens of each species in each zone is given in brackets.

General subdivisions

The stage and sub-stage subdivision of the Lower Triassic and their definitions were already discussed in the foreword (Ware & Bucher 2018; this volume). Tozer (1965) subdivided the Griesbachian and the Dienerian into two parts (lower and upper) each, a scheme which has been followed by every author since. Because of the paucity of Griesbachian faunas in the Salt Range, this twofold subdivision cannot be assessed here. However, for the Dienerian, the much higher resolution subdivisions obtained in this study leads to a threefold subdivision of the Dienerian (lower, middle and upper). The lower Dienerian is characterized by the occurrence of the genera *Gyronites* and *Ussuridiscus*, the middle Dienerian by the co-occurrence of the genera *Ambites* and *Mullericeras*, and the upper Dienerian by the appearance of Paranoritidae and Sagecerataceae.

Griesbachian ammonoid faunas

Griesbachian ammonoids are very rare and generally poorly preserved in the Salt Range. Moreover, Griesbachian ammonoids from other regions of the Northern Indian Margin have not been investigated in detail since the work of Diener (1897) and von Krafft and Diener (1909). A new biozonation for the Griesbachian of Spiti Valley (Indian Himalayas) has been attempted by Krystyn *et al.* (2004), but without explanation or illustration of the taxonomic definitions, thus making this scheme difficult to apply or to test in other areas. As a consequence, the three

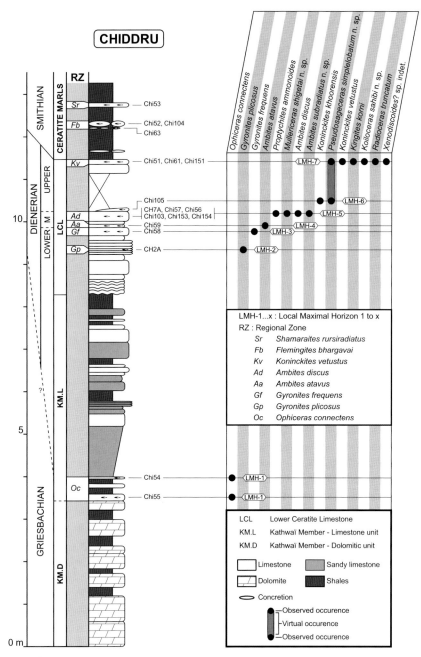

Fig. 6. Composite section of Chiddru: lithostratigraphy, ammonoid occurrences and biostratigraphy. Faunal content of CHI63, CHI52, CHI104 and CHI53: see Brühwiler *et al.* (2012).

Griesbachian regional biozones described here cannot be correlated with confidence.

Hypophiceras cf. *H. gracile* Regional Zone

Co-occurring species. – *Hypophiceras* cf. *H. gracile* (*n* = 1).

Occurrence in the investigated sections. – This zone is here based on a single specimen found in the dolomitic unit of the Kathwai Member in Nammal Nala.

Correlation. – Although species assignment remains uncertain, identification at the genus level of this single specimen is very robust. The genus *Hypophiceras* is known in the Arctic (Siberia, Arctic Canada and NE Greenland), where it occurs together with lower Griesbachian taxa. However, Kummel (1970) reported one specimen from the Kathwai Member of Kathwai, which may be conspecific with ours, and correlated this bed on the basis of lithology with the *Ophiceras connectens* Zone of Chiddru (upper Griesbachian). In the absence of additional material, and

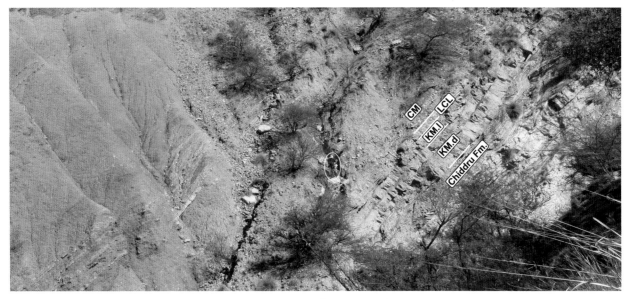

Fig. 7. Section about 1 km south-east of the village of Amb (N32°29′48.1″, E71°56′20.6″). Scale indicated by a geologist (circled) in the middle of the view. KM.d: Kathwai Member, dolomitic unit; KM.l: Kathwai Member, limestone unit; LCL: Lower Ceratite Limestone; CM: Ceratite Marls.

considering that this correlation is exclusively based on lithology, this zone is here kept separate from the subsequent *Ophiceras connectens* regional Zone. Additional work on the Griesbachian of the Tethys is necessary to decipher whether the genus *Hypophiceras* is restricted to the lower Griesbachian or if it ranges up into the upper Griesbachian. This zone corresponds to the horizon MH-G1 in Romano *et al.* (2013).

Ophiceras connectens Regional Zone

Co-occurring species. – *Ophiceras connectens* (*n* = 13).

Occurrence in the investigated sections. – This zone has been identified in Nammal Nala (base of the Lower Ceratite Limestone) and in Chiddru (base of limestone unit of the Kathwai Member) and only yields poorly preserved specimens of *Ophiceras connectens*.

Correlation. – This species was considered as a synonym of *Ophiceras tibeticum* by Waterhouse (1994), who thus treated this fauna as an equivalent of the Himalayan *Ophiceras tibeticum* zone of late Griesbachian age. Acceptance or rejection of this treatment requires revision of the taxonomy of Ophiceratidae and of the Himalayan Griesbachian faunal succession. This zone corresponds to the horizon MH-G2 in Romano *et al.* (2013).

Ophiceras sakuntala Regional Zone

Co-occurring species. – *Ophiceras sakuntala* (*n* = 4).

Occurrence in the investigated sections. – This regional zone has only been recognized in the condensed layer at the base of the Lower Ceratite Limestone in Nammal Nala.

Correlation. – *Ophiceras sakuntala* also occurs in Shalshal Cliff in the central Himalayas, where it was originally described by Diener (1897). This condensed layer corresponds to the horizon MH-G3 in Romano *et al.* (2013).

Remarks. – Because this layer yields both late Griesbachian and earliest Dienerian taxa (LMH-3, Fig. 4), the natural association of species belonging to this regional zone is obscured.

Early Dienerian ammonoid faunas

Early Dienerian ammonoids from the Salt Range are abundant and usually quite well preserved, but can be difficult to mechanically separate from the matrix. They are all found in the Lower Ceratite Limestone, an interval characterized by low accumulation and sedimentation rate. The lower Dienerian biostratigraphical record must be treated with caution with respect to potential hiatuses and condensed deposition. The Salt Range nevertheless provides the most

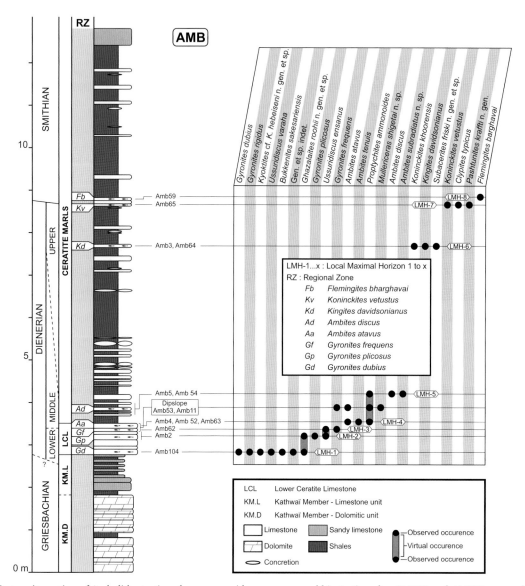

Fig. 8. Composite section of Amb: lithostratigraphy, ammonoid occurrences and biostratigraphy. AMB53 and AMB11 were collected on a dipslope formed by the two uppermost beds of the Lower Ceratite Limestone, without any further distinction.

expanded and diverse faunal succession for the early Dienerian worldwide. A likely case of condensation is found at the base of the Lower Ceratite Limestone in Nammal Nala, where *Gyronites dubius* of earliest Dienerian age occurs together with the late Griesbachian *Ophiceras sakuntala* (LMH-3, Fig. 4).

The lower Dienerian is best characterized by the genus *Gyronites*. In the Salt Range, the genus *Ussuridiscus* is also restricted to the lower Dienerian, but in Primorye, its type species *Ussuridiscus varaha* is found in four consecutive beds belonging to three different zones (Shigeta & Zakharov 2009), ranging from the upper Griesbachian to the middle Dienerian, whereas it is restricted to a single zone in the Salt Range. Its presence in the upper Griesbachian is uncertain, as its occurrence in this substage is based

on bed 1009 of Shigeta and Zakharov (2009, fig. 15), which yielded a poorly preserved specimen assigned to *Lytophiceras*? sp. indet. However, the presence of *Ussuridiscus varaha* in the middle Dienerian is confirmed by its association in bed 1013 of Shigeta & Zakharov (2009) with '*Ambitoides*' *fuliginatus* (here re-assigned to *Mullericeras*), *Proptychites ammonoides* and '*Gyronites*' (here re-assigned to *Ambites* on the basis of the bottleneck shape of the venter). The *Gyronites subdharmus* Zone, which Shigeta and Zakharov (2009) assigned to the upper Griesbachian, contains the genus *Gyronites* and therefore likely correlates with the lower Dienerian as described here. However, *Gyronites subdharmus* is absent in the Salt Range, so a correlation at the species level cannot be made. Krystyn *et al.* (2004) placed their

Fig. 9. Section of Munta Nala, about 2.5 km West of the village of Wargal (N32°27'07″, E72°01'56.7″; Kummel 1966; Kummel & Teichert 1966, 1970,). Thickness of Kathwai Member and Lower Ceratite Limestone amounts to ca. 6 m. KM.d: Kathwai Member, dolomitic unit; KM.l: Kathwai Member, limestone unit; LCL: Lower Ceratite Limestone; CM: Ceratite Marls.

'*Pleurogyronites*' beds from Spiti into the upper Griesbachian. These beds also yield *Gyronites* and thus better correlate with our lower Dienerian. Additional material from these layers in Spiti is described in the companion paper of this volume (Ware *et al.* 2018). In north-western Guangxi (South China), Brühwiler *et al.* (2008) found some specimens initially ascribed to '*Koninckites*' cf. '*K.*' *timorense* but which were synonymized with *Ussuridiscus varaha* by Shigeta and Zakharov (2009; an assignment which we partially confirm here), so their *Proptychites candidus* beds may partially correlate with the lower Dienerian as defined here (whereas the original *Proptychites candidus* Zone of Tozer is here considered as middle Dienerian; see below). The genus *Gyronites* has never been reported outside of the Tethys, but conodonts from the *Bukkenites strigatus* Zone of the Canadian Arctic indicate that the topmost part of this zone may actually already be Dienerian (Orchard 2008) and thus may in part correspond to our lower Dienerian.

Gyronites dubius Regional Zone

Co-occurring species. – {*Bukkenites sakesarensis* (*n* = 21)}, *Ghazalaites roohii* (*n* = 26), {*Gyronites dubius* (*n* = 27)}, {*Gyronites rigidus* (*n* = 1)}, *Kyoktites* cf. *K. hebeiseni* (*n* = 2) and {*Ussuridiscus varaha* (*n* = 7)}.

Occurrence in the investigated sections. – This regional zone is here based on the association found at the base of the Lower Ceratite Limestone in Amb. It also occurs at the base of the Lower Ceratite Limestone in Nammal Nala, but it is there included in a condensed layer (LMH-3, Fig. 4).

Correlation. – *Gyronites dubius* and *Gyronites rigidus* also occur in several localities in the Indian Himalayas, where they were first described by Diener (1897) and von Krafft and Diener (1909), indicating further extension of this fauna to the Northern Indian Margin. This regional zone corresponds to UA-zone DI-1 in Ware *et al.* (2015) and partially to the condensed horizon MH-G3 in Romano *et al.* (2013).

Gyronites plicosus Regional Zone

Co-occurring species. – *Ghazalaites roohii* (*n* = 19), {*Gyronites plicosus* (*n* = 85)}, {*Gyronites sitala* (*n* = 2)}, *Proptychites oldhamianus* (*n* = 2), {*Proptychites wargalensis* (*n* = 3)}, *Ussuridiscus ensanus* (*n* = 8), {*Ussuridiscus ornatus* (*n* = 3)}, {*Ussuridiscus ventriosus* (*n* = 1)}.

Occurrence in the investigated sections. – This regional zone has been recognized in every investigated section, in several beds in the middle of the Lower Ceratite Limestone, except in Chiddru where it occurs in the lower third of the Lower Ceratite Limestone.

Correlation. – *Gyronites sitala*, *Ussuridiscus ensanus* and some species here synonymized with *Gyronites plicosus* and *Proptychites oldhamianus* described by Diener (1897) and von Krafft and Diener (1909) occur in several localities in the Indian Himalayas,

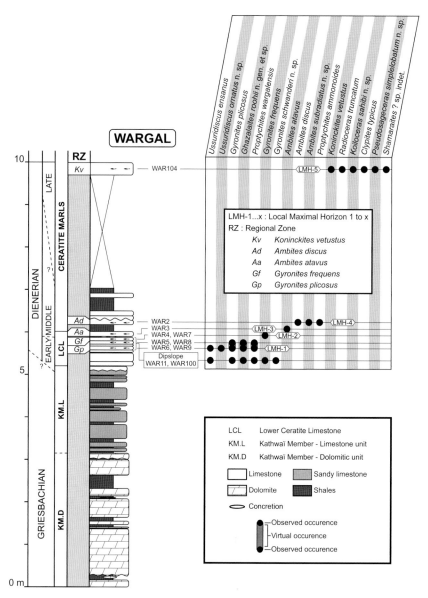

Fig. 10. Composite section of Wargal: lithostratigraphy, ammonoid occurrences and biostratigraphy. WAR11 and WAR100 were collected on a dipslope formed by several beds in the middle part of the Lower Ceratite Limestone, without any further precision.

indicating the broader extension of this fauna throughout the Northern Indian Margin. This regional zone corresponds to horizon MH-D1 (Romano *et al.* 2013) and to the UA-zone DI-2 (Ware *et al.* 2015).

Remarks. – *Gyronites sitala*, *Proptychites oldhamianus*, *Ussuridiscus ornatus* and *Ussuridiscus ventriosus* were only found in Nammal (LMH-4, Fig. 4), whereas *Proptychites wargalensis* has only been found in Wargal (LMH-1, Fig. 10), thus forming two distinct local maximal horizons. These two local maximal horizons are here lumped, as this distinction may be the result of the scarcity of some of their respective characteristic species. Moreover, these two

local maximal horizons have not been found in sequence anywhere, thus providing an additional argument to merge them.

Gyronites frequens Regional Zone

Co-occurring species. – *Ghazalaites roohii* ($n = 1$), {*Gyronites frequens* ($n = 90$)}, {*Gyronites schwanderi* ($n = 2$)}, *Proptychites oldhamianus* ($n = 1$), *Ussuridiscus ensanus* ($n = 4$).

Occurrence in the investigated sections. – This regional zone occurs in every investigated section: in the two topmost beds of the Lower Ceratite Limestone in Nammal Nala, in the penultimate bed of the

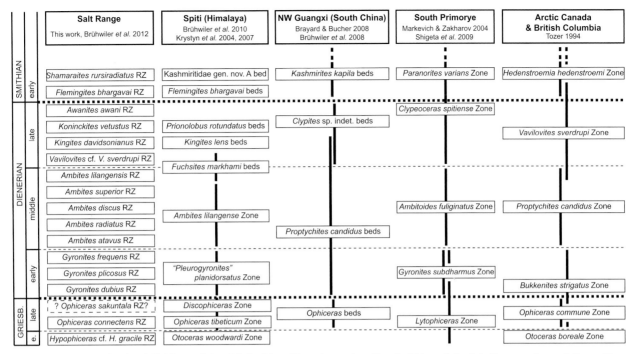

Fig. 11. Biostratigraphical subdivisions of the Griesbachian, Dienerian and earliest Smithian of the Salt Range and correlation with zonations of other regions. Thick vertical black bars indicate uncertainty intervals for correlations. The *O. sakuntala* beds, indicated with question marks and a dashed frame, correspond to a horizon identified only in the condensed horizon of Nammal Nala (LMH-3, Fig. 4). See text for details (RZ: Regional Zone).

Lower Ceratite Limestone in Amb and Wargal, and in one bed in the middle of the Lower Ceratite Limestone in Chiddru.

Correlation. – Some species described by von Krafft & Diener (1909) are here synonymized with *Gyronites frequens*, thus indicating the extension of this regional zone in the Indian Himalayas. Some specimens described by Wang & He (1976) and Waterhouse (1996) may also belong to *Gyronites frequens*, indicating that this regional zone may also expand to Nepal and Tibet. This regional zone corresponds to the horizon MH-D2 in Romano *et al.* (2013) and to UA-zone DI-3 of Ware *et al.* (2015).

Remarks. – In Nammal Nala, this regional zone is represented by two distinct local maximal horizons (LMH-5 and LMH-6, Fig. 4). They are only differentiated by two rare species (*Gyronites schwanderi* and *Ghazalaites roohii*) and are thus lumped here.

Middle Dienerian ammonoid faunas

Depending on the areas, middle Dienerian ammonoids occur either in the Lower Ceratite Limestone or at the base of the Ceratite Marls. The middle Dienerian is here defined by the occurrence of

Ambites, a genus which is generally very abundant and with a worldwide distribution. The genus *Mullericeras* is mostly restricted to the middle Dienerian, except for one species (*Mullericeras spitiense*) which extends up into the late Dienerian. In Spiti, *Ambites* is very abundant in the lower half of the 'Ambites beds', in the *Ambites lilangense* zone and at the base of the *Fuchsites markhami* beds (Brühwiler *et al.* 2010a). Several species of *Ambites* were found in different sections in north-western Guangxi (Brühwiler *et al.* 2008), which were all considered as belonging to the *Proptychites candidus* beds. In this work, the authors did not provide any more precise subdivisions because of the scattered occurrence of the faunas, but a comparison with the faunas from the Salt Range should allow the refinement of the biozonation for this area. In South Primorye, *Ambites* is present in bed 1013 of Shigeta & Zakharov (2009), corresponding to their '*Ambitoides*' *fuliginatus* Zone.

Correlation with the Dienerian biozonation established by Tozer (1965, 1994) is not clear. The faunas from north-eastern British Columbia described by Tozer (1963, 1994) are different from those described by Tozer (1961, 1994) from Arctic Canada. Hence correlation between these two basins is uncertain. What Tozer referred to as *Proptychites candidus* in British Columbia, on

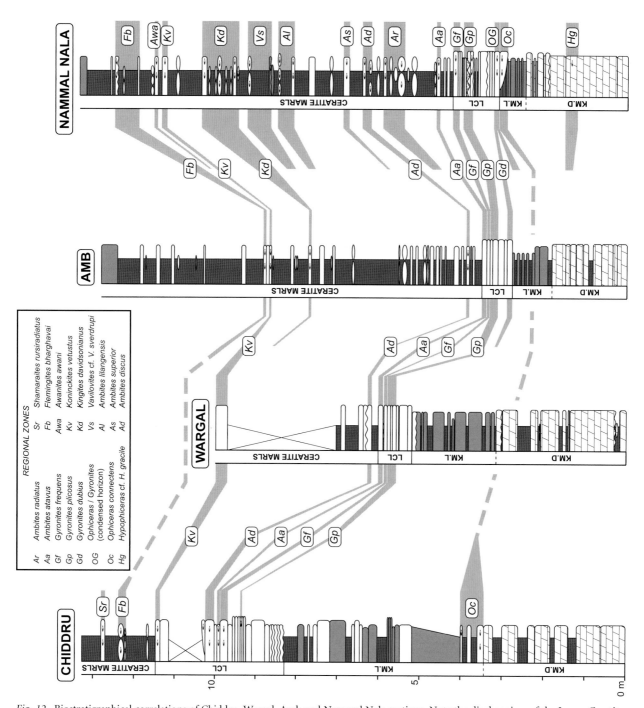

Fig. 12. Biostratigraphical correlations of Chiddru, Wargal, Amb and Nammal Nala sections. Note the diachronism of the Lower Ceratite Limestone between the different sections.

which his correlation with Arctic Canada is based, is in our opinion not conspecific with the species originally described by Tozer (1961) in Arctic Canada.

Moreover, several species clearly belonging to the genus *Ambites* occur in British Columbia, indicating that the Candidus Zone of this region is middle Dienerian. However, Tozer assigned only one species (*Ambites ferruginus*) from Arctic Canada to *Ambites*. The latter differs from all other *Ambites* species by its involute coiling and unusually thick trapezoidal whorl section. Hence, the correlation of the Candidus Zone of Arctic Canada with our middle Dienerian cannot be confirmed. The zonation of Tozer (1994) is based on scattered occurrences of these

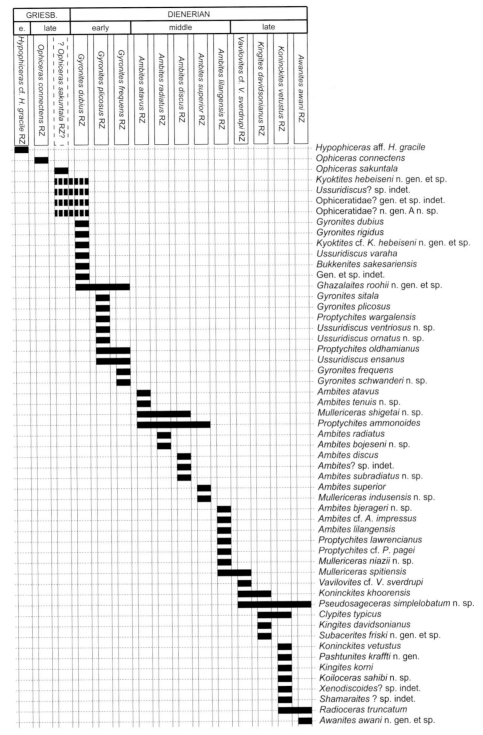

Fig. 13. Range chart showing the biostratigraphical distribution of Griesbachian and Dienerian ammonoid species in the Salt Range. Dashed thick lines correspond to species of uncertain age found in the condensed horizon of Nammal Nala (RZ: Regional Zone).

faunas, often without superpositional information. Additional investigations in these areas would probably allow the construction of a more detailed biozonation and more accurate correlations.

Ambites atavus Regional Zone

Co-occurring species. – {*Ambites atavus* ($n = 74$)}, {*Ambites tenuis* ($n = 5$)}, *Mullericeras shigetai* ($n = 1$), *Proptychites ammonoides* ($n = 1$).

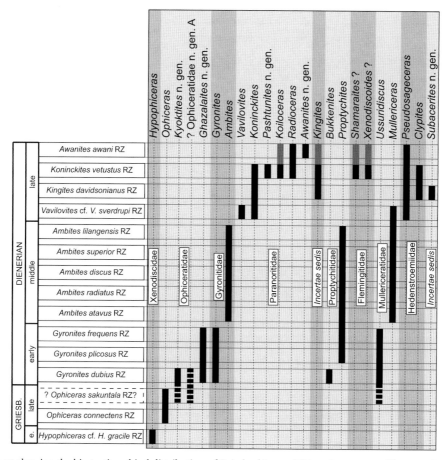

Fig. 14. Range chart showing the biostratigraphical distribution of Griesbachian and Dienerian ammonoid genera (grouped by families) in the Salt Range. Dashed thick lines correspond to genera of uncertain age occurring in the condensed horizon of Nammal Nala. Grey thick lines correspond to virtual occurrences (RZ: Regional Zone).

Occurrence in the investigated sections. – This regional zone has been documented in every studied locality: at the base of the Ceratite Marls in Nammal Nala, in the topmost bed of the Lower Ceratite Limestone in Amb and Wargal, and in a bed in the middle of the Lower Ceratite Limestone in Chiddru.

Correlation. – *Ambites atavus* has so far only been recorded from the Salt Range. This regional zone corresponds to the horizon MH-D3 in Romano *et al.* (2013) and to UA-zone DI-4 in Ware *et al.* (2015).

Ambites radiatus Regional Zone

Co-occurring species. – {*Ambites bojeseni* (*n* = 18)}, {*Ambites radiatus* (*n* = 104)}, *Mullericeras shigetai* (*n* = 3), *Proptychites ammonoides* (*n* = 2).

Occurrence in the investigated sections. – This regional zone has only been recognized in Nammal Nala, in a group of nodular beds about 1 m above the base of the Ceratite Marls.

Correlation. – *Ambites radiatus* also occurs in Jinya in north-western Guangxi, where it was originally described by Brühwiler *et al.* (2008), extending this fauna to the South China block. This regional zone corresponds to the horizon MH-D4 by Romano *et al.* (2013) and to UA-zone DI-5 of Ware *et al.* (2015).

Ambites discus Regional Zone

Co-occurring species. – {*Ambites discus* (*n* = 163)}, {*Ambites subradiatus* (*n* = 8)}, *Mullericeras shigetai* (*n* = 5), *Proptychites ammonoides* (*n* = 34).

Occurrence in the investigated sections. – This regional zone has been recorded in every studied locality: in a group of beds about 2 m above the base of the Ceratite Marls in Nammal Nala, at the base of the Ceratite Marls in Amb and Wargal, and in a bed in the middle of the Lower Ceratite Limestone in Chiddru.

Correlation. – Some species described by Diener (1897) and von Krafft and Diener (1909) from several localities in the Indian Himalayas are here synonymized with *Ambites discus*, thus enlarging the distribution of this regional zone to this region. Some specimens described by Wang & He (1976) and Mu *et al.* (2007) could possibly belong to *Ambites discus*, thus suggesting the presence of the *Ambites discus* Zone in Tibet and north-western Guangxi, respectively. One species from the Candelaria Hills (Nevada: *Ambites* aff. *A. radiatus*; Ware *et al.* 2011) is here synonymized with *Ambites subradiatus*, thus indicating the correlation of the *Ambites discus* Zone with the Candelaria Formation in Nevada. The *Ambites discus* regional Zone corresponds to the horizon MH-D5 in Romano *et al.* (2013) and to UA-zone DI-6 in Ware *et al.* (2015).

Ambites superior Regional Zone

Co-occurring species. – {*Ambites superior* (*n* = 48)}, {*Mullericeras indusense* (*n* = 6)}, *Proptychites ammonoides* (*n* = 4).

Occurrence in the investigated sections. – This regional zone has only been recognized in Nammal Nala area, in a bed about 3 m above the base of the Ceratite Marls.

Correlation. – One specimen described by von Krafft and Diener (1909) may belong to *Ambites superior*, indicating that this regional zone may also expand to the Indian Himalayas. This regional zone corresponds to the horizon MH-D6 in Romano *et al.* (2013) and to UA-zone DI-7 in Ware *et al.* (2015).

Ambites lilangensis Regional Zone

Co-occurring species. – {*Ambites bjerageri* (*n* = 32)}, {*Ambites lilangensis* (*n* = 64)}, {*Mullericeras niazii* (*n* = 11)}, *Mullericeras spitiense* (*n* = 2), {*Proptychites lawrencianus* (*n* = 35)}, {*Proptychites* cf. *P. pagei* (*n* = 5)}.

Occurrence in the investigated sections. – This regional zone has only been identified in Nammal Nala, in a bed about 4 m above the base of the Ceratite Marls.

Correlation. – One species, originally ascribed to *Gyronites frequens* by Brühwiler *et al.* (2008) and here synonymized with *Ambites bjerageri* n. sp., occurs in Shanggan in north-west Guangxi, indicating the correlation of this regional zone to South China. *Ambites lilangensis* and its synonyms have been found in various localities worldwide: in Spiti Valley in India (von Krafft & Diener 1909), in British

Columbia (Tozer 1994), in Nepal (Waterhouse 1996) and in Nevada (Ware *et al.* 2011). Some specimens described by Wang & He (1976) may also belong to *Ambites lilangensis*, indicating that this regional zone may also be present in Tibet. This regional zone corresponds to the horizon MH-D7 in Romano *et al.* (2013) and to UA-zone DI-8 in Ware *et al.* (2015).

Late Dienerian ammonoid faunas

Upper Dienerian ammonoids occur in the upper part of the lower third of the Ceratite Marls in Nammal, Amb and Wargal, and in the second half of the Lower Ceratite Limestone in Chiddru. The upper Dienerian substage is here defined by the first occurrence of Paranoritidae, as exemplified by the typically Dienerian genera *Koninckites*, *Vavilovites* and *Awanites*. Sagerataceae also appear in this interval, and the rare genus *Clypites* has so far only been documented in this time interval. The upper Dienerian substage can easily be identified in Spiti, where Paranoritidae occur in abundance in the upper part of the 'Ambites beds' of Brühwiler *et al.* (2010a), at the top of the 'Fuchsites markhami beds', in the 'Kingites lens beds' and in the 'Prionolobus rotundatus beds'. A species described as *Proptychites* sp. indet. by Brühwiler *et al.* (2008) from Yuping section is here synonymized with *Koninckites khoorensis*, thus indicating the presence of the base of the upper Dienerian in the 'Proptychites candidus beds' of north-western Guangxi. In the same area, the '*Clypites* sp. indet. beds' *sensu* Brayard & Bucher (2008) from Waili section are also most likely late Dienerian in age as this genus is so far only known in this interval. In Primorye (Shigeta & Zakharov 2009), one species originally ascribed to *Clypeoceras spitiense* is here synonymized with *Clypites typicus*, a Sagerataceae which is so far only known from the late Dienerian. *Ambitoides orientalis*, with which it co-occurs, has a suture line typical of Paranoritidae, so the *Clypeoceras timorense* zone of Primorye is most likely upper Dienerian. The genus *Vavilovites* is here considered as a representative of Paranoritidae, with a typical range in the upper Dienerian. Therefore, the *Vavilovites sverdrupi* Zone of Tozer (1994) correlates at least in part with the upper Dienerian substage. However, correlation of the three subzones distinguished in the *Vavilovites sverdrupi* Zone is difficult. The first subzone is based on a single characteristic species (*Koninckites dimidiatus* Tozer 1994) for which no equivalent is known anywhere else. The second subzone contains the type species of *Vavilovites*, and thus clearly belongs to the upper

Dienerian. The third one is characterized by two species belonging to genera which do not occur in the Salt Range: *Heibergites heibergensis* and '*Kingites*' *discoidalis*. The assignment of the latter to *Kingites* is here rejected (see taxonomic descriptions) and no correlations can be made on the basis of these two taxa. Tozer (1994, p. 24) also describes some species as being derived 'from beds in the Toad Formation that closely follow Subzone 2', among which *Flemingites reticulatus* and *Xenodiscoides scapulatus* are undoubtedly of Smithian affinity. This fauna may thus correlate with the lowermost Smithian as defined by Brühwiler *et al.* (2010b). In Eastern Verkhoyansk (Siberia; Dagys & Ermakova 1996), the Dienerian Stage is divided into three zones. The first one is a clear correlative of the *Vavilovites sverdrupi* Zone, and is thus upper Dienerian. The second zone is subdivided into two subzones characterized by *Vavilovites subtriangularis* and *Vavilovites umbonatus*, respectively. The assignment of these two species to the genus *Vavilovites* is uncertain because of their different whorl sections. Other co-occurring species are also referred to as *Vavilovites* and are again restricted to this region. Therefore, no robust correlations can be established between the Subtriangularis Zone of Eastern Verkhoyansk and the Salt Range. The third zone is characterized by *Kingites*? *korostolevi*, a species whose assignment to *Kingites* is here rejected (see taxonomic descriptions). It also contains *Sakhaitoides allarense*, *Sakhaitoides verkhoyanicum* and *Episageceras antiquuum*, three species for which no equivalent are known anywhere else, thus making its assignment to the upper Dienerian uncertain.

Vavilovites cf. *V. sverdrupi* Regional Zone

Co-occurring species. – {*Koninckites khoorensis* (*n* = 54), *Mullericeras spitiense* (*n* = 11)}, *Pseudosageceras simplelobatum* (*n* = 1), {*Vavilovites* cf. *V. sverdrupi* (*n* = 5)}.

Occurrence in the investigated sections. – This regional zone has only been recognized in Nammal, in two consecutive beds ca. 4 m above the base of the Ceratite Marls.

Correlation. – The two specimens assigned to *Vavilovites* sp. indet. from floated blocks in the Candelaria Hills (Nevada; Ware *et al.* 2011) are here synonymized with *Vavilovites* cf. *V. sverdrupi*, thus documenting this regional zone in Nevada. This regional zone corresponds to the horizon MH-D8 in Romano *et al.* (2013) and to UA-zone DI-9 of Ware *et al.* (2015).

Remarks. – The two beds of Nammal assigned to this regional zone correspond to two different local maximal horizons (LMH-12 and LMH-13; Fig. 4). The first one contains only thirteen specimens belonging to *Vavilovites* cf. *V. sverdrupi* and *Mullericeras spitiense*. The second local maximal horizon contains 58 specimens, including 54 *Koninckites khoorensis*, three *Vavilovites* cf. *V. sverdrupi* and one *Pseudosageceras simplelobatum*. Because of the small sample size of the first horizon and the scarcity of *Mullericeras spitiense* (the species which distinguishes the first local maximal horizon from the second one), these two local maximal horizons are lumped together.

Kingites davidsonianus Regional Zone

Co-occurring species. – {*Clypites typicus* (*n* = 11), *Koninckites khoorensis* (*n* = 776)}, {*Kingites davidsonianus* (*n* = 75)}, *Pseudosageceras simplelobatum* (*n* = 50), {*Subacerites friski* (*n* = 1)}.

Occurrence in the investigated sections. – This regional zone occurs in three successive beds between 5 and 6 m above the base of the Ceratite Marls in Nammal. It has also been identified in Amb, in a bed about 4 m above the top of Lower Ceratite Limestone.

Correlation. – In Spiti valley, the *Kingites lens* beds (Brühwiler *et al.* 2010a) are an exact correlative of this regional zone. This regional zone corresponds to the horizon MH-D9 in Romano *et al.* (2013) and to UA-zone DI-10 of Ware *et al.* (2015).

Remarks. – In Chiddru, a single isolated specimen of *Koninckites khoorensis* has been found ca. 1 m below the top of the Lower Ceratite Limestone (LMH-6, Fig. 6), without further associated species. It may belong to this regional zone or to the previous one.

Koninckites vetustus Regional Zone

Co-occurring species. – *Clypites typicus* (*n* = 11), *Kingites korni* (*n* = 14), {*Koiloceras sahibi* (*n* = 7)}, {*Koninckites vetustus* (*n* = 189)}, {*Pashtunites kraffti* (*n* = 11)}, *Pseudosageceras simplelobatum* (*n* = 16), *Radioceras truncatum* (*n* = 9).

Occurrence in the investigated sections. – The *Koninckites vetustus* Zone has been identified in every studied locality: in Nammal Nala ca. 7 m above the base of the Ceratite Marls, in Amb ca. 5 m above the base of the Ceratite Marls, in Wargal ca. 4 m above the base of the Ceratite Marls, and in Chiddru at the top of the Lower Ceratite Limestone.

Correlation. – In Spiti, the *Prionolobus rotundatus* bed (Brühwiler *et al.* 2010a) is identical to the *Koninckites vetustus* regional Zone. This regional zone corresponds to the horizon MH-D10 in Romano *et al.* (2013) and to UA-zone DI-11 of Ware *et al.* (2015).

Awanites awani Regional Zone

Co-occurring species. – {*Awanites awani* (n = 22)}, *Kingites korni* (n = 0, virtual occurrence only), *Pseudosageceras simplelobatum* (n = 2), *Radioceras truncatum* (n = 1).

Occurrence in the investigated sections. – The *Awanites awani* Zone has only been identified in Nammal Nala, in a bed situated just above the *Koninckites vetustus* bed.

Correlation. – This regional zone corresponds to the horizon MH-D11 in Romano *et al.* (2013) and to UA-zone DI-12 in Ware *et al.* (2015).

Remarks. – No specimen of *Kingites korni* could be recovered from these beds. However, it is present in the underlying zone, and some specimens described by Brühwiler *et al.* (2012) in the early Smithian of the Salt Range are here synonymized with this species. It is therefore considered as virtually present (i.e. present in the underlying and overlying beds) in this regional zone.

Conclusion

Abundant and well preserved Dienerian ammonoid faunas were sampled in detailed bedrock controlled Lower Triassic successions in Nammal Nala, Chiddru, Amb and Wargal in the Salt Range. This led to a thorough revision of the taxonomy of Dienerian ammonoids, with emphasis on their intraspecific variability and ontogeny. Most Dienerian ammonoid species were previously insufficiently known because of insufficient sample size and/or poor preservation. Griesbachian ammonoids are much rarer in the Salt Range, and will need further investigation in other areas of the Tethys to refine their taxonomy and biostratigraphy.

This new data enabled us to construct a detailed biostratigraphical scheme for the Dienerian of the Salt Range, with a total of 12 successive ammonoid zones. Today, this is the most comprehensive Dienerian ammonoid record known worldwide, confirming that the Salt Range represents a key area for the study of the Early Triassic biotic recovery and its ammonoid zonation. Detailed studies such as this one are a necessary prerequisite for analyses of ammonoid diversity dynamics and phylogeny.

Systematic palaeontology

By David Ware *and* Hugo Bucher

Classification

The suprageneric classification used here is mostly based on the classification established by Spath (1934), with some modifications which are discussed in the text. Our classification diverges from more recent ones (e.g. Tozer 1994; Dagys & Ermakova 1996; Shigeta & Zakharov 2009), in which Gyronitidae are treated as a synonym of Meekoceratidae and Paranoritidae as a synonym of Proptychitidae, two points which are rejected here.

Waterhouse (1994) provided a completely different classification which was exclusively based on the suture lines. However, his approach did not take into account the intraspecific variability of the suture line, and as a consequence, extreme variants belonging to a single species were assigned to different families in his classification. According to the very narrow definitions of his classification and considering the asymmetry of suture lines (here illustrated by some specimens for which the suture line could be drawn from both sides), most of our specimens would be placed in different families depending on which side of the specimen the suture line is obtained from. Hence, the classification of Waterhouse (1994) is not further considered here.

Our classification is essentially based on abundant material, and as most taxa considered here were initially described by Waagen (1895) and Spath (1934) from the Salt Range, emended diagnoses can be provided at both genus and species levels. At the family level, emended diagnoses are given only for two families (Gyronitidae and Mullericeratidae), whose respective ranges are restricted to the Dienerian. For other families, a more comprehensive study including taxa from other regions and time intervals would be necessary to establish emended diagnoses.

The population approach: intraspecific variability and convergences

As already underlined by many authors (e.g. Tozer 1994; Brayard & Bucher 2008; Monnet *et al.* 2010), ammonoid species usually exhibit a wide range of intraspecific variability, so the taxonomy of ammonoids must be based on a population approach. This variability mostly follows the 'Buckman's Law of Covariation' of Westerman (1966), ranging from involute, compressed shells with subdued ornamentation and a highly frilled suture line to more evolute, depressed, coarsely ornamented shells with a simpler suture line. This is particularly important for Dienerian ammonoids. They have a very low morphological disparity (Brosse *et al.* 2013), and as already noticed by Brayard & Bucher (2008) for Smithian ammonoids and Monnet & Bucher (2005) for Anisian ammonoids, convergences of shell shapes of compressed end-member variants from different lineages are very frequent. Such convergences can only be detected by the analysis of large samples within a highly resolved time frame. As differences between compressed end-members of different taxa can be small, and intraspecific variability is generally large, it is necessary to work with large populations to decipher which characteristics are of intra- or interspecific significance. Several clear examples of convergences are here given. For example, the convergence between *Koninckites khoorensis* and Proptychitidae, between some very evolute species of *Ambites* and smooth forms of *Gyronites*, or between involute tabulate Paranoritidae and Mullericeratidae. In these cases, differences between shell shapes of different taxa are very small, and not always visible on each specimen of the population. For example, *Ambites bjerageri* n. sp. has a sub-serpenticonic shape and a tabulate venter, thus superficially resembling *Gyronites*, but some individuals have a clear bottleneck shaped venter and weak spiral ridges on their flanks, two characteristics which are typical of the genus *Ambites*. These two characteristics are however not present on every specimen, so the affinities of this species can only be deciphered by working at the population level. Moreover, some of these characteristic traits can be detected only on well preserved specimens. Thus, good preservation is also a prerequisite to establish a reliable taxonomy, as even slight weathering, corrosion or distortion of the specimen can be sufficient to alter diagnostic traits.

Ideally, the intraspecific variability should be quantified for every trait of the shell. Considering that ornamentation and whorl section are the result of a highly integrated growth module whose accurate shape is difficult to quantify (e.g. Urdy *et al.* 2010a, b; Erlich *et al.* 2016), only the four classic geometrical parameters of the ammonoid shell are here statistically treated: the diameter (D), whorl height (H), whorl width (W) and umbilical diameter (U), and the corresponding ratios of the latter three parameters with the diameter (H/D, W/D and U/D). Providing that these parameters are available for at least five specimens within each species, H, W, U and the corresponding ratios are first plotted against the diameter to illustrate variability and potential allometric growth. These plots allow detecting phases of allometric growths, as well as changes in allometric rates. Normality of the three relative parameters within the population is then tested for the growth phase being the closest to isometric growth (i.e., usually excluding the juveniles, which are always more evolute than adults, and in some cases the largest specimens, when they show a clear umbilical egression). Providing that at least 30 specimens (excluding juveniles and forms with an umbilical egression at maturity) were available within a species, normality was tested using the Lilliefors (1967) test: the label 'normal' on each histogram indicates that the test cannot reject the null hypothesis of normality at a confidence interval of 95%. They are otherwise labelled 'not normal'. To compare different closely related species, boxplots (e.g. Monnet & Bucher 2005) for each of these three relative parameters are provided. These display the 25th, 50th (median) and 75th percentiles of the range of measures covered by 99% of the specimens from a normally distributed sample. Outliers represent specimens not falling within the normal distribution. Histograms and boxplots were calculated only on a certain range of size corresponding to the phase of growth the closest to isometry, excluding juveniles and forms with a strong umbilical egression at maturity, to avoid problems of growth allometries.

Suture lines

As a consequence of the comparatively low morphological disparity of Dienerian ammonoids, most authors focused on the suture lines to establish their taxonomy and phylogeny. The most extreme example is Waterhouse (1994), who established a completely new taxonomy for Triassic ammonoids exclusively based on suture lines and underplayed the characters of the shell. The problem is that the morphology of the suture line is not an independent trait, as it covaries with the shape of the whorl section (as already noticed by Spath 1919a). Within the frame of intraspecific variation, involute and compressed variants have more elements and more indentations on the lobes and auxiliary series than evolute and depressed variants. This covariation is most likely the result of the morphogenesis of the septum, and has been tentatively explained by the means of viscous fingering model by Checa &

Garcia-Ruiz (1996, p. 272) through a domain effect, i.e. the number of elements depends on the 'distance between opposite elements of the septal suture'. This can be illustrated in our material by the development of the auxiliary series. In involute and compressed forms, this distance is very short. It corresponds to the distance between the flank of the whorl and the flank of the overlapped part of the preceding whorl, which imparts a long auxiliary series with differentiated elements. In similarly involute but more depressed variants, the auxiliary series is also long but with less differentiated elements. In evolute shells, the overlap with the preceding whorl is much smaller, resulting in an auxiliary series which is very short with less differentiated elements. As a consequence, the presence or absence of an auxiliary lobe is not necessarily of taxonomic significance, as it is partly the product of covariation with the shell shape. Hence, it may appear or disappear during ontogeny as a result of allometric growth. This is especially important in our case, as many authors have considered this characteristic as extremely important. For example, Tozer (1994) divided Dienerian ammonoids into two groups mostly based on this characteristic: Meekoceratidae, without auxiliary lobe, and Proptychitidae, with an auxiliary lobe. He therefore considered Paranoritidae as a synonym of Proptychitidae, and Gyronitidae as a synonym of Meekoceratidae, a view we do not endorse because an auxiliary lobe can appear or disappear within a clade without having any phylogenetic significance.

Another problem with suture lines is that their variability has never been taken into account. The suture line has often been analysed in a very typological way. However, as seen previously, the pattern of the suture line covaries with the shape of the whorl section. It is of course difficult to quantify this variability, but here we tried to give an overview of it by drawing, when possible, suture lines of different variants (e.g. involute and evolute, with a vertical and with an oblique umbilical wall, etc.). This showed that the suture line is, within a single population, highly variable, especially the external dorsal half (i.e. the auxiliary series and third lateral saddle). We also noticed a general asymmetry of the suture line. Here, three complete external suture lines have been drawn as example, and all of them are strongly asymmetrical, especially concerning the auxiliary series and the ventral lobe, two parameters which are usually considered as of primary importance for ammonoid classification. These specimens do not show any asymmetry in their shell shape, so they cannot be considered as pathological. In *Koninckites vetustus* (Pl. 15, fig. 20) and in *Proptychites lawrencianus* (Pl. 23, fig. 4), an auxiliary lobe is clearly individualized on the right side, but not on the left side. As a consequence, according to Tozer's classification, their right side is typical of Proptychitidae, while their left side is typical of Meekoceratidae. Another example is given for a specimen of *Clypites typicus* (Pl. 31, fig. 18), which has a clearly individualized adventitious saddle on the right side, not on the left side. Yacobucci & Manship (2011) noticed a similarly strong asymmetry of the suture line in Cretaceous ammonites and Carboniferous goniatites, and interpreted this as a possible consequence of soft part asymmetry. Actually, this asymmetry is better explained as being the product of a septum plane which is not exactly perpendicular to the coiling plane (i.e. a line connecting homologous elements on both sides of the shell is not perpendicular to the coiling plane). The intraspecific variability and asymmetry of suture lines were already noticed by Spath (1919b), who concluded that classification of ammonoids could not be solely based on the characteristics of the suture lines. These examples show that the shape of the suture line cannot always be directly treated as a character of higher taxonomic significance. It must be treated with the same weight, including its intraspecific variability and a certain plasticity, as all other traits of the shell.

In the present work, all species have a ceratitic suture line. To facilitate comparison between the different suture lines illustrated, those from the right side were mirrored to match the orientation of those from the left side.

Ontogeny and growth allometry

Early Triassic ammonoids generally show important growth allometries, with significant morphological changes between juvenile and adult stages (e.g. Brayard & Bucher 2008; Brühwiler *et al.* 2012). As a consequence, some authors placed juvenile specimens in different taxa (even sometimes different families) than the adult ones. For example, juveniles of *Koninckites khoorensis* (here considered as belonging to Paranoritidae) were placed in the genus *Dinarites* (family Dinaritidae) by Waagen (1895), and in the new genus *Prejuvenites* (family Melagathiceratidae) by Waterhouse (1996; see description of *Koninckites khoorensis* below). Numerous juvenile specimens were found, and in order to associate these with the corresponding adults, the ontogeny of some key taxa is here studied, providing that enough specimens were available. These growth allometries were quantified by cutting some specimens perpendicularly to the coiling plane, and grinding them until the middle of the protoconch was reached.

These sections allow the measurement of the four geometric parameters of ammonoids previously mentioned every half whorl, and thus to construct the growth trajectories of these specimens and to plot them on the same graphs as the rest of the population. The change in whorl section shape (appearance/disappearance of the tabulate venter, modification of the umbilical wall or change of convexity of the flanks) could also be analysed visually on these polished sections.

Ten specimens were cut, including *Ambites discus* (*n* = 1), *Ambites superior* (*n* = 1), *Koninckites khoorensis* (*n* = 3), *Koninckites Vetustus* (*n* = 3), *Proptychites lawrencianus* (*n* = 1), *Kingites davidsonianus* (*n* = 1) and *Pseudosageceras simplelobatum* (*n* = 1). They all follow the same four stages of growth: (1) the neanic stage, until a diameter of 5–10 mm, where the shell becomes more evolute; (2) the juvenile stage, where the shell becomes more involute and compressed, until a diameter of 20–30 mm; (3) the submature stage, where the shell almost keeps the same proportions or becomes more evolute; (4) the adult stage, where umbilical egression finally occurs. This last stage is not present in every species and could not be clearly seen on the growth trajectories as, first, the specimens cut did not have their complete body chamber, and second because this umbilical egression often occurs very rapidly, in less than half a whorl, and thus the resolution obtained from sections is not sufficient to capture this change. In fact, it was only detected by visual examination of the complete specimens. The study of these growth allometries is very important, as the evolution of the different lineages are largely the result of developmental heterochronies on forms with a strong allometric growth, for example a minor change in the slope of allometry of the submature stage or of the diameter at which a specimen changes growth stage can result in major differences in adult morphology. It also allows a better understanding of the intraspecific variability within each species, which is mostly the result of different allometric slopes and/or of change of growth stage. It also explains the larger intraspecific variability of juveniles compared with adult forms, already noticed by some authors (e.g. Monnet *et al.* 2012). A good example is given here with the growth trajectories of *Koninckites vetustus* (Fig. 36). In this species, the submature stage is reached at a smaller diameter in evolute variants than in involute variants. During the submature stage, shell proportions of evolute variants remain approximately the same or become slowly more involute, while involute variants become slowly more evolute. These convergences of ontogenetic trajectories lead to more homogenous morphologies towards maturity.

Systematic descriptions

Synonymy lists and taxa in open nomenclature are annotated following the recommendations of Matthews (1973) and Bengtson (1988). The proportions of the shells are quantified using the four classic geometrical parameters of the ammonoid shell: diameter (D), whorl height (H), whorl width (W) and umbilical diameter (U). Locality numbers are reported on measured sections (Figs 4, 6, 8, 10).

Repository. – Figured and measured specimens are abbreviated PIMUZ (Paläontologisches Institut und Museum der Universität Zürich) and GSI (Geological Survey of India, Calcutta).

Class Cephalopoda Cuvier, 1797

Subclass Ammonoidea Agassiz, 1847

Order Ceratitida Hyatt, 1884

Superfamily Xenodiscaceae Frech, 1902

Family Xenodiscidae Frech, 1902

Genus *Hypophiceras* Trümpy, 1969

Type species. – *Glyptophiceras triviale* Spath, 1935.

***Hypophiceras* aff. *H. gracile* (Spath, 1930)**

Plate 1, figures 1–3

? 1930 *Glyptophiceras gracile* n. gen., n. sp. Spath, pp. 34–36, pl. 7, figs 3, 4, 5 (holotype), 6, pl. 8, fig. 10.

? 1970 *Glyptophiceras* sp. indet. Kummel, pl. 1, fig. 1.

Material. – One specimen.

Description. – Very evolute sub-serpenticonic shell with weakly compressed whorl section. Flanks with maximal width at the umbilical shoulder, slightly converging towards the broadly rounded venter, without forming any ventro-lateral shoulder. Umbilical wall high, slightly oblique, delimited by a narrowly rounded umbilical shoulder. Flanks ornamented by distant thick blunt radial ribs in the

Table 1. Measurements of Induan ammonoids from the Salt Range for which less than 5 specimens were measurable.

Genus	Species	Specimen number	Section	Age	D	H	W	U
Ambites	cf. *impressus*	PIMUZ30349	Nammal	*Al*	36.6	14.2	9.4	11.6
Ambites	*tenuis*	PIMUZ30309	Amb	*Aa*	28.6	10.2	5.6	11.1
Ambites	*tenuis*	PIMUZ30308	Amb	*Aa*	39.9	14.1	7.2	15.9
Ambites?	sp. indet.	PIMUZ30326	Nammal	*Ad*	38.7	18.8	8.7	7.1
Ambites?	sp. indet.	PIMUZ30327	Nammal	*Ad*	37.3	20.3	8.1	5
Ambites?	sp. indet.	PIMUZ30325	Nammal	*Ad*	41	21.5	8	5.8
Ambites?	sp. indet.	PIMUZ30328	Nammal	*Ad*	24	12.6	5.6	3.6
Ambites?	sp. indet.	PIMUZ30329	Nammal	*Ad*	27	14.8	6.7	3.5
Gen. Indet.	sp. indet.	PIMUZ30511	Amb	*Gd*	13.7	8.3	3.4	1
Gyronites	*schwanderi*	PIMUZ30279	Wargal	*Gd*	41.4	15	10.8	16.3
Gyronites	*schwanderi*	PIMUZ30280	Wargal	*Gf*	49.4	18	10.7	18.4
Gyronites	*schwanderi*	PIMUZ30281	Nammal	*Gf*	66.5	19.3	13.8	31
Gyronites	*schwanderi*	PIMUZ30282	Nammal	*Gf*	70.8	25.1	15.2	29
Gyronites	*rigidus*	PIMUZ30267	Amb	*Gd*	19	6.8	4.4	7.6
Gyronites	*sitala*	PIMUZ30277	Nammal	*Gp*	33.6	10.4	8.4	14.5
Gyronites	*sitala*	PIMUZ30278	Nammal	*Gp*	41.2	11.1	9	21
Hypophiceras	aff. *gracile*	PIMUZ30236	Nammal	*Hg*	30.6	9.3	7.7	13.6
Kyoktites	cf. *hebeiseni*	PIMUZ30245	Amb	*Gd*	40	17.5	8.3	9.4
Kyoktites	*hebeiseni*	PIMUZ30244	Nammal	OG	30.5	14.6	6.8	6.2
Kyoktites	*hebeiseni*	PIMUZ30243	Nammal	OG	49.9	23.9	11.6	10.2
Mullericeras	*indusense*	PIMUZ30461	Nammal	*As*	33.4	16.7	7.3	6.1
Mullericeras	*indusense*	PIMUZ30463	Nammal	*As*	33.9	16.7	7.4	6.7
Mullericeras	*indusense*	PIMUZ30459	Nammal	*As*	35.7	17.5	7.5	7
Mullericeras	*indusense*	PIMUZ30460	Nammal	*As*	44	21.4	9.6	8.4
Mullericeras	*spitiense*	PIMUZ30452	Nammal	*Al*	35	21.1	9.2	0
?Ophiceratidae gen. et sp. indet.		PIMUZ30242	Nammal	OG	18.8	10	4.8	2.5
Ophiceras	*sakuntala*	PIMUZ30241	Nammal	OG	41.9	16.3	9.2	12.2
Ophiceras	*sakuntala*	PIMUZ30240	Nammal	OG	58.1	22	13.6	16.4
Ophiceras	*connectens*	PIMUZ30237	Chiddru	*Oc*	43.9	16.1	11.1	14.2
Ophiceras	*connectens*	PIMUZ30238	Nammal	*Oc*	45.6	16.3	10.5	16.4
Ophiceras	*connectens*	PIMUZ30239	Nammal	*Oc*	63.9	22.8	15.2	24.9
Proptychites	cf. *pagei*	PIMUZ30450	Nammal	*Al*	55.6	35.4	15	0
Proptychites	cf. *pagei*	PIMUZ30451	Nammal	*Al*	83.2	52.9	19.9	0
Proptychites	cf. *pagei*	PIMUZ30448	Nammal	*Al*	100.9	63.4	30.2	0
Proptychites	cf. *pagei*	PIMUZ30449	Nammal	*Al*	90.1	57.8	26.8	0
Proptychites	*oldhamianus*	PIMUZ30440	Nammal	*Gp*	86.6	44	31	16.7
Proptychites	*oldhamianus*	PIMUZ30441	Nammal	*Gp*	39.4	19.2	14.5	7.8
Proptychites	*oldhamianus*	PIMUZ30439	Nammal	*Gf*	66.9	36.1	24	9.6
Proptychites	*wargalensis*	PIMUZ30438	Wargal	*Gp*	53.4	25.5	16.6	11.6
Shamaraites?	sp. indet.	PIMUZ30426	Wargal	*Kv*	12.5	4.7	3.3	5
Subacerites	*friski*	PIMUZ30510	Amb	*Kd*	40.4	24.8	9	0
Ussuridiscus	*ornatus*	PIMUZ30481	Nammal	*Gp*	33.8	17.1	7.2	4.4
Ussuridiscus	*ornatus*	PIMUZ30479	Wargal	*Gp*	26	13.8	5.6	2.5
Ussuridiscus	*ornatus*	PIMUZ30480	Wargal	*Gp*	40.7	21.2	9.1	4.2
Ussuridiscus	*ventriosus*	PIMUZ30482	Nammal	*Gp*	39.4	21.4	10.3	4.1
Ussuridiscus?	sp. indet.	PIMUZ30484	Nammal	OG	19	11.8	4.3	1.2
Ussuridiscus?	sp. indet.	PIMUZ30483	Nammal	OG	29.4	16.1	5.6	2.4
Vavilovites	cf. *sverdrupi*	PIMUZ30357	Nammal	*Vs*	39.5	19.8	12.1	7.1
Vavilovites	cf. *sverdrupi*	PIMUZ30359	Nammal	*Vs*	45.7	23.4	12.8	6.7
Vavilovites	cf. *sverdrupi*	PIMUZ30358	Nammal	*Vs*	60	31.8	16.3	8.4
Xenodiscoides?	sp. indet.	PIMUZ30425	Chiddru	*Kv*	22.5	8.9	5.8	7.7

D, Diameter; H, whorl height; W, whorl width; U, umbilical diameter.
Age abbreviations as in Figure 12.

inner whorls, becoming progressively thinner and denser on the last whorl. Suture line not preserved.

Measurements. – See Table 1.

Discussion. – This specimen differs from the specimens originally described by Spath (1930) by its less evolute coiling. It is otherwise very close to this species. As we have only one specimen available, a definitive species assignment is not possible.

Griesbachian Xenodiscidae are in need of a revision, as these forms have been split into different genera and species without taking intraspecific variability and covariation into account. The specimen figured by Kummel (1970), which also comes from the dolomitic unit of the Kathwai member, seems very close to ours, but is too incomplete to be firmly identified. In the Boreal Realm, the genus *Hypophiceras* occurs only in the lower Griesbachian, so this single specimen and that of Kummel (1970) would suggest that this bed is equally early Griesbachian in age. However, Griesbachian Xenodiscidae are very poorly known in the Tethys, and an in-depth study of the Tethyan Griesbachian ammonoids is needed prior giving a firm conclusion about the age of this specimen.

Occurrence. – Nammal Nala, sample Nam361.

Superfamily Meekocerataceae Waagen, 1895
Family Ophiceratidae Arthaber, 1911

Genus *Ophiceras* Griesbach, 1880

Type species. – *Ophiceras tibeticum* Griesbach, 1880.

Ophiceras connectens Schindewolf, 1954

Plate 1, figures 4–13

1954 *Ophiceras connectens* n. sp. Schindewolf, pp. 178–180, fig. 4, pl. 6, fig. 4 (holotype).

1970 *Ophiceras connectens* Schindewolf, 1954; Kummel, pp. 189–191, fig. 2B–D, pl. 1, figs 2 (holotype), 3–9.

Material. – 15 specimens.

Description. – Evolute (U/D ≈ 36%) *Ophiceras* with a slightly compressed whorl section (W/D ≈ 24%, W/H ≈ 67%) and a broadly rounded venter without any trace of ventro-lateral shoulders. Flanks nearly flat with maximal thickness at the umbilical shoulder, slowly converging towards the venter. Umbilical wall oblique, individualized by a rounded shoulder. No ornamentation visible. Suture line poorly preserved, with rather elongated saddles having a rounded tip and lobes having very fine indentations at their base.

Measurements. – See Table 1.

Discussion. – Our largest specimen is more evolute than the two smaller ones. It is also the case for the holotype, which is larger and more evolute than the other specimens figured by Kummel (1970). These differences are most likely the result of a growth allometry, but not enough well-preserved specimens are available to demonstrate it. The suture line was visible on only one specimen. It has deeper lobes and saddles than

the holotype and other specimens illustrated by Kummel (1970). However, as shown here for other species, the depth of lobes can be subject to substantial intraspecific variability, so this difference is here not considered as important. Kummel (1970) suggested that this species could be a synonym of some species from the Himalayas described by Diener (1897), but a detailed revision of the Himalayan Ophiceratidae would be first necessary. Waterhouse (1994) actually placed *Ophiceras connectens* in synonymy with *Ophiceras tibeticum*, and considered Kummel's specimens to belong to another species, *Lytophiceras chamunda* (Diener 1897). However, we follow Kummel's approach by keeping the species name *Ophiceras connectens* for our specimens, until a revision of the faunas from the central Himalayas becomes available.

Occurrence. – Chiddru, samples Chi54 (*n* = 4) and Chi55 (*n* = 1); Nammal Nala, samples Nam376 (*n* = 6), Nam390 (*n* = 4).

Ophiceras sakuntala Diener, 1897

Plate 1, figures 14–18

1897 *Ophiceras sakuntala* n. sp. Diener, pp. 114–118, pl. 10, figs 1 (lectotype), 2–7; pl. 11, figs 1, 2, 4.

Material. – 3 specimens.

Description. – Rather involute (U/D ≈ 29%) *Ophiceras* with a compressed whorl section (W/D ≈ 23%, W/H ≈ 59%) and a narrowly rounded venter without any trace of ventro-lateral shoulders. Flanks slightly convex with maximal thickness at the umbilical shoulder, strongly converging towards the venter. Umbilical wall very oblique, individualized by an indistinct rounded shoulder. No ornamentation visible. Suture line with relatively short and rounded saddles, lobes with very fine denticules at their base, auxiliary series short with hardly discernible indentations.

Measurements. – See Table 1.

Discussion. – The two specimens described here are in perfect agreement with the specimens originally described by Diener (1897). Considering that only two specimens are here available, no detailed study of their ontogeny and intraspecific variability could be done and hence, no synonymy list is provided. An extended synonymy list for this species was given by Bando (1981), but it was partially contradicted by Tozer (1994). As for the other species of *Ophiceras* described here, a revision of Himalayan Ophiceratidae is necessary.

Occurrence. – Nammal Nala, sample Nam391.

Genus *Kyoktites* n. gen.

Derivation of name. – After the Kyokti Valley (Himachal Pradesh, India), where the holotype of *Meekoceras kyokticum* von Krafft, 1909 comes from.

Type species. – *Kyoktites hebeiseni* n. sp.

Composition of the genus. – *Kyoktites hebeiseni* n. sp. and *Meekoceras kyokticum* von Krafft, 1909.

Diagnosis. – Moderately involute, compressed sub-platyconic Ophiceratidae with a narrowly rounded venter and a shallow umbilicus with a vertical umbilical wall.

Occurrence. – ?Latest Griesbachian to earliest Dienerian of Spiti valley (India) and Salt Range.

Discussion. – This genus is very close to other involute Ophiceratidae genera such as *Discophiceras* or *Ghazalaites*, but it differs by its more platyconic shape with more compressed whorl section and flatter flanks. Waterhouse (1996) created the genus *Khangsaria* in which he included the species *Meekoceras kyokyticum* von Krafft, 1909. However, the type species of this genus, *Khangsaria galteensis* Waterhouse, 1996, is based on material which poor preservation precludes any identification at the species level. Therefore, the genus *Khangsaria* is considered here as a *nomen dubium*.

Kyoktites hebeiseni n. sp.

Plate 2, figures 1–4

Derivation of name. – Named after Markus Hebeisen (PIMUZ).

Holotype. – Specimen PIMUZ30243 (Pl. 2, figs 1–4).

Type locality. – Nammal Nala, Salt Range, Pakistan.

Type horizon. – Sample Nam391, condensed horizon from the base of Lower Ceratite Limestone.

Diagnosis. – Rather involute platyconic *Kyoktites* with sub-parallel flanks.

Material. – One specimen.

Description. – Rather involute (U/D = 20%) compressed (W/D = 23%, W/H = 47%) platyconic shell with a rounded venter. Flanks with maximal thickness at mid-height, the flanks being only very slightly convex or sub-parallel. Umbilical wall low and vertical, bordered by a rounded shoulder. Some very low thin sigmoidal folds occur on the inner half of flanks. Suture line simple, with a narrow and shallow ventral lobe, elongated rounded first and second lateral saddle, a deep first lateral lobe and a shallow second lateral lobe both with a few small indentions at their bases, and a very broad but short sub-rectangular third lateral saddle. Auxiliary series short, not well enough preserved to be described.

Measurements. – See Table 1.

Discussion. – This species is very close to *Kyoktites kyokticum* (von Krafft, 1909), from which it differs by its more parallel flanks. As the two only specimens studied here come from a condensed bed containing both typical Griesbachian and typical Dienerian faunas, its precise age cannot be defined here.

Occurrence. – Nammal Nala, sample Nam391.

Kyoktites cf. *H. hebeiseni* n. sp.

Plate 2, figures 5–8

Material. – Two specimens.

Description. – Moderately involute (U/D = 23.5%) very compressed (W/D = 21%, W/H = 47%) sub-platyconic shell with a narrowly rounded venter. Flanks convex with maximal thickness at inner third. Umbilical wall very low and vertical, poorly individualized by a rounded shoulder. No ornamentation visible. Suture line very simple with short rounded lateral saddles. Lateral lobes with very small indentations at their base. Third lateral saddle broader than high and broadly rounded. Auxiliary series short with numerous small indentations.

Measurements. – See Table 1.

Discussion. – This specimen is very close to the holotype of *Kyoktites hebeiseni* n. sp. but differs by its more evolute and compressed whorls, its more convex flanks and its suture line with shallower lobes and shorter saddles, the third lateral saddle being broadly rounded. Considering the small amount of material available for study, it cannot be excluded that these two species could represent variants of a single species.

Occurrence. – Amb, sample Amb104.

Genus *Ghazalaites* n. gen.

Derivation of name. – Named after Ghazala Roohi (Pakistan Museum of Natural History).

Type species. – *Ghazalaites roohii* n. sp. (Pl. 2, figs 12–24).

Composition of the genus. – Type species only.

Diagnosis. – Compressed shell with a narrowly rounded venter and a pronounced allometric growth. Inner whorls are very involute, the umbilicus being almost occluded, and a strong umbilical egression starts on the last whorl of the adult phragmocone. No differentiated umbilical wall. Mature ornamentation consists of broad and proverse wavy

folds on flanks. Suture line characterized by an asymmetrical third lateral saddle merging with the auxiliary series without clear boundary.

Occurrence. – Early Dienerian of the Salt Range.

Discussion. – This genus is close to *Discophiceras* and *Kyoktites*, from which it differs by its more involute inner whorls followed by a stronger umbilical egression, and by its suture line with the characteristic shape of its third lateral saddle. It also differs from *Kyoktites* by its convex flanks and asymmetric third lateral saddle. The resemblance between these three genera suggests a close phylogenetic relationship.

Ghazalaites roohii n. sp.

Plate 2, figures 12–24; Figure 15

v ? 1978   *Lytophiceras* sp. ind. Guex, pl. 1, fig. 4.

Derivation of name. – Named after Ghazala Roohi (Pakistan Museum of Natural History).

Holotype. – Specimen PIMUZ30248 (Pl. 2, figs 15–17).

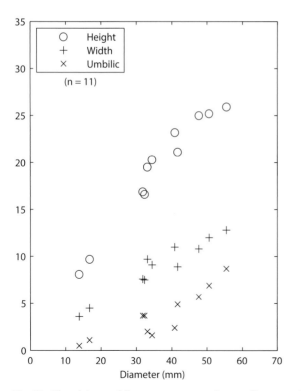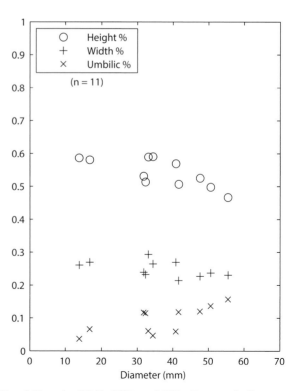

Fig. 15. Ghazalaites roohii n. gen. et n. sp. Scatter diagrams of H, W and U, and of H/D, W/D, and U/D (D = conch diameter, H = whorl height, W = whorl width, U = umbilical diameter).

Type locality. – Nammal Nala, Salt Range, Pakistan.

Type horizon. – Sample Nam332 (middle Lower Ceratite Limestone), *Gyronites plicosus* Regional Zone, lower Dienerian.

Diagnosis. – As for the genus.

Material. – 47 specimens.

Description. – Compressed (W/D ≈ 25%, W/H ≈ 46%) discoidal shell with a rounded venter showing a strong umbilical egression at maturity. Inner whorls are very involute (U/D ≈ 5%), with maximal width at inner quarter of the flanks. From the point of maximal thickness, the flanks converge towards the narrowly rounded venter, forming a slightly convex gentle slope. The flanks bend rapidly towards the umbilicus, joining the preceding whorl at a right angle but without individualizing any umbilical wall. Flanks are smooth. Near maturity, at a point which varies between one whorl before the end of the phragmocone and the beginning of the body chamber, the coiling becomes more evolute. On the largest measured specimen, at a diameter of 55.5 mm, U/D = 16%. At the same stage, the point of maximal thickness shifts towards the inner third of flanks, forming an indistinct shoulder grading into the very oblique umbilical wall. Some low, broad, prorsiradiate sigmoidal folds following the shape of the growth lines appear progressively and become stronger towards the aperture. Suture line with shallow lobes and short saddles. Ventral lobe divided by a shallow ventral saddle into two small lateral branches with three small indentations at their base. Lateral lobes shallow with a few small indentations at their base. First two lateral saddles broad and rounded. Third lateral saddle very characteristic, asymmetrical, without any individualized flank on its dorsal side. The auxiliary series is thus differentiated from the third lateral saddle only by the progressive appearance of small indentations. The auxiliary series is long, with many very small regularly spaced indentations, and becomes shorter when the egression starts.

Measurements. – See Figure 15.

Discussion. – Guex (1978) illustrated a small poorly preserved, broken and distorted specimen coming from the same bed as the ones here studied and with which it is probably conspecific. However, its poor preservation precludes any further comparison.

Occurrence. – Amb, samples Amb2 ($n = 2$) and Amb104 ($n = 26$); Nammal Nala, samples Nam320 ($n = 2$), Nam331 ($n = 1$), Nam332 ($n = 1$), Nam335a ($n = 2$), Nam377 ($n = 5$) and Nam76 ($n = 1$); Wargal, samples War5 ($n = 1$), War6 ($n = 2$), War8 ($n = 1$), War9 ($n = 2$) and War100 ($n = 1$).

Ophiceratidae? n. gen. A n. sp. A

Plate 2, figures 9–11

Material. – One, poorly preserved, specimen.

Description. – Very involute and compressed platyconic shell with broadly rounded venter and parallel flanks. Umbilicus not clearly visible, very narrow. Maximal thickness apparently at outer third of the flanks. Suture line very simple, with extremely shallow lobes and short saddles, lobes having very few small denticules at their base. Third lateral saddle extremely broad and flat. Auxiliary series long, with numerous small indentations.

Measurements. – Specimen too distorted and poorly preserved to be measured. Maximal diameter can be estimated at about 6 cm, U/D ≈ 10%, and W/H ≈ 45%.

Discussion. – The simple suture line and broadly rounded venter indicate that this specimen most likely belongs to Ophiceratidae. However, its very involute platyconic shape and its peculiar suture line are very unusual, so it likely belongs to a new genus and species. This specimen is nevertheless too poorly preserved to be used as a holotype and to provide a detailed diagnosis. Therefore, it is here let in open nomenclature. A better preserved specimen would be necessary to erect a new genus and species for this form.

Occurrence. – Nammal Nala, sample Nam391.

Ophiceratidae? gen. et sp. indet.

Plate 1, figures 19–21

Material. – One, poorly preserved, specimen.

Description. – Very involute (U/D = 13%) discoidal small shell with a narrowly rounded venter. Flanks slightly convex with maximal width at inner third. No ornamentation. Suture line only partially visible, very simple, with shallow lobes and rounded short saddles, lobes having very small denticules at their base, and the third lateral saddle being broader than high.

Measurements. – See Table 1.

Discussion. – This single juvenile specimen, with its simple suture line and its rounded venter, most probably belongs to Ophiceratidae, possibly to one of the two co-occurring ophiceratid species, *Ophiceras sakuntala* or *Kyoktites hebeiseni*. However, considering that only a couple of specimens belonging to these two species have been found, their intraspecific variability and ontogeny could not be studied here. Hence, the taxonomic assignment of this small specimen remains uncertain.

Occurrence. – Nammal Nala, sample Nam391.

Family Gyronitidae Waagen, 1895

Type genus. – *Gyronites* Waagen, 1895.

Composition of the family. – *Gyronites* Waagen, 1895 and *Ambites* Waagen, 1895.

Emended diagnosis. – Compressed shells of variable involution with a tabulate venter delimitated by sharp, angular shoulders which may fade out on adult body chamber. Suture line ceratitic with relatively broad lobes and saddles, lobes having small indentations at their base, saddles being well rounded. Auxiliary series short, generally with just a few small indentations and no auxiliary lobe.

Discussion. – This family, which was emended by Spath (1934), has been considered by most subsequent authors (e.g. Tozer 1994; Shigeta & Zakharov 2009) as a synonym of Meekoceratidae Waagen, 1895; a family which was then considered as having a very broad scope. However, Brühwiler *et al.* (2012) indicated that *Meekoceras*, the type genus of Meekoceratidae, was actually very close to Prionitidae Hyatt, 1900 and that Meekoceratidae is a largely polyphyletic taxon, the superficial and occasional similarity between Gyronitidae and some representatives of Prionitidae being the result of convergence. This opinion is here followed.

After a careful analysis of our faunas, Gyronitidae are here restricted to two genera, and thus appear in the early Dienerian and disappear at the end of the middle Dienerian, and are succeeded by Paranoritidae in the late Dienerian. Actually, considering the synonymy of the different genera included by Spath

(1934) in Gyronitidae, our understanding of this family and his are identical: the genus *Gyrophiceras* is based on a single, obviously pathological specimen which most probably belongs to *Gyronites*; *Gyrolecanites* is based on a species which is here considered as belonging to *Ambites*; the type species of *Prionolobus* (*Prionolobus atavus* Waagen, 1895) is considered as belonging to the genus *Ambites*. Spath also included the genus *Catalecanites* within Gyronitidae, but only with doubts, mentioning that it could also correspond to a simplified offshoot of Flemingitidae. This genus and its type species, based on a fragment illustrated by Diener (1897) has never been used by any subsequent author, and is here considered as a *nomen dubium*. One problem in Spath's concept of this family concerns the genus *Prionolobus*. Although the type species of this genus is here considered as belonging to *Ambites*, the other species included in it by Waagen, as well as the additional species of Spath, all belong to different genera, mostly *Koninckites* and *Paranorites*, which are both (in agreement with Spath) included in Paranoritidae. Hence, Gyronitidae were considered to range up to the early Smithian in previous work, a distribution which must now be shortened to the early and middle Dienerian.

Tozer (1965) based his definition of the Dienerian on the first appearance of Gyronitidae (which he synonymized in later works with Meekoceratidae). In Nammal Nala, a few specimens of *Gyronites dubius* were found associated with the typical Griesbachian species *Ophiceras sakuntala*. However, this association could be the result of condensation (see the biostratigraphy section), so it does not contradict Tozer's definition for the base of the Dienerian. The hypothetical presence of *Gyronites* in the Griesbachian would require confirmation from a section without any sign of condensation.

Genus *Gyronites* Waagen, 1895

Type species. – *Gyronites frequens* Waagen, 1895.

Composition of the genus. – *Gyronites frequens* Waagen, 1895; *Meekoceras dubium* von Krafft, 1909; *Danubites rigidus* Diener, 1897; *Gyronites plicosus* Waagen, 1895; *Danubites sitala* Diener, 1897; *Gyronites subdharmus* Kiparisova, 1961 and *Gyronites schwanderi* n. sp.

Emended diagnosis. – Compressed, evolute, sub-platyconic to serpenticonic small sized Gyronitidae (maximal diameter of ca. 7 cm) with a tabulate venter, becoming occasionally rounded on the adult body chamber. Ornamentation varying from nearly smooth forms to variants with strong blunt radial ribs, which tend to disappear on the adult body chamber. Suture line very simple, lobes with few small indentations, second lateral saddle much larger than the others while the third one is much smaller, auxiliary series very short, sometimes nearly absent, with few and very small indentations.

Occurrence. – Early Dienerian of the Salt Range (Pakistan), Central Himalayas (India, Nepal), South Tibet (China), South Primorye (Russia).

Discussion. – The nearly smooth variants of this genus can be mistaken for evolute species of *Ambites*, from which it differs by its venter, tabulate but not bottleneck shaped, simpler suture line with fewer, smaller denticulations on the lobes and a less developed auxiliary series, and by the absence of spiral ridges on the flanks. Sub-serpenticonic ribbed variants resemble some variants of Griesbachian Xenodiscidae (e.g. *Tompophiceras, Hypophiceras*) and early Smithian Kashmiritidae, but the presence of a clearly tabulate venter in their inner whorls allows differentiating them clearly from these two families. Tozer (1994) created the new genus *Pleurogyronites* based on one specimen from the Dienerian of British Columbia. He considered this specimen as very close to *Gyronites*, differing only by the presence of ribs. According to our emended definition of the genus

Gyronites, the presence or absence of ribs is not of generic significance, and falls within the normal range of intraspecific variability. Moreover, the type species of *Gyronites* was originally described by Waagen (1895) as perfectly smooth, but he based his diagnosis on weathered specimens. Additionally, it is shown here that *Gyronites frequens* varies from nearly smooth forms to forms with strong, blunt radial ribs. It is therefore likely that *Pleurogyronites* is a synonym of *Gyronites*. Waterhouse (1996) created the genus *Wangyikangia* for the species *Gyronites nangaensis* Waagen, 1895; but this species is here shown to be a thinly ribbed variant of *Gyronites frequens* (see below), hence the genus erected by Waterhouse is a junior synonym of *Gyronites*. He also erected the genus *Himoceras* for the species *Meekoceras dubium* von Krafft, 1909, differing from *Gyronites* by its sub-trapezoidal whorl section and its very simple suture line which resembles the species *Ophiceras tibeticum* Griesbach, 1880. These differences are here not considered as justifying its separation from other species of *Gyronites*, hence the genus *Himoceras* is also a junior synonym of *Gyronites*. The similarities between *Gyronites dubius* and *Ophiceras tibeticum* indicate that *Gyronites* most probably derives from *Ophiceras*.

Gyronites frequens Waagen, 1895

Plate 3, figures 1–5; Plate 4, figures 1–17;
Figures 16–18

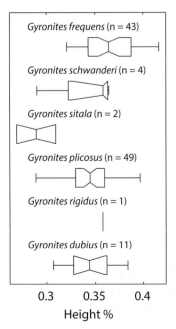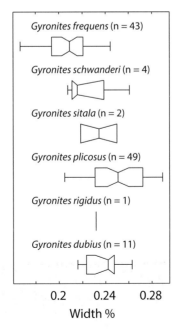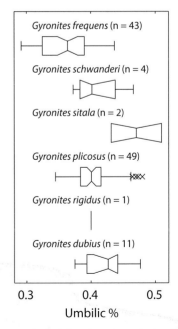

Fig. 16. Boxplots for the different species of *Gyronites* from the Salt Range. Because of allometric growth, these boxplots were calculated excluding specimens smaller than 20 mm in diameter, except for *G. frequens* which shows a strong allometry and for which specimens smaller than 30 mm in diameter have been excluded. Abbreviations as in Figure 15.

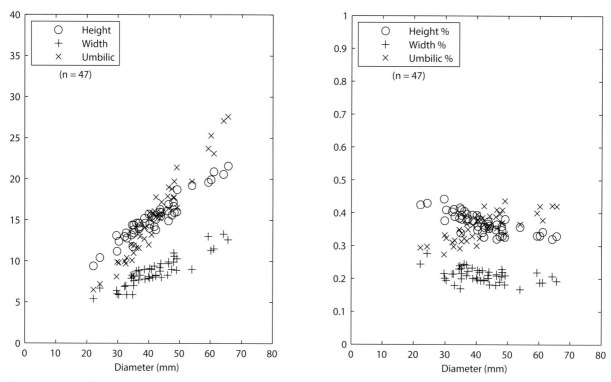

Fig. 17. *Gyronites frequens* Waagen, 1895. Scatter diagrams of H, W and U, and of H/D, W/D, and U/D (abbreviations as in Fig. 15).

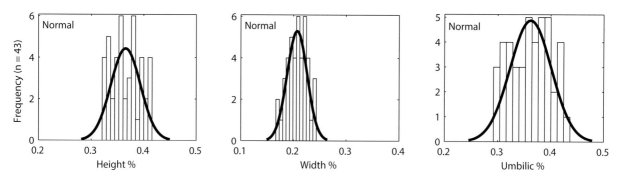

Fig. 18. *Gyronites frequens* Waagen, 1895. Histograms of H/D, W/D, and U/D (abbreviations as in Fig. 15). Because of allometric growth, specimens smaller than 30 mm in diameter have been excluded.

1895   *Gyronites frequens* n. gen., n. sp. Waagen, pp. 292–294, pl. 38, figs 1, 2 (lectotype), 3, 4, pl. 40, fig. 4.

1895   *Gyronites nangaensis* n. gen., n. sp. Waagen, pp. 297, 298, pl. 37, fig. 5 (holotype).

1895   *Lecanites psilogyrus* n. sp. Waagen, pp. 280, 281, pl. 39, fig. 5 (holotype).

1895   *Lecanites undatus* n. sp. Waagen, pp. 281, 282, pl. 38, figs 1 (lectotype), 2.

1895   *Prionolobus compressus* n. gen., n. sp. Waagen, pp. 313–315, pl. 35, fig. 3 (holotype).

1895   *Prionolobus plicatus* n. gen., n. sp. Waagen, pp. 315, 316, pl. 35, fig. 2 (holotype).

1895   *Prionolobus plicatilis* n. gen., n. sp. Waagen, pp. 318, 319, pl. 36, fig. 1 (holotype).

? 1909   *Xenodiscus lilangensis* n. sp. von Krafft, pp. 97–99, pl. 25, figs 6–10.

1909   *Xenodiscus khoorensis* n. sp. von Krafft, p. 88.

1934   *Gyronites frequens* Waagen, 1895; Spath, pp. 91, 92, fig. 19 [cop. Waagen, 1895].

? 1976   *Gyronites psilogyrus* Waagen, 1895;
     Wang & He, p. 274, fig. 7a, pl. 1, figs
     9, 10.

? 1976   *Prionolobus plicatilis* Waagen, 1895;
     Wang & He, p. 275, fig. 8a, pl. 3, figs
     13–15.

v 1978   *Gyronites frequens* Waagen, 1895;
     Guex, pl. 1, fig. 3.

v 1978   *Gyronites undatus* Waagen, 1895;
     Guex, pl. 8, fig. 3.

? 1996   *Gyronites frequens* Waagen, 1895;
     Waterhouse, pp. 33, 34, text-fig. 4A,
     pl. 1, figs 1–4.

? 1996   *Gyronites planissimus* Koken, 1934;
     Waterhouse, pp. 34, 35, text-fig. 4A,
     pl. 1, figs 5, 8.

? 1996   *Gyronites spiralis* n. sp. Waterhouse, pp.
     35, 36, text-fig. 4A, pl. 1, figs 6, 7, 9, 10.

non v 2008   *Gyronites frequens* Waagen, 1895;
     Brühwiler, Brayard, Bucher & Guo-
     dun, p. 1168, pl. 5, figs 7, 8 (= *Ambites*
     bjerageri n. sp.).

Emended diagnosis. – Compressed, moderately evolute *Gyronites* with a strong allometric growth. Flanks convex, most compressed variants being weakly ribbed and with a low, poorly differentiated oblique umbilical wall, depressed variants having sub-radial stronger ribs and a high sub-vertical umbilical wall individualized by a narrowly rounded shoulder.

Material. – 111 specimens.

Description. – Compressed (W/D ≈ 21%, W/H ≈ 57%) evolute sub-platyconic shells with a strong growth allometry, becoming more evolute with growth. Until a diameter of ca. 4 cm, the relative whorl height exceeds the relative umbilical width (*i.e.* for D < 4 cm, 34.6 < H/D < 44.3% and 27.4 < U/D < 39.1%), and above a diameter of 4 cm, the umbilical width becomes larger than the whorl height (i.e. for D > 4 cm, 32.1 < H/D < 38.1% and 32.3 < U/D < 43.7%). Venter tabulate with sharp ventro-lateral shoulders at all stages of growth. Moderately to very compressed whorl section, with convex flanks and maximum width generally at inner third of the flanks, except on the most robust variants where it is situated just above the umbilical wall. Umbilical wall undifferentiated on compressed

variants, the flanks sloping progressively towards the umbilicus without forming any shoulder and reaching the preceding whorl at a slightly obtuse angle. On robust variants, the umbilical wall is nearly vertical, well separated from the flanks by a narrowly rounded umbilical shoulder. Ornamentation is variable, varying as a function of shell geometry. Compressed variants are nearly smooth, with faint slightly sigmoid folds following the trajectory of the growth lines, while robust variants have weak sub-radial blunt ribs which rise near the umbilical shoulder, reach their maximal strength around inner third of the flank and fade towards the venter. These ribs or folds appear at a diameter of ca. 2 cm, reach their maximal strength at a diameter of ca. 4 cm, become more numerous and thinner with growth and tend to disappear on the adult body chamber. Suture line with relatively deep lobes and elongated saddles compared with other species of *Gyronites*.

Measurements. – See Figures 16–18.

Discussion. – The type series of *Gyronites frequens*, which come from the lowest bed of the Lower Ceratite Limestone of the hills West of Khoora, are all weathered inner moulds. Waagen (1895) stated in his report (p. 293) that although abundant, there are 'only a few localities where specimens of it can be detached from the rock'. Waagen considered this species as being perfectly smooth, but this is most probably the consequence of the weathering of his material, as even compressed variants actually have folds. Two characters lead Waagen to split this species into 7 species: the ornamentation and the suture line. As the ornamentation of this species is quite variable, Waagen separated smooth variants from variants with low broad folds and variants with blunt ribs. It is here shown that these variants form a continuum, and therefore all belong to the same species. Waagen also gave much importance to the indentations of the suture line, and placed specimens with a goniatitic suture line in the genus *Lecanites*, specimens with a goniatitic ventral lobe and a few, small indentations on the lateral lobes in the genus *Gyronites*, and specimens with a few indentations on the lateral branches of the ventral lobe and strong denticulations in the lateral lobes and the auxiliary series in the genus *Prionolobus*. Hence, several specimens with a nearly identical shell morphology were separated into different species belonging to different genera based on their suture line. This is especially the case, for example, of *Lecanites psilogyrus* and *Gyronites frequens* for smooth variants, of *Lecanites undatus* and *Prionolobus plicatus* for variants with low broad folds, and of *Gyronites nangaensis* and *Prionolobus plicatilis* for variants with small, numerous ribs. *Prionolobus compressus* corresponds to a robust variant of *Gyronites frequens*. Considering that, as stated by Noetling (1901) and underlined again by Kummel (1966), most of Waagen's type specimens were poorly preserved and strongly weathered, these differences of suture line are most probably the reflect of the weathering stage of the specimen rather than a genuine morphological difference.

The species *Xenodiscus lilangensis* von Krafft, 1909 may be a synonym of *Gyronites frequens*, corresponding to robust variants of this species, but the ornamentation seems to be stronger than that of all our specimens, and the type specimens are strongly distorted, making comparisons with our material difficult. Wang & He (1976) figured two specimens which they ascribe to a species here considered as synonyms of *Gyronites frequens*. These are very poorly preserved and thus cannot be clearly identified. Guex (1978) collected two specimens which can both be ascribed to *Gyronites frequens* without any doubt. The specimens ascribed to *Gyronites frequens*, *Gyronites planissimus* and *Gyronites spiralis* by Waterhouse (1996) are very close to our material included in *Gyronites frequens*, but are too poorly preserved to be synonymized with certainty. The specimens described by Brühwiler *et al.* (2008) as *Gyronites frequens* are actually placed in the genus *Ambites* (see below).

Occurrence. – Amb, samples Amb11 ($n = 13$), Amb53 ($n = 3$) and Amb62 ($n = 22$); Chiddru, sample Chi58 ($n = 1$); Nammal Nala, samples Nam76 ($n = 21$), Nam339 ($n = 18$) and Nam354 ($n = 8$); Wargal, samples War4 ($n = 12$), War7 ($n = 8$) and War100 ($n = 5$).

Gyronites dubius (von Krafft, 1909)

Plate 4, figures 18–35; Figures 16, 19

? *non* 1897   *Meekoceras* sp. ind. ex. aff. *plicatile* (Waagen, 1895); Diener, pp. 137, 138, pl. 15, fig. 6.

1909   *Meekoceras dubium* n. sp. von Krafft, pp. 50, 51, pl. 24, figs 6–10, 11 (lectotype), 12–14.

Lectotype. – Designated by Waterhouse (1996) as specimen GSI9499 illustrated by von Krafft & Diener, 1909, pl. 24, fig. 11.

Diagnosis. – *Gyronites* characterized by its sub-trapezoidal thick whorl cross-section, with high vertical

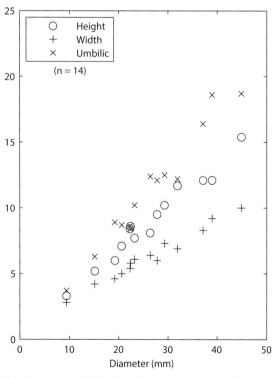
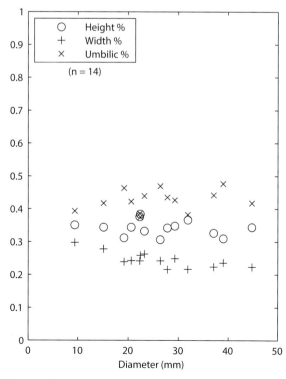

Fig. 19. *Gyronites dubius* (von Krafft, 1909). Scatter diagrams of H, W and U, and of H/D, W/D, and U/D (abbreviations as in Fig. 15).

umbilical wall and nearly flat flanks converging towards the tabulate venter.

Material. – 29 specimens.

Description. – Very evolute (U/D ≈ 42%) shell with a thick whorl section (W/D ≈ 24%, W/H ≈ 72%). Venter tabulate with sharp ventro-lateral shoulders. Umbilical wall very high and vertical, delimited by a narrowly rounded shoulder. Flanks flat to slightly convex with maximal width at the umbilical shoulder, converging towards the venter, giving the whorl section a sub-trapezoidal outline. Ornamentation varying from very vague radial folds on the most compressed variants to relatively strong radial ribs on robust variants. Folds and ribs do not cross the venter, which is perfectly smooth. Ornamentation fades on the adult body chamber. Suture line typical of *Gyronites* with finely denticulated lobes, a very large second lateral saddle, and a small third lateral saddle. Auxiliary series poorly preserved, with or without very small indentations.

Measurements. – See Figures 16, 19.

Discussion. – This species can easily be differentiated from all other congeneric species by its very characteristic sub-trapezoidal whorl cross-section. As already noticed by von Krafft & Diener, 1909 and by Waterhouse (1996), this species is very close to *Ophiceras tibeticum* Griesbach, 1880; differing only by its tabulate venter with sharp shoulders. This similarity may indicate a direct phylogenetic link between these two species. von Krafft & Diener (1909) also indicated that the specimen ascribed by Diener (1897) to *Meekoceras* sp. ind. ex. aff. *plicatile* is a synonym of the present species. Actually, according to the drawing and the measurements provided by Diener, this small specimen (D = 19 mm) has a more compressed whorl section than the specimens of similar size examined here. Its whorl section is less trapezoidal than our specimens. Additional material would be necessary to analyse the intraspecific variability of the juveniles of this species in order to validate any synonymy. Waterhouse (1996) created *Himoceras* as a new genus based on this species, which in our opinion is not necessary because this species shows all the characteristics of the genus *Gyronites* as revised here. Waterhouse (1996, p. 58) also designated 'the specimen GSI 9499 (von Krafft & Diener, 1909; pl. 24, fig. 1)' as lectotype for this species. However, the figure he designated corresponds to the species *Xenodiscus nivalis* Diener, 1897, not to *Meekoceras dubium*. This specimen corresponds actually to

the one figured in von Krafft & Diener (1909, pl. 24, fig. 11).

Occurrence. – Amb, sample Amb104 (*n* = 27); Nammal Nala, sample Nam391 (*n* = 2).

Gyronites rigidus (Diener, 1897)

Plate 5, figure 1–3; Figure 16

1897   *Danubites rigidus* n. sp. Diener, pp. 36, 37, pl. 15, figs 4 (lectotype), 5.

? 1909   *Xenodiscus* cf. *plicosus* (Waagen, 1895); von Krafft & Diener, pp. 101, 102, pl. 25, fig. 4.

Lectotype. – Here designated as the specimen GSI6023 (Diener, 1897, pl. 15, fig. 4).

Diagnosis. – Evolute, moderately compressed sub-platyconic *Gyronites* of small size with a thick tabulate venter delimited by very sharp ventro-lateral shoulders. Flanks convex with thin, very slightly sigmoid ribs which disappear on the outer third of the flanks.

Material. – One specimen.

Description. – Evolute (U/D = 40%) small shell (D = 19 mm) with a moderately compressed whorl section (W/D = 23%, W/H = 65%). Flanks convex with maximal width at mid-flanks, merging the thick tabulate venter almost at a right angle, thus imparting very sharp-shaped ventro-lateral shoulders. Upper flanks grading into indistinct umbilical wall. Ornamentation consisting of thin and blunt ribs following the shape of the growth lines, reaching their maximal strength at mid-flanks and fading towards the perfectly smooth venter. Suture line only partially visible and too recrystallized to be drawn and described.

Measurements. – See Figure 16 and Table 1.

Discussion. – This single specimen can be easily differentiated from the associated *Gyronites dubius* by its whorl cross section and ornamentation. It can also be distinguished from all other species of *Gyronites* by its more evolute shape at similar size, the absence of individualized umbilical wall, its thin ribs and its very sharp ventro-lateral shoulders. It is nearly identical to the smallest of the two specimens of *Danubites rigidus* illustrated by Diener (1897, pl.

15, fig. 5), differing only by its slightly less numerous ribs (9 ribs on the last half whorl preserved for our specimen, against 12 for Diener's specimen). This small difference is here interpreted as intraspecific variability. The other specimen described by Diener, here designated as lectotype of this species, is slightly larger (D = 30 mm) and has even more numerous ribs, a probable consequence of its larger size. It has 15 ribs on its last half whorl, but only 9 (as our specimen) on its penultimate half whorl. The ribs of this specimen tend to disappear at the end of its body chamber, indicating that it is probably an adult specimen. The specimen identified as *Xenodiscus* cf. *plicosus* by von Krafft & Diener, 1909 differs from the specimens described here by its thinner whorl section and more distant ribs. This difference may represent intraspecific variability, but additional material would be necessary to confirm it.

Occurrence. – Amb, sample Amb104.

Gyronites plicosus Waagen, 1895

Plate 5, figures 4–28; Figures 16, 20, 21

1895 *Gyronites plicosus* n. gen., n. sp. Waagen, pp. 298–300, pl. 38, fig. 11 (holotype).

1895 *Prionolobus buchianus* (de Koninck, 1863) n. gen.; Waagen, pp. 320–322, pl. 35, fig. 5.

1897 *Danubites* sp. ind. ex. aff. *planidorsato* n. sp. Diener, pp. 35, 36, pl. 15, figs 3, 7.

? 1905 *Celtites radiosus* Koken, pl. 22, fig. 1.

? 1905 *Celtites fortis* Koken, pl. 22, fig. 2.

1909 *Xenodiscus rotula* (Waagen, 1895); von Krafft & Diener pp. 93–97, pl. 23, figs 4, 5, pl. 25, fig. 11, pl. 27, figs 4, 5.

v 1978 *Hypophiceras plicosum* (Waagen, 1895); Guex, pl. 1, figs 1, 2, 5.

v ? 1978 *Gyronites plicatilis* (Waagen, 1895); Guex, pl. 5, fig. 9.

Diagnosis. – Relatively thick, moderately to very evolute *Gyronites* with weak to strong ribs that fade on the mature body chamber. Compressed variants have weak ribs, a poorly differentiated oblique umbilical wall and a thick tabulate venter with sharp shoulders extending to the end of the adult body chamber. Robust variants have coarse ribs, a high sub-vertical umbilical wall individualized by a narrowly rounded shoulder, and a tabulate venter on the inner whorls, becoming rounded at maturity.

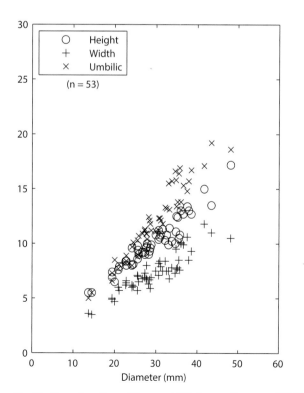

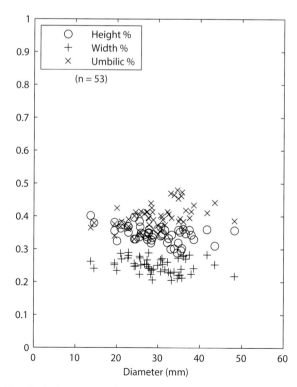

Fig. 20. Gyronites plicosus Waagen, 1895. Scatter diagrams of H, W and U, and of H/D, W/D, and U/D (abbreviations as in Fig. 15).

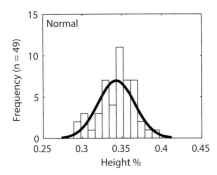 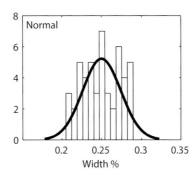 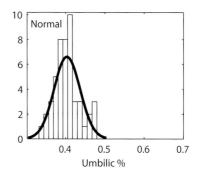

Fig. 21. Gyronites plicosus Waagen, 1895. Histograms of H/D, W/D, and U/D (abbreviations as in Fig. 15). Because of allometric growth, specimens smaller than 20 mm in diameter have been excluded from this analysis.

Suture line with a very short, sometimes almost absent auxiliary series.

Material. – 98 specimens.

Description. – Evolute (U/D ≈ 40%) shell with a rather thick whorl section (W/D ≈ 25%, W/H ≈ 73%), the shell becoming slightly more evolute with growth. The whorl section and ornamentation are highly variable. Compressed specimens have a whorl section with maximal width at inner third and convex flanks. Their venter is tabulate, delimited by sharp ventro-lateral shoulders, even on the adult body chamber. Their oblique umbilical wall is low, poorly individualized by a broadly rounded shoulder. Their ornamentation consists of numerous thin, sigmoid and blunt ribs (up to 18 on the last half whorl of the phragmocone). On thick variants, the maximal width is situated just above the umbilical wall and the flanks are nearly flat and converging towards the venter. Their venter is tabulate on inner whorls, delimited by narrowly rounded shoulders, and becomes broadly rounded on the last whorl. Their umbilical wall is high, almost vertical with a narrowly rounded shoulder. Their ornamentation consists of few (down to 6 on the last half whorl of the phragmocone) radial, thick and blunt ribs. In all variants, ribs become weaker towards the smooth venter, and fade on the body chamber. Suture line with a very broad and asymmetrical second lateral saddle, the first lateral lobe being much deeper than the second one. The third lateral saddle is very small with a flattened tip, and situated so close to the umbilical seam that there is no space left for the auxiliary series.

Measurements. – See Figures 16, 20, 21.

Discussion. – The illustration of the type specimen of *Gyronites plicosus* (Waagen, 1895; pl. 38, fig. 11) appears to be more compressed than the specimens assigned here to this species. Waagen (1895, p. 299)

measured a whorl thickness of 7 mm, but mentions that 'this latter measurement is only approximate as the specimen is firmly embedded in the rock'. As this specimen is otherwise identical to our thick variants, this measurement is most likely underevaluated, and thus the drawing of the specimen he provided probably should have a thicker whorl section. Waagen (1895) did not illustrate the suture line, but the description he gives is also in agreement with our material. The same reasoning applies to the specimen that Waagen called *Prionolobus buchianus* (de Koninck, 1863), which is strongly weathered. It should be noted here that the original species of de Koninck is poorly illustrated and accompanied by a vague description. It should be considered as a *nomen dubium* unless the original specimen is refigured and described. *Danubites planidorsatus* Diener, 1897 differs from our specimens by its much thinner whorl section. But the two specimens Diener (1897) referred to as *Danubites* sp. ind. ex. aff. *planidorsato* are very similar to our specimens, and most likely conspecific. Noetling (1905) considered two specimens from the Lower Ceratite Limestone as belonging to the genus *Celtites*. These specimens most likely belong to *Gyronites plicosus*, but the illustrations in Noetling's work are insufficient to confirm it. The specimens assigned to *Xenodiscus rotula* Waagen, 1895 by von Krafft and Diener (1909) correspond to robust variants of *Gyronites plicosus*. The original species *Gyronites rotula* of Waagen (1895) comes from the Ceratite Sandstones and is thus early Smithian in age, and most probably belongs to the genus *Kashmirites*. Guex (1978) assigned this species to the genus *Hypophiceras*, probably because of its ornamentation and its rounded venter on outer whorl. A re-examination of his material showed that his specimens, like ours, do have a tabulate venter at least at the beginning of their last whorl, indicating that his specimens are clearly conspecific with ours and belong to the genus *Gyronites*. The specimen assigned to *Gyronites plicatilis* by Guex (1978) is an incomplete, poorly preserved fragment of *Gyronites* which may correspond to a compressed variant of *Gyronites plicosus*, but is too incomplete to allow identification at the species level. Robust variants of *Gyronites plicosus* are actually very close to *Gyronites sitala*, differing only from the latter by their greater involution and their more angular whorl section. They also resemble inner whorls of *Gyronites schwanderi* n. sp., from which they differ by their more rounded venter. Small and smooth variants of *Gyronites plicosus* resemble *Gyronites rigidus* (see above), but are more involute.

Occurrence. – Amb, sample Amb2 ($n = 10$); Chiddru, sample CH2A ($n = 2$); Nammal Nala, samples Nam320 ($n = 3$), Nam331 ($n = 10$), Nam332 ($n = 2$), Nam335a ($n = 6$) and Nam377 ($n = 37$); Wargal, samples War5 ($n = 3$), War6 ($n = 4$), War9 ($n = 8$) and War100 ($n = 13$).

Gyronites sitala (Diener, 1897)

Plate 5, figures 29–33; Figure 16

1897 *Danubites sitala* n. sp. Diener, pp. 49–50, pl. 15, figs 12 (lectotype), 13.

Lectotype. – Here designated as the specimen GSI6031 (Diener, 1897, pl. 15, fig. 12).

Diagnosis. – Very evolute and serpenticonic *Gyronites* with a sub-rounded whorl-section and a very narrowly tabulate venter.

Material. – Two specimens.

Description. – Very evolute (U/D $\approx 47\%$) serpenticonic shell with a sub-rounded, slightly compressed whorl section (W/D $\approx 23\%$, W/H $\approx 81\%$). Flanks strongly convex with maximal width at mid-flanks. No differentiated umbilical wall, the flanks sloping progressively towards the umbilicus and merging the umbilical seam at a right angle. Venter tabulate and very narrow (its width represents about a third of the whorl width), the flanks joining the venter at a very obtuse angle, the ventro-lateral shoulders being thus angular but only visible with fringing light. Inner whorls with widely spaced and blunt radial ribs, becoming thinner, more numerous and sigmoid on the last whorl. Suture line not well enough preserved to be drawn and described.

Measurements. – See Figure 16 and Table 1.

Discussion. – In the original description of this species, Diener (1897) mentioned that the venter was rounded. However, von Krafft (1909) mentioned that the two specimens, on which Diener based this species, are 'too fragmentary and weather-worn to deserve the introduction of a new specific name'. He nevertheless recognized that it 'is probably an independent species', a statement which is in contradiction with his previous sentence. The original specimens are probably too weathered to see the very obtuse angular ventro-lateral shoulders, but they otherwise agree perfectly with our specimens. We therefore consider this species name as valid. With their serpenticonic shape and strong ribbing, their morphology is somewhat reminiscent of that of Griesbachian Xenodiscidae such as *Hypophiceras* or *Tompophiceras* and that of Smithian Kashmiritidae. However, they differ clearly from these taxa in having a narrow tabulate venter.

Occurrence. – Nammal Nala, samples Nam331 ($n = 1$) and Nam377 ($n = 1$).

Gyronites schwanderi n. sp.

Plate 5, figures 34–38; Figure 16

Derivation of name. – Named after Hans-Jörg Schwander (Faculty of Science, University of Zürich).

Holotype. – Specimen PIMUZ30281 (Pl. 5, figs 36–38).

Type locality. – Nammal Nala, Salt Range, Pakistan.

Type horizon. – Sample Nam378 (topmost Lower Ceratite Limestone), *Gyronites frequens* Regional Zone, lower Dienerian.

Diagnosis. – *Gyronites* with a sub-trapezoidal whorl section and strong, blunt ribs which disappear completely on the body chamber. Venter tabulate with sharp ventro-lateral shoulders. Umbilical wall steep and oblique, outlined by a narrowly rounded shoulder.

Material. – 4 specimens.

Description. – Evolute (U/D $\approx 41\%$) shell with a sub-trapezoidal, moderately compressed whorl section (W/D $\approx 22\%$, W/H $\approx 66\%$). Venter narrowly tabulate with sharp ventro-lateral shoulders becoming narrowly rounded at the end of the adult body chamber. Flanks nearly flat with maximal width at the umbilical shoulder, converging towards the venter and imparting the whorl section its sub-trapezoidal section. Umbilical wall steep and oblique, delimited by a narrowly rounded shoulder, and becoming less steep on the adult body chamber. Inner whorls with widely spaced and blunt ribs, which disappear completely on the body chamber. Suture line not well enough preserved to be drawn and described.

Measurements. – See Figure 16 and Table 1.

Discussion. – This species differs from the co-occurring species *Gyronites frequens* by its more evolute coiling, its thicker whorl section, its sub-trapezoidal whorl section and its stronger ribbed inner whorls. It resembles robust variants of *Gyronites plicosus* but differs from them by its larger size, its more angular whorl section with a sharper umbilical shoulder, and its narrower tabulate venter which persists until the end of the body chamber. Some specimens assigned to two species described by von Krafft and Diener (1909) are quite close to *Gyronites schwanderi*. The first ones, misidentified as *Xenodiscoides radians* (Waagen, 1895), differ by their more convex flanks. The original species of Waagen comes from the Ceratite Sandstones and is, as in the case of *Gyronites rotula* previously discussed, a Smithian Kashmiritidae. The second species, *Ophiceras obtusoangulatum* Diener, 1909, differs by its narrower venter and less angular umbilical shoulders. In both cases, the differences with our specimens are not very important and could conceivably reflect intraspecific variability. However, with only four specimens in our collections, this hypothesis cannot be tested and we therefore prefer to erect a new species.

Occurrence. – Nammal Nala, sample Nam378 (*n* = 2); Wargal, sample War100 (*n* = 2).

Genus *Ambites* Waagen, 1895

Type species. – *Ambites discus* Waagen, 1895.

Emended diagnosis. – Compressed, moderately involute, platyconic to serpenticonic Gyronitidae with a tabulate venter. Ventro-lateral shoulders protruding over the flanks, usually underlined by a very thin strigation and by a slight concavity on the flanks, giving the ventral part of the whorl section a bottlenecked outline. Ornamentation varying from nearly smooth forms to variants with two spiral ridges and sigmoid ribs or folds on their flanks. Suture line simple, lobes with numerous small indentations, auxiliary series short to long with small indentations and generally no differentiated auxiliary lobe.

Composition of the genus. – *Ambites discus* Waagen, 1895; *Prionolobus atavus* Waagen, 1895; *Gyronites superior* Waagen, 1895; *Lecanites impressus* Waagen, 1895; *Meekoceras lilangense* von Krafft, 1909, *Pleurambites frechi* Tozer, 1994, *Pleurambites radiatus* Brühwiler, Brayard, Bucher & Guodun, 2008, *Ambites tenuis* n. sp., *Ambites bojeseni* n. sp., *Ambites subradiatus* n. sp. and *Ambites bjerageri* n. sp.

Occurrence. – Middle Dienerian of the Salt Range (Pakistan), Central Himalayas (India, Nepal), South Tibet and Guangxi (China), South Primorye (Russia), British Columbia (Canada), Nevada (USA).

Discussion. – The most characteristic trait of this genus is its bottlenecked venter, although within each species there are always specimens which do not show it and it waxes and wanes during ontogeny. This peculiar trait is never seen among representatives of *Gyronites* or of Paranoritidae. It also differs from *Gyronites* by the occasional presence of two low spiral ridges on flanks. Its suture line has more rounded lobes and saddles, with more numerous indentations in the lobes and a longer auxiliary series with more denticulations than that of *Gyronites*. It differs from *Koninckites* by its rounder umbilical shoulder, more evolute adult shell and suture line without auxiliary lobe and with more numerous but smaller indentations. Some species have very involute inner whorls (e.g. *Ambites discus*) which can be very difficult to differentiate from co-occurring juvenile specimens of *Mullericeras*. Only some small specimens can be clearly assigned to one or the other genus. The most evolute ones and the ones which have an already clearly visible 'bottlenecked venter' belong to *Ambites*, whereas the most involute ones with a very broad third lateral saddle clearly belong to *Mullericeras*. However, a few intermediate specimens cannot be clearly identified.

This genus was originally defined by Waagen (1895) as having a goniatitic suture line. Some confusion arose from this original description, leading him and other authors to create new genera for species which had clearly ceratitic suture lines. However, von Krafft & Diener (1909, p. 48) noticed on the type specimen of *Ambites discus* some 'very delicate, indistinct denticulations on two or three succeeding principal lateral lobes'. Additional specimens from the type locality (lowest bed of the Ceratite Marls in Amb) all do show these small denticulations, even at very small diameters (ca. 1 cm). Therefore, the suture line of this genus is never goniatitic, but simply ceratitic with numerous small indentations in the lobes. As a consequence, the genus *Ambitoides* created by Shigeta & Zakharov (2009), based on a species of Tozer (1994)

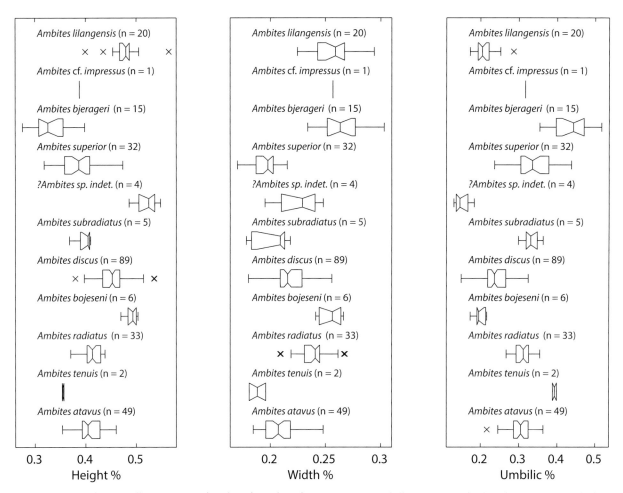

Fig. 22. Boxplots for the different species of *Ambites* from the Salt Range. Because of allometric growth, these boxplots were calculated excluding specimens smaller than 25 mm in diameter. Abbreviations as in Figure 15.

originally assigned to *Ambites* (*Ambites fuliginatus* Tozer, 1994) and explicitly defined by Shigeta & Zakharov (2009) as being distinguished from the genus *Ambites* because of its ceratitic suture line is here considered as a synonym of *Ambites* (note that we here reject the assignment of their specimens to Tozer's species, and consider them as possibly belonging to a new species belonging to *Mullericeras*; see below).

Because of their ceratitic suture line, many species belonging to *Ambites* were assigned to the genus *Prionolobus*. The latter is here considered as a synonym of *Ambites* (although not all species originally assigned to *Prionolobus* actually belong to *Ambites*), as its type species (*Prionolobus atavus* Waagen, 1895) clearly belongs to *Ambites* (see below). Spath (1934) erected the genus *Gyrolecanites* for *Lecanites impressus* Waagen (1895). According to Waagen's drawing and description, this species has a very clearly bottlenecked shaped venter, and the holotype is, according to Spath (1934), strongly weathered and still embedded in the rock.

Its apparently very simple suture line is then probably a consequence of the weathering, and its bottleneck shaped venter clearly places this species within *Ambites*, the genus *Gyrolecanites* becoming then a junior synonym of *Ambites*. Tozer (1994) introduced the name *Pleurambites* for shells with thicker whorl section and stronger ornamentation. This is a typical case of the first Buckman law of covariation, and some species described here do range from nearly smooth to rather strongly ribbed variants. The presence of ribs therefore does not indicate that it is a separate species, and the genus *Pleurambites* is consequently treated here as a synonym of *Ambites*. Finally, Waterhouse (1996) introduced the genus *Lilangia* based on the species *Meekoceras lilangense* von Krafft, 1909. This species, with its strongly bottleneck shaped venter clearly belongs to *Ambites*, and the genus *Lilangia* is also a synonym of *Ambites*. Being extremely poorly preserved, new species of *Lilangia* established by Waterhouse (1996) cannot be clearly identified.

Ambites discus Waagen, 1895

Plate 6, figures 1–31; Plate 7, figures 1–17;
Figures 22–24

1895 *Ambites discus* n. sp. Waagen, pp. 152–154, pl. 21, figs 4, 5 (lectotype).

1895 *Ambites magnumbilicatus* n. sp. Waagen, pp. 154, 155, pl. 21, fig. 6 (holotype).

1895 *Koninckites impressus* n. sp. Waagen, pp. 263–265, pl. 35, fig. 6 (holotype).

1897 *Meekoceras hodgsoni* n. sp. Diener, pp. 133–135, pl. 6, fig. 1 (holotype).

? 1897 *Koninckites vidharba* n. sp. Diener, pp. 139–150, pl. 7, fig. 9 (holotype).

1905 *Ophiceras discus* (Waagen, 1895); Noetling, pl. 13: unnumbered text-fig. in footnote 2 (lectotype).

non 1905 *Ophiceras discus* (Waagen, 1895); Noetling, pl. 13, fig. 4.

p 1909 *Meekoceras lilangense* n. sp. von Krafft, pl. 14, figs 1, 2.

p 1909 *Meekoceras hodgsoni* Diener, 1897; von Krafft & Diener, pp. 26–28, pl. 2, fig. 9 (holotype).

p? 1909 *Meekoceras hodgsoni* Diener, 1897; von Krafft & Diener, pl. 3, fig. 2.

non 1909 *Meekoceras hodgsoni* Diener, 1897; von Krafft & Diener, pl. 30, fig. 1.

1909 *Meekoceras* cf. *discus* (Waagen, 1895); von Krafft & Diener, pp. 47–50, pl. 6, fig. 2.

1934 *Prionolobus impressus* (Waagen, 1895); Spath, pp. 100, 101, fig. 22 [cop. Waagen, 1895].

1934 *Ambites discus* Waagen, 1895; Spath, pp. 102, 103, fig. 23 (lectotype) [cop. Waagen, 1895].

? 1976 *Prionolobus tulungensis* n. sp. Wang & He, p. 277, fig. 8d, pl. 2, figs 13–15 (holotype).

1985 *Prionolobus impressus* (Waagen, 1895); Pakistani-Japanese Research group, pl. 12, fig. 1.

? 2007 *Prionolobus impressus* (Waagen, 1895); Mu, Zakharov, Li & Shen, p. 869, figs 12.4, 13.1–13.3

Emended diagnosis. – Relatively involute (U/D ≈ 24%) compressed (W/D ≈ 22%, W/H ≈ 49%) *Ambites* with very involute internal whorls and usually strongly convex flanks with two strong spiral ridges, the venter being then quite narrow and the umbilical wall very low but vertical with a narrowly rounded shoulder.

Material. – 164 specimens.

Description. – Moderately involute, compressed platyconic shell with a relatively thin tabulate venter (the venter amounts to 30–35% of the whorl width). This species shows a very strong growth allometry. The shell becomes very involute up to a diameter of about 25 mm, where U/D varies between 10 and 20%. From this diameter on, the umbilicus opens slowly, reaching between 25 and 35% of the diameter for the largest specimens found. As these largest specimens do not have their body chamber preserved, the umbilicus probably becomes even larger at maturity. The intensity of this allometry is highly variable, some specimens reaching adult size with a still strong involution and compressed whorl section, whereas others already nearly have an adult morphology at a diameter of 30 mm. The ornamentation and whorl section shape varies according to this allometry and the first Buckman law of covariation. The flanks are divided dorso-ventrally into three parts by two low spiral ridges which are absent on the most involute immature specimens and well visible on the most evolute specimens and on adults. The maximum width is situated at the inner third of the whorl on involute juveniles, and at mid-flanks, between the two spiral ridges on more evolute forms. On adults and sub-adults, the whorl section shape is very variable. The flanks of the most involute forms are sub-parallel on their inner and middle third, and bend progressively on their outer third. On other variants, the whorl section is occasionally sub-octagonal, with both inner and outer third of the flanks (clearly delineated by the two spiral ridges) forming a straight or slightly concave slope towards the umbilical wall and the venter, respectively. Some other variants have a sub-hexagonal whorl section, the inner third of the flanks being sub-parallel and the umbilical wall relatively high (see Pl. 6, figs 17, 18). Some low sigmoid ridges on inner whorls of the most involute adult forms almost disappear on outer whorls, whereas they are stronger on the most evolute variant, sometimes becoming sub-angular and proverse on the inner and outer third of the flanks, and radial between the two spiral ridges. As already stated previously, the suture line is clearly ceratitic.

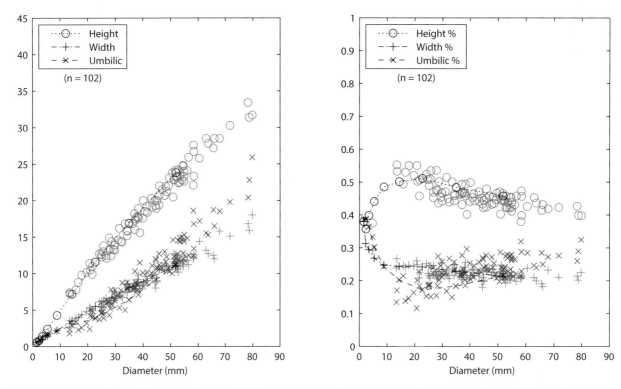

Fig. 23. Ambites discus Waagen, 1895. Scatter diagrams of H, W and U, and of H/D, W/D, and U/D (abbreviations as in Fig. 15). Ontogenetic trajectories obtained from sectioned specimen illustrated on Plate 6, figures 17, 18 are shown in black.

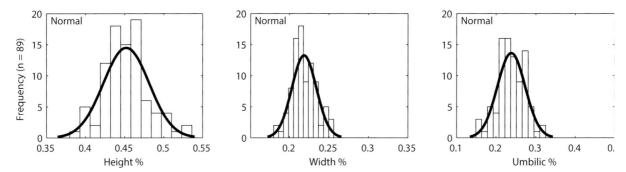

Fig. 24. Ambites discus Waagen, 1895. Histograms of H/D, W/D, and U/D (abbreviations as in Fig. 15). Because of allometric growth, specimens smaller than 25 mm in diameter have been excluded.

It bears very small indentations already at small diameter (less than 1 cm), where its second lateral saddle is bent spirally and its third lateral saddle is sub-rectangular (broader than high) and hardly differentiated from the auxiliary series which bears a few minor indentations. At larger diameters, the second lateral saddle becomes more straight, the two first lateral saddles and the two lateral lobes being well rounded and of almost even size. The length of lobes and saddles is very variable. Some specimens have lobes and saddles much longer than broad, having then a ventral lobe divided by a narrow elongated ventral saddle, a third lateral saddle quite elongated and slightly bent towards the umbilicus, and an auxiliary series with numerous indentations.

Others have lobes and saddles nearly as long as broad, having then a ventral lobe divided by a short ventral saddle, a third lateral saddle broader than high with a well rounded to sub-rectangular outline, and an auxiliary series with a few minor indentations. The auxiliary series is always quite long, without any clearly differentiated auxiliary lobe.

Measurements. – See Figures 22–24.

Discussion. – This species is clearly differentiated from other species of *Ambites* by its compressed whorl section with a relatively thin venter combined with a rather strong involution, even at adult size. Waagen (1895) already noticed the strong allometric growth of this species. The lectotype corresponds to a small involute variant, the paratype is larger and more evolute, both agreeing with our specimens. He described a second species of *Ambites*, *Ambites magnumbilicatus*, which he considered as difficult to differentiate from the type species, and is here, in agreement with von Krafft and Diener (1909), considered as a synonym.

The species *Koninckites impressus* Waagen, 1895 is very similar to our specimens in every respect except for its suture line, which has a clearly differentiated auxiliary lobe. This distinction was used by Waagen (1895, p. 264) for a distinct assignment to *Koninckites*. It is otherwise very similar to the variants with a concave inner third of the whorl, and according to his original description, also has a bottleneck shaped venter. As discussed previously, the portion of the suture line near the umbilicus (*i.e.* the third lateral saddle and the auxiliary series) can be highly variable within one species, and within a species with a simple but long auxiliary series like *Ambites superior* (see below), it is possible that a few individuals do have an auxiliary lobe whereas the majority do not. This species is therefore here considered as an extreme variant (in terms of suture line) of *Ambites discus*.

Noetling (1905) illustrated a cast of the holotype of *Ambites discus* together with an emended diagnosis of the species in a footnote of his plate 13. However, the specimen he ascribed to this species (Neotling 1905, pl. 13, fig. 4) is clearly different from this species. With its very involute tabulate shell and parallel flanks, Noetling's specimen most probably belongs to Mullericeratidae. The type of *Meekoceras hodgsoni* Diener, 1897 is clearly identical to those of our specimens that exhibit a rather high, vertical umbilical wall. Among the two other specimens attributed to this species by von Krafft and Diener (1909), the type of *Koninckites vidharba* Diener (1897, pl. 7, fig. 9, re-figured in von Krafft and Diener (1909), pl. 3, fig. 2) could correspond to the involute inner whorls of *Ambites discus*, but identification at this diameter, especially for these compressed, involute and tabulate forms is very difficult if not impossible without a clear stratigraphical control or the associated variants. The second specimen illustrated by von Krafft and Diener (1909), pl. 30, fig. 1) is very involute, and may belong to *Ussuridiscus*. von Krafft and Diener (1909) attributed two specimens to *Meekoceras lilangense* which are more compressed than the type of this species and are most probably conspecific with *Ambites discus*. Specimens referred to as *Prionolobus compressus* by the Pakistani-Japanese Research Group (1985) clearly correspond to this species. Two specimens from

China, one described by Wang & He (1976) as *Prionolobus tulungensis* and one described by Mu *et al.* (2007) as *Prionolobus impressus*, agree in proportion with *Ambites discus*, but their poor state of preservation prevents further comparison.

Occurrence. – Amb, samples Amb5 ($n = 2$) and Amb54 ($n = 10$); Chiddru, samples CH7A ($n = 17$), Chi56 ($n = 11$), Chi57 ($n = 5$), Chi103 ($n = 2$), Chi153 ($n = 5$) and Chi154 ($n = 2$); Nammal Nala, samples Nam50 ($n = 17$), Nam52 ($n = 18$), Nam54 ($n = 1$), Nam303 ($n = 1$), Nam364 ($n = 14$), Nam370 ($n = 6$), Nam380 ($n = 12$), Nam382 ($n = 3$), Nam383 ($n = 12$), Nam400 ($n = 1$), Nam526 ($n = 9$) and Nam527 ($n = 6$); Wargal, sample War2 ($n = 10$).

Ambites atavus (Waagen, 1895)

Plate 8, figures 1–26; Plate 9, figures 1–2; Figures 22, 25, 26

> 1895 *Prionolobus atavus* n. sp. Waagen, pp. 309, 310, pl. 34, fig. 4 (lectotype), pl. 35, fig. 4.
>
> 1895 *Gyronites evolvens* n. sp. Waagen, pp. 295–297, pl. 35, fig. 7 (holotype).
>
> *non* 1905 *Prionolobus atavus* Waagen, 1895; Noetling, pl. 22, fig. 7.
>
> 1934 *Gyronites planissimus* n. sp. Koken, pp. 92, 93, pl. 8, fig. 4 (holotype).
>
> *non* 1976 *Gyronites evolvens* Waagen, 1895; Wang & He, pp. 273, 274, fig. 7b–c, pl. 3, figs 6–12.
>
> *non* 2010a *Mudiceras planissimus* (Koken, 1934); Brühwiler, Ware, Bucher, Krystyn & Goudemand, p. 732, fig. 13.
>
> *non* 2012 *Mudiceras planissimus* (Koken, 1934); Brühwiler & Bucher, pp. 47–51, fig. 28Q–U.

Emended diagnosis. – Relatively evolute (U/D ≈ 30%) compressed (W/D ≈ 21%, W/H ≈ 51%) *Ambites* with usually weak ornamentation. Spiral ridges very weak but always present, even at small size (for D around 3 cm). Umbilical wall poorly differentiated on evolute variants, very low and vertical on the most involute ones. Suture line with a broad third lateral saddle with a flattened tip.

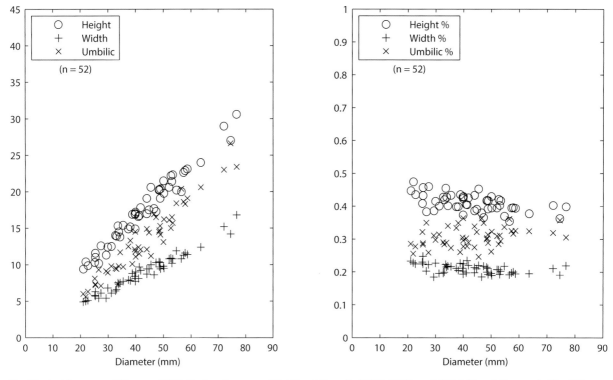

Fig. 25. *Ambites atavus* (Waagen, 1895). Scatter diagrams of H, W and U, and of H/D, W/D, and U/D (abbreviations as in Fig. 15).

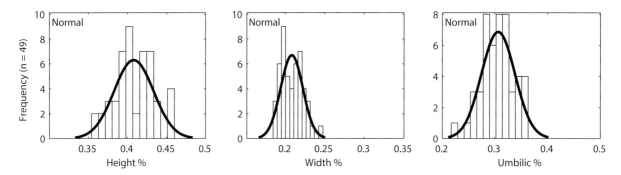

Fig. 26. *Ambites atavus* (Waagen, 1895). Histograms of H/D, W/D, and U/D (abbreviations as in Fig. 15). Because of allometric growth, specimens smaller than 25 mm in diameter have been excluded.

Material. – 80 specimens.

Description. – Evolute, compressed, discoidal to sub-platyconic shell with a thin tabulate venter (the venter representing between 30 and 35% of the whorl width). Growth slightly allometric, juveniles tending to be slightly more involute and thicker than the adults. Flanks usually strongly convex with maximum width at the inner third of whorl height, at the level of the inner spiral ridge. In some involute, compressed variants, the segment of the flanks comprised between the two spiral ridges is sub-parallel. From the maximum width, the flanks form a nearly flat slope towards the umbilicus, where they abruptly bend before joining the preceding whorl, thus forming a low vertical umbilical wall poorly individualized by a broadly rounded edge (as described by Waagen (1895) for the type specimen). Suture line simple, with lobes generally narrower than saddles. Lateral lobes with numerous thin indentations which are either aligned in an almost straight line, or along a broadly rounded curve. The two-first lateral saddles are relatively broad and rounded, the second one being larger than the first one and slightly bent towards the umbilicus. The third lateral saddle is broader than high, generally sub-rectangular but occasionally well rounded. Auxiliary series short with very small indentations and occasional weakly developed auxiliary lobe (see Pl. 8, figs 9, 23).

Measurements. – See Figures 22, 25, 26.

Discussion. – This species is very close to *Ambites superior* (Waagen, 1895), from which it differs mainly by its suture line. *Ambites atavus* has a very broad, sometimes sub-rectangular third lateral saddle, and has fewer indentations on the lateral lobes than *Ambites superior*. Its umbilical wall is also usually less differentiated. Additionally, no variants of *Ambites atavus* have an ornamentation as robust as that of *Ambites superior*, and *Ambites atavus* never becomes as evolute as some variants of *Ambites superior* (see Fig. 22). *Ambites atavus* is generally more evolute than *Ambites discus*, but involute variants of the former are difficult to distinguish from the latter species, differing only by their suture line with less elongated saddles and a second lateral saddle wider than the two adjacent lobes (whereas they are of even size in *Ambites discus*). However, as *Ambites discus* never becomes as evolute as *Ambites atavus*, the two species can easily be distinguished as long as the number of individuals from the same sample is large enough. *Ambites atavus* and *Ambites tenuis*, which are the oldest representatives of the genus and share an identical stratigraphical distribution, are morphologically the nearest *Ambites* species to the youngest *Gyronites*, i.e. *Gyronites frequens*. However, *Ambites atavus* clearly differs from *Gyronites frequens* by its bottleneck shaped venter and the presence of two spiral ridges on the flanks. Waagen (1895) erected two different species (*Prionolobus atavus* and *Gyronites evolvens*) for the material here included in *Ambites atavus*. We choose the name *Ambites atavus* as the valid name because the type specimen has the most typical morphology for this species. Incidentally, *Ambites atavus* was the type species of the genus *Prionolobus* and was used as such by many previous authors, whereas *Gyronites evolvens* has seldom been used. The lectotype of *Ambites atavus* (designated by Spath, 1934) is a weathered phragmocone, which explains why Waagen did not notice its bottleneck shaped venter and the two spiral ridges which are barely visible on these evolute, relatively thick variants of *Ambites atavus*. The second species, *Gyronites evolvens*, corresponds to a more involute variant of *Ambites atavus*. The material referred to as *Gyronites evolvens* by Wang & He (1976) is too poorly preserved to be clearly identified, but these specimens are more involute and thicker than any variant of *Ambites atavus*. Koken (1934) created the species *Gyronites planissimus* for a large specimen coming from the Lower Ceratite Limestone of Wargal. The holotype is quite weathered and its ornamentation is not visible anymore. Although larger than any of our specimens, its morphology agrees with variants of *Ambites atavus* with compressed whorl section and sub-parallel flanks, and its suture line with a relatively broad third lateral saddle is, as far as can be seen from his photo (Spath 1934; pl. 8, fig. 4), nearly identical to the one shown here in Plate 8, figures 13–16. Brühwiler *et al.* (2010a, 2012) misidentified some early Smithian specimens as *Gyronites planissimus*, and erected the new genus *Mudiceras* for these specimens. Although their overall shape is similar to the original specimen described by Spath (1934), their suture line is very different, typical of Paranoritidae, with a poorly individualized auxiliary lobe. These early Smithian specimens thus belong to an unnamed new species.

Occurrence. – Amb, samples Amb4 ($n = 3$), Amb11 ($n = 3$), Amb52 ($n = 27$), Amb53 ($n = 3$) and Amb63 ($n = 19$); Chiddru, sample Chi59 ($n = 5$); Nammal Nala, samples Nam60 ($n = 10$) and Nam524 ($n = 2$); Wargal, sample War3 ($n = 8$).

Ambites tenuis n. sp.

Plate 9, figures 3–10; Figure 22

Derivation of name. – From the Latin '*tenuis*', meaning thin.

Holotype. – Specimen PIMUZ30308 (Pl. 9, figs 3–7).

Type locality. – Amb, Salt Range, Pakistan.

Type horizon. – Sample Amb63 (topmost Lower Ceratite Limestone), *Ambites atavus* Regional Zone, middle Dienerian.

Diagnosis. – Unusually widely umbilicated and compressed *Ambites* (U/D $\approx$ 39%, W/D $\approx$ 19%, W/H $\approx$ 53%), with thin, sub-radial ribs.

Material. – 5 specimens.

Description. – Very evolute and compressed discoidal shell with a relatively broad tabulate venter (the venter amounting to 45–50% of the whorl width). Flanks convex with maximal width at midflanks. On the holotype, flanks are slightly concave just above the venter, and have two very indistinct spiral ridges that form a swelling at the intersections with the ribs. The flanks slope gently towards the umbilicus and bend just before reaching it, forming an undifferentiated vertical umbilical wall without shoulder. Suture line very simple, with rounded lateral saddles broader than lateral lobes, with a few very thin indentations at the bottom of the lobes. Auxiliary series short with five small indentations.

Measurements. – See Figure 22 and Table 1.

Discussion. – This species is based on only two specimens found together with *Ambites atavus* in Amb. Although some specimens of similar small size of *Ambites atavus* are close to these two specimens, they are still slightly less evolute and do not show the strong ornamentation of *Ambites tenuis*. The very evolute platyconic shape and simple suture line of *Ambites tenuis* compares well with that of some ribbed forms of *Gyronites*, but the presence of a slight concavity underlying the venter and of the two spiral ridges, although visible only on the largest specimen, allow to place them within the genus *Ambites*.

Occurrence. – Amb, sample Amb63.

Ambites radiatus (Brühwiler, Brayard, Bucher & Guodun, 2008)

Plate 9, figures 11–28; Figures 22, 27, 28

v 2008   *Pleurambites radiatus* n. sp. Brühwiler,
              Brayard, Bucher & Guodun, p. 1168, pl.
              5, figs 1 (holotype), 2, 3.

Emended diagnosis. – Evolute (U/D ≈ 32%) platyconic *Ambites* of small size (maximal observed diameter of 41.7 mm on specimens with nearly complete body chamber) with relatively thick (W/H ≈ 59%, W/D ≈ 24%) sub-rectangular whorl-section and almost no or little allometric growth.

Material. – 104 specimens.

Description. – Very evolute platyconic to sub-serpenticonic shell with a relatively thick whorl section and thick tabulate venter (the venter amounting to 45–55% of the whorl width). Flanks weakly convex with maximal width at mid-flanks. Umbilical wall vertical and relatively high, with a narrowly rounded shoulder. The broad tabulate venter, the high vertical umbilical wall and the weakly convex flanks confer a

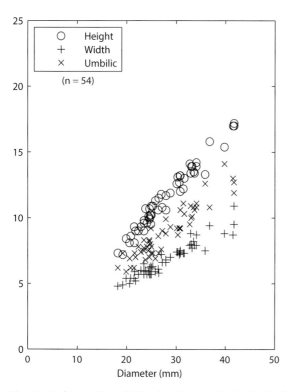
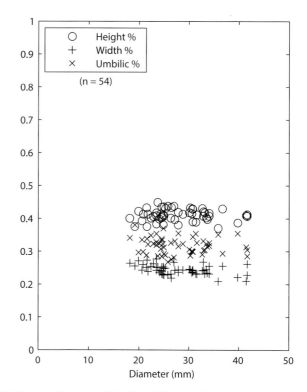

Fig. 27. Ambites radiatus (Brühwiler, Brayard, Bucher & Guodun, 2008). Scatter diagrams of H, W and U, and of H/D, W/D, and U/D (abbreviations as in Fig. 15).

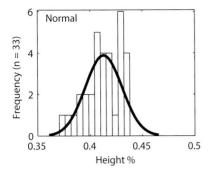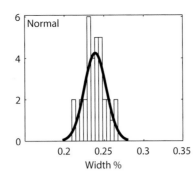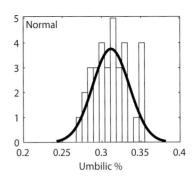

Fig. 28. *Ambites radiatus* (Brühwiler, Brayard, Bucher & Guodun, 2008). Histograms of H/D, W/D, and U/D (abbreviations as in Fig. 15). Because of allometric growth, specimens smaller than 25 mm in diameter have been excluded.

sub-rectangular shape to the whorl-section. With further growth, the whorl-section tends to become slightly more compressed and the tabulate venter becomes slightly convex, but still with clear, slightly protruding ventro-lateral shoulders. Strength of ornamentation variable, ranging from smooth shells to slightly sigmoid folds or ribs and with the usual two spiral ridges. On the most heavily ornamented variants, the ribs tend to cross the venter, giving the shell a wavy outline in lateral view. Suture line quite simple, with numerous very thin indentations in the lobes. The lateral branches of the ventral lobe are very broad, with numerous small indentations at their base. Lateral lobes and saddles of nearly equivalent size, relatively elongated. The third lateral saddle is as broad as high, with its top slightly rounded. Auxiliary series with numerous regularly spaced indentations, without differentiated elements.

Measurements. – See Figures 22, 27, 28.

Discussion. – With its small size, evolute coiling and sub-rectangular rather thick whorl section, this species can easily be distinguished from other species of *Ambites*. This species was originally based on only three specimens, including two very small juveniles. Our new, more abundant material allows a better understanding of its characteristics, and we therefore provide here an emended diagnosis. Compared to other specimens, the holotype is a strongly ornamented variant.

Occurrence. – Nammal Nala, samples Nam72 (*n* = 10), Nam81 (*n* = 2), Nam379 (*n* = 2), Nam381 (*n* = 57), Nam384 (*n* = 11), Nam520 (*n* = 18) and Nam525 (*n* = 4).

Ambites bojeseni n. sp.

Plate 9, figures 29–31; Plate 10, figures 1–11;
Figures 22, 29

Derivation of name. – Named after Dr. Jørgen Bojesen-Koefoed (Geological Survey of Denmark and Greenland, GEUS, Denmark).

Holotype. – Specimen PIMUZ30318 (Pl. 10, figs 1–3).

Type locality. – Nammal Nala, Salt Range, Pakistan.

Type horizon. – Sample Nam384 (lower Ceratite Marls, ca. 1.5 m above base), *Ambites radiatus* Regional Zone, middle Dienerian.

Diagnosis. – Rather involute (U/D ≈ 20% for D > 25 mm) and thick (W/H ≈ 52%, W/D ≈ 25% for D > 25 mm) *Ambites*, becoming more evolute with size (U/D estimated at 25% on the holotype), with very weak ornamentation and a simple suture line with deep, nearly symmetrical lateral lobes and saddles, the lobes bearing numerous very small indentations at their base.

Material. – 18 specimens.

Description. – Involute and thick platyconic shell with a broad tabulate venter, the venter amounting to ca. 50% of the whorl thickness. Despite the small number of specimens, a slight allometry could be detected, the shell becoming more evolute with growth. Flanks reaching their maximum thickness at mid-flanks, sub-parallel between the umbilical shoulder and the point of maximum width from where they gently slope towards the venter, thus forming a faint spiral ridge. Umbilical wall vertical and high, poorly differentiated in small specimens but well delineated by a sharp shoulder on larger specimens. Ornamentation nearly absent apart from very low and slightly sigmoid folds. Suture line simple, with narrow and deep lateral lobes. Lobes with numerous small indentations at their base. Ventral

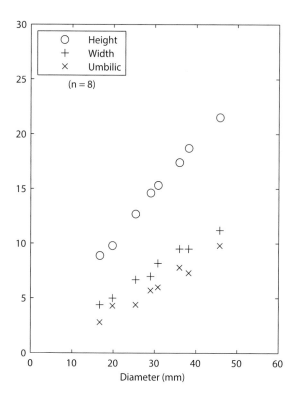
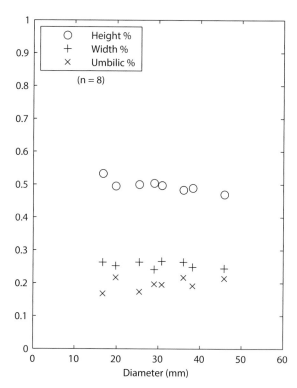

Fig. 29. *Ambites bojeseni* n. sp. Scatter diagrams of H, W and U, and of H/D, W/D, and U/D (abbreviations as in Fig. 15).

lobe large, divided by a deep ventral saddle into two large branches. The two-first lateral saddles are elongated with rounded top and are approximately of equal size. The third lateral saddle has a flattened top and is thus sub-rectangular, as broad as high on juveniles, broader than high on larger specimens. Auxiliary series short, with numerous regularly spaced small indentations.

Measurements. – See Figures 22, 29.

Discussion. – This species is very close to *Ambites lilangensis*, with which it shares similar proportions. It mainly differs by its suture line, with deeper, narrower lobes and more elongated saddles, a broader ventral lobe, lobes with more numerous indentations at their base, a more simple auxiliary series without any tendency to have a differentiated auxiliary lobe, and an almost symmetrical third lateral saddle with a sub-rectangular outline. It also tends to have less convex flanks, with less ornamentation, and its venter is always clearly tabulate and flat, whereas some specimens of *Ambites lilangensis* have a slightly convex venter. It is clearly differentiated from all other species of *Ambites* by its more involute coiling and thicker whorl thickness.

Occurrence. – Nammal Nala, samples Nam72 ($n = 2$), Nam381 ($n = 8$), Nam384 ($n = 3$),

Nam520 ($n = 2$), Nam521 ($n = 1$) and Nam525 ($n = 2$).

Ambites subradiatus n. sp.

Plate 10, figures 12–19; Figures 22, 30

v 2011   *Ambites* aff. *radiatus* (Brühwiler, Brayard, Bucher & Guodun); Ware, Jenks, Hautmann & Bucher, pp. 165–168, figs 7, 8b.

Derivation of name. – Refers to the species name of *Ambites radiatus*, with the prefix 'sub', in reference to its close affinity with this species.

Holotype. – Specimen PIMUZ30321 (Pl. 10, figs 12–14).

Type locality. – Amb, Salt Range, Pakistan.

Type horizon. – Sample Amb54 (lowermost bed of the Ceratite Marls), *Ambites discus* Regional Zone, middle Dienerian.

Diagnosis. – Very evolute (U/D $\approx$ 32%) platyconic *Ambites* with rather compressed whorl-section (W/H $\approx$ 52%, W/D $\approx$ 21%) and almost no detectable growth allometry.

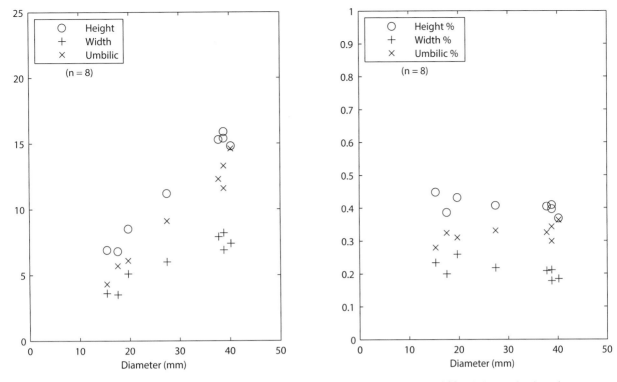

Fig. 30. *Ambites subradiatus* n. sp. Scatter diagrams of H, W and U, and of H/D, W/D, and U/D (abbreviations as in Fig. 15).

Material. – 8 specimens.

Description. – Very evolute thin sub-platyconic shell with a pronounced bottleneck shaped venter. Flanks convex with a low and variable umbilical wall, ranging from variants without umbilical wall, the flanks joining the preceding whorl at an obtuse angle, to variants with a low umbilical wall delineated by a rounded shoulder. Flanks almost smooth, with very weak slightly sigmoid folds and two weak spiral ridges. Suture line simple, with numerous small indentations in lateral lobes. The lateral branches of the ventral lobe are very thin, with only three indentations at their base. The first two lateral saddles are of similar size, moderately elongated, with well rounded tips. The first lateral lobe is almost twice deeper than second lateral lobe, but has a comparable width. The third lateral saddle is much smaller than the two-first ones and also rounded. Auxiliary series with a few, irregularly spaced indentations.

Measurements. – See Figures 22, 30.

Discussion. – This species is very close to *Ambites radiatus*, from which it differs mainly by its thinner whorl section, lower umbilical wall, more convex flanks and its more strongly marked bottleneck shaped venter. Its suture line differs from that of *Ambites radiatus* by its smaller third lateral saddle

and its ventral lobe with thinner lateral branches. No robust variant such as the holotype of *Ambites radiatus* has been found. However, only very few specimens are available and it cannot be completely excluded that this difference is due to the small number of specimens available. On the other hand, specimens from Nevada described by Ware *et al.* (2011) are strictly identical to those described here, thus supporting a distinction at the species level.

Occurrence. – Amb, sample Amb 54 (*n* = 3); Chiddru, sample Chi56 (*n* = 1); Nammal Nala, samples Nam50 (*n* = 1), Nam52 (*n* = 1) and Nam380 (*n* = 1); Wargal, sample War2 (*n* = 1).

Ambites? sp. indet.

Plate 10, figures 20–25; Figure 22

Material. – Four specimens.

Description. – Very involute (U/D ≈ 15%) compressed (W/H ≈ 43%) platyconic shell with a relatively thin tabulate venter. Flanks slightly convex with maximal width at inner third. Umbilical wall vertical, individualized by a narrowly rounded shoulder. Flanks nearly smooth with weak and slightly sigmoid folds. Suture line simple with lobes bearing only few very small denticulations. Ventral lobe divided into two branches by a very shallow ventral saddle, the branches being short with only two or three small indentations at their base. The third lateral saddle is sub-rectangular (wider than deep), much smaller than the two others. Auxiliary series long, with numerous regularly spaced small indentations, without auxiliary lobe.

Measurements. – See Figure 22 and Table 1.

Discussion. – These four specimens closely resemble juveniles (i.e. at a diameter of ca. 2 cm) of the co-occurring species *Ambites discus*, and could therefore represent extreme variants of this species which keep a juvenile morphology up to 3 or even 4 cm in diameter. However, they plot separately from *Ambites discus* of similar size (not illustrated in figures). Moreover, if we include them in the histograms of *Ambites discus*, the histogram (Fig. 24) corresponding to the relative whorl height (H/D) does not show anymore a normal distribution, these four specimens forming a small peak at the extreme right end of the distribution (for H/D between 53 and 55%). We therefore decided to keep them separate, but their identification at the species level remains difficult because

of the small number of available specimens. They closely resemble juveniles of *Ambites discus*, but they lack the typical bottleneck shaped venter of this genus. They also closely resemble *Mullericeras niazii* n. sp. (see below), but their suture line is much more simple than this species, and they lack the large ventral lobe with large branches with many indentations typical of *Mullericeras*. They most probably belong to Gyronitidae, but additional material would be necessary to decide whether they belong to a distinct species or genus.

Occurrence. – Nammal Nala, samples Nam380 (n = 1), Nam382 (n = 2) and Nam526 (n = 1).

Ambites superior (Waagen, 1895)

Plate 10, figures 26–33; Plate 11, figures 1–17; Plate 12, figures 1–3; Figures 22, 31, 32

    1895 *Gyronites superior* n. gen., n. sp. Waagen, pp. 294, 295, pl. 37, fig. 6 (holotype).

? 1895 *Prionolobus ovalis* n. gen., n. sp. Waagen, pp. 316, 317, pl. 35, fig. 1 (holotype).

p ? 1909 *Meekoceras disciforme* n. sp. von Krafft, pp. 45–47, pl. 3, figs 5, 6.

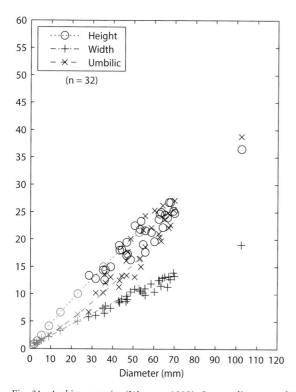
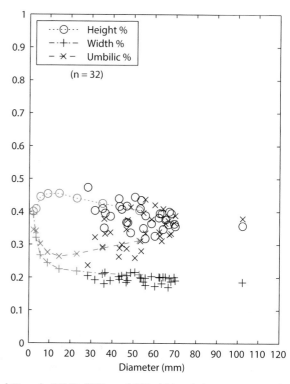

Fig. 31. Ambites superior (Waagen, 1895). Scatter diagrams of H, W and U, and of H/D, W/D, and U/D (abbreviations as in Fig. 15). Ontogenetic trajectories obtained from sectioned specimen illustrated on Plate 10, figures 32, 33 are shown in black.

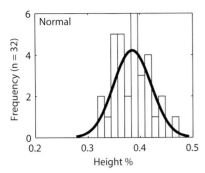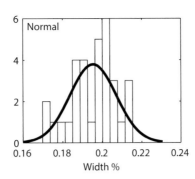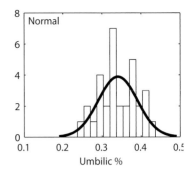

Fig. 32. Ambites superior (Waagen, 1895). Histograms of H/D, W/D, and U/D (abbreviations as in Fig. 15). Because of allometric growth, specimens smaller than 25 mm in diameter have been excluded.

Emended diagnosis. – Very evolute (U/D ≈ 34%) compressed (W/H ≈ 51%, W/D ≈ 20%) *Ambites* with a suture line with very deep lobes and elongated saddles, the third lateral saddle being much smaller than the others.

Material. – 48 specimens.

Description. – Very evolute compressed sub-platyconic shell with usually a pronounced bottleneck shaped venter. Evolute variants with convex flanks and maximum width at inner third of the whorl. Involute variants with more parallel flanks, the part of the flanks situated between the two lateral spiral ridges being parallel, with maximal width at mid-flanks. Umbilical wall varying from vertical with a very narrowly rounded umbilical shoulder to gently curved and sloping gradually towards the umbilical seam without any clear shoulder. Flanks with two relatively strong spiral ridges and slightly sigmoid folds whose strength is very variable. Some specimens are almost smooth, while others bear strong ribs. Growth slightly allometric, juveniles being more involute and less compressed than larger specimens. The ornamentation tends to disappear on very large specimens. Suture line with deep, well indented lobes. The lateral branches of the ventral lobe are usually broad with many indentations. The three lateral saddles are very elongated with well rounded tips, the second one being the largest, the third one being very small. Auxiliary series varying from a few regularly spaced small indentations to a clearly individualized auxiliary lobe followed by a few deep indentations.

Measurements. – See Figures 22, 31, 32.

Discussion. – This species is clearly differentiated from other representatives of *Ambites* by its very evolute, compressed shell and its suture line with very deep lobes. The holotype corresponds to a moderately evolute variant, with slightly convex flanks. *Prionolobus ovalis* Waagen, 1895, which is based on one single specimen from the same bed as the holotype of *Ambites superior* (topmost Lower Ceratite Limestone of Khoora) differs from the latter by its venter, described by Waagen as 'indistinctly flattened', a characteristic which does not apply to any specimens of *Ambites superior*. It is otherwise very similar and it cannot be excluded that this difference in the ventral shape is the result of a shell repair. Some specimens of *Ambites discus* in our samples clearly display this type of shell repair, where the bottleneck shaped venter is replaced by a sub-rounded venter in the subsequent part of the whorl. This assumption can only been confirmed by a close examination of the type specimen, which leaves the synonymy of this species uncertain. Waterhouse (1996) based his new genus *Ovaliconchia* on *Prionolobus ovalis*, which opens the possibility that the genus erected by Waterhouse may ultimately become a junior synonym of *Ambites*. Two specimens illustrated by von Krafft and Diener (1909) and assigned to his new species *Meekoceras disciforme* are quite close to our specimens. However, they do not have a clearly visible bottlenecked venter and their suture line does not have the characteristic deep lobes and elongated saddles, but these specimens are, as stated by the author, strongly weathered, which may explain these differences. However, their much more compressed whorl section distinguishes them from the two other illustrated specimens which were designated by von Krafft as the types of this species. It should be noted here that Spath (1934, p. 103) designated the specimen figured by von Krafft (1909, pl. 3, fig. 6), as the lectotype of *Meekoceras disciforme*, whereas von Krafft (1909) designated two other specimens as types of this species. This conflicts with the code of zoological nomenclature, which clearly indicates (article 74.2) that a lectotype must be chosen among the syntypes. The designation of Spath is therefore not valid.

Occurrence. – Nammal Nala, samples Nam53 ($n = 30$), Nam302 ($n = 5$), Nam308 ($n = 10$), and Nam371 ($n = 3$).

Ambites lilangensis (von Krafft, 1909)

Plate 12, figures 4–19; Plate 13, figures 1–6; Figures 22, 33

p 1909 *Meekoceras lilangense* n. sp. von Krafft, pp. 23–25, pl. 1, figs 1, 2 (lectotype), 3, 5–7.

non p 1909 *Meekoceras lilangense* n. sp. von Krafft, pl. 14, figs 1, 2.

1909 *Meekoceras lingtiense* n. sp. von Krafft, pp. 25, 26, pl. 2, fig. 1 (holotype).

1934 *Prionolobus lilangensis* (von Krafft, 1909); Spath, pp. 101, 102, pl. 4, fig. 4.

1967 *Prionolobus* sp. cf. *P. lilangense* (von Krafft, 1909); Tozer, pp. 18, 73.

1970 *Prionolobus lilangense* (von Krafft, 1909); Tozer, pl. 16, fig. 6.

? 1976 *Prionolobus lilangensis* (von Krafft, 1909); Wang & He, p. 276, text-fig. 8b, pl. 3, figs 4, 5.

p ? 1976 *Gyronites evolvens* Waagen, 1895; Wang & He, pp. 273, 274, text-fig. 7c, pl. 3, figs 11, 12.

p 1994 *Ambites fuliginatus* n. sp. Tozer, p. 67, fig. 15a, pl. 13, figs 5, 7, pl. 14, fig. 8 (holotype).

non p 1994 *Ambites fuliginatus* n. sp. Tozer, pl. 13, fig. 4.

1996 *Lilangia lilangensis* (von Krafft, 1909); Waterhouse, pp. 36, 60, 61, text-fig. 7G, pl. 5, figs 2–7, 9, 10, pl. 6, fig. 1.

v 2011 *Ambites lilangensis* (von Krafft, 1909); Ware, Jenks, Hautmann & Bucher, pp. 164, 165, figs 4, 5.

Material. – 64 specimens.

Description. – Involute ($U/D \approx 21\%$) and thick ($W/H \approx 54\%$, $W/D \approx 26\%$) discoidal to sub-

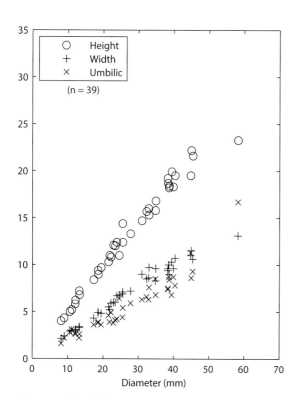
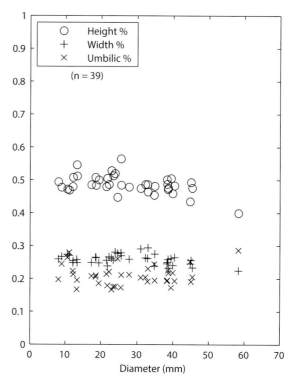

Fig. 33. Ambites lilangensis (von Krafft, 1909). Scatter diagrams of H, W and U, and of H/D, W/D, and U/D (abbreviations as in Fig. 15).

platyconic shell with a broad tabulate venter, the venter representing about 50% of the whorl width. Coiling becoming more evolute with growth. Venter tabulate to slightly convex with strong bottleneck shaped edges. Flanks with maximum thickness at inner third of the whorl height on juvenile stage and at mid-flanks on sub-adult stage. Flanks slightly to strongly convex, showing occasionally two marked spiral ridges which impart the whorl section a compressed sub-octagonal shape. Umbilical wall vertical and high, with a sharp shoulder. Ornamentation varying from nearly absent to rather strong sigmoid folds following the trajectory of the growth lines and crossing the venter. Suture line with generally asymmetrical, broad and short lateral saddles bent towards the umbilicus. Ventral lobe narrow, with narrow lateral branches bearing a few (3 or 4) deep indentation at their base. Lateral lobes with a low rounded base bearing few rather strong indentations. Auxiliary series with irregularly spaced, relatively deep indentations, occasionally with a small and clearly individualized auxiliary lobe.

Measurements. – See Figures 22, 33.

Discussion. – Because no large specimens of this species were found in the Salt Range, and because its type locality is in the Spiti Valley, no emended diagnosis is provided here. As discussed previously, this species closely resembles *Ambites bojeseni*, from which it differs mainly by its suture line. Among the syntypes of *Ambites lilangensis*, von Krafft (1909) figured two specimens which do not match with our sample, both having a much thinner whorl section and one being much more evolute. These two specimens are here assigned to *Ambites discus* (see above). As a consequence of this revised identification, their assumption that this species was found in five different layers within the '*Meekoceras* beds' of Lalung can be questioned. This species has been documented in a single regional zone by us in the Salt Range. Von Krafft, (1909, p. 26) distinguished *Meekoceras lingtiense* from this species on the basis of its 'delicate striæ, which cross the external part', which are 'directly opposed to the concentric external striæ [...] of *Meekoceras lilangense*'. The presence of these very delicate 'concentric external striæ' is strongly influenced by the preservation. It is visible only when the shell is exceptionally well preserved, and has not been observed in our specimens, but this feature is most probably present in every species of *Ambites* and can therefore not be considered as diagnostic. The specimen figured by Wang & He (1976) is too poorly preserved to validate its identification. As previously

mentioned, one of their specimens assigned to *Gyronites evolvens* agrees in proportions with *Ambites lilangensis*, but is also too poorly preserved for a reliable identification. Tozer (1967, 1970, 1994) initially ascribed some specimens from British Columbia to this species, but finally decided to erect a distinct species for these (*Ambites fuliginatus*) on the basis of their 'appreciably more conspicuous ribs and striae on the body chamber'. It is shown here that this is part of the intraspecific variability of this species, and actually some specimens he did not illustrate (D. Ware, pers. obs., 2011) are strictly identical to ours. One of the specimens he ascribed to this new species (*in* Tozer 1994; pl. 13, fig. 4) is distinctly more involute and thus may belong to the genus *Mullericeras* (see below). The specimens described by Waterhouse (1996) are reasonably well preserved and we agree with their assignment to *Ambites lilangensis*. However, as already discussed, erecting the new genus *Lilangia* for this species is not justified in our view.

Occurrence. – Nammal Nala, samples Nam100 ($n = 20$), Nam344 ($n = 14$), Nam345 ($n = 1$), Nam501 ($n = 9$) and Nam503 ($n = 20$).

Ambites bjerageri n. sp.

Plate 13, figures 10–26; Figures 22, 34

v 2008 *Gyronites frequens* Waagen, 1895; Brühwiler, Brayard, Bucher & Guodin, p. 1168, pl. 5, figs 7, 8.

Derivation of name. – Named after Dr. Morten Bjerager (Geological Survey of Denmark and Greenland, GEUS, Denmark).

Holotype. – Specimen PIMUZ30351 (Pl. 13, figs 12–15).

Type locality. – Nammal, Salt Range, Pakistan.

Type horizon. – Sample Nam100 (4 m above base of Ceratite Marls), *Ambites lilangensis* Regional Zone, middle Dienerian.

Diagnosis. – Small-sized sub-serpenticonic to serpenticonic *Ambites* (maximal diameter estimated at ca. 45 mm).

Material. – 34 specimens.

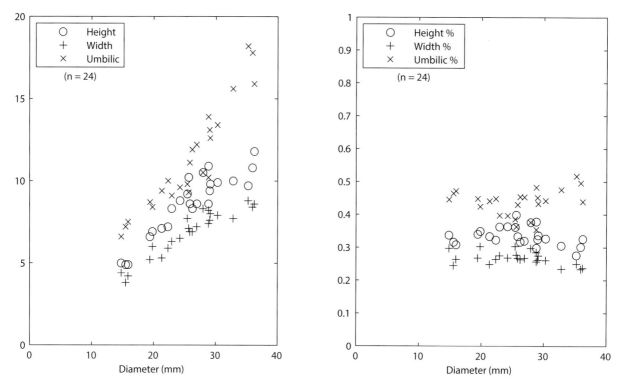

Fig. 34. Ambites *bjerageri* n. sp. Scatter diagrams of H, W and U, and of H/D, W/D, and U/D (abbreviations as in Fig. 15).

Description. – Very evolute (U/D ≈ 44%) sub-ser-penticonic to serpenticonic shell with an almost isometric whorl section (W/H ≈ 80%, varying between 70 and 90%). Flanks very convex with maximum width at mid-flanks. The tabulate ven-ter sometimes tends to disappear on adult speci-mens, and every intermediate exist between forms with a thick tabulate, bottleneck shaped venter, giving the whorl section a sub-rectangular shape, and forms without any clearly differentiated ven-ter and a sub-rounded whorl-section. Umbilical wall undifferentiated on forms with a sub-rounded whorl-section, oblique to sub-vertical on forms with a sub-rectangular whorl-section, vaguely differentiated by a broadly rounded shoulder. Ornamentation almost absent, consist-ing of very weak, slightly sigmoid folds following the trajectory of the growth lines and two occa-sional very weak spiral ridges. Suture line simple with rounded lateral saddles. The branches of the ventral lobe are very thin, with two or three small indentations at their base. The two lateral lobes are of nearly equal size, with numerous small indentations at their base. The second lat-eral saddle is much larger than the others, and the third lateral saddle is very small. Auxiliary series very short, with a few regularly spaced indentations.

Measurements. – See Figures 22, 34.

Discussion. – With its sub-serpenticonic shape, this species clearly differs from all other species of *Ambites*. The presence of variants with a clear *Ambites* like whorl section (sub-rectangular with a clearly marked bottleneck shaped venter) indicates a clear affinity with *Ambites*. The specimens identified by Brühwiler *et al.* (2008) as *Gyronites frequens* are clearly identical to our specimens with a broad tabu-late bottleneck shaped venter. Shigeta & Zakharov (2009) synonymized these specimens with *Gyronites subdharmus* Kiparisova (1961), an assignment we do not endorse because *Gyronites subdharmus* lacks the characteristic bottleneck shaped venter, includes variants with a strong ornamentation and has a sim-pler suture line with fewer indentations at the base of the lobes.

Occurrence. – Nammal Nala, samples Nam92 (*n* = 1), Nam100 (*n* = 10), Nam300 (*n* = 2), Nam344 (*n* = 8), Nam501 (*n* = 4), and Nam503 (*n* = 9).

Ambites cf. *A. impressus* (Waagen, 1895)

Plate 13, figures 7–9; Figure 22

? 1895 *Lecanites impressus* n. sp. Waagen, pp. 286, 287, pl. 37, figs 7 (lectotype, here designated), 8.

? 1934 *Gyrolecanites impressus* (Waagen, 1895); Spath, pp. 95, 96, fig. 21 [cop. Waagen, 1895].

Material. – One specimen.

Description. – Moderately evolute (U/D = 31.7%) relatively thick (W/H = 66.2%, W/D = 25.7%) discoidal shell. Strongly convex flanks with maximum whorl width at inner third. Umbilical wall high, sub-vertical, poorly individualized by a broadly rounded umbilical shoulder. Flanks nearly smooth, with very weak and slightly sigmoid folds. Suture line not visible.

Measurements. – See Figure 22 and Table 1.

Discussion. – This specimen has proportions, which are exactly intermediate between the two co-occurring species of *Ambites* (*Ambites bjerageri* and *Ambites lilangensis*). It cannot be excluded that it represents an extreme variant of one of these two species. It closely resembles the two specimens ascribed to *Lecanites impressus* by Waagen (1895), whose intraspecific variability cannot be assessed. Waagen also indicated that the type specimens were heavily weathered, and still mostly embedded in the matrix, but that their goniatitic suture line was probably not entirely due to this weathering. However, Waagen clearly described its bottleneck shaped venter, which is clearly visible on his illustrations. This species therefore clearly belongs to the genus *Ambites*, and its supposed very simple goniatitic suture line without auxiliary series is most likely a preservation artefact.

Occurrence. – Nammal Nala, sample Nam344.

Family Paranoritidae Spath, 1930

Type genus. – *Paranorites* Waagen, 1895.

Composition of the family. – *Paranorites* Waagen, 1895; *Koninckites* Waagen, 1895; *Vavilovites* Tozer, 1971; *Radioceras* Waterhouse, 1996; *Nanningites* Brayard & Bucher, 2008; *Wailiceras* Brayard & Bucher, 2008; *Urdyceras* Brayard & Bucher, 2008; *Vercherites* Brühwiler, Ware, Bucher, Krystyn & Goudemand, 2010a, *Koiloceras* Brühwiler & Bucher, 2012, *Pashtunites* n. gen. and *Awanites* n. gen.

Discussion. – Since its introduction and discussion by Spath (1930, 1934), this family has been considered as a synonym of Proptychitidae (e.g. Arkell *et al.* 1957; Tozer 1961, 1994; Brühwiler *et al.* 2012). However, typical representative of the Proptychitidae, such as the genera *Proptychites*, *Pseudoproptychites*, *Bukkenites* or *Pseudaspidites* all have a broadly rounded venter without any distinct shoulder, inflated inner whorls, no spiral ridges on the flanks and a suture line with usually very deep lobes and elongated saddles, the saddles tending to be slightly phylloid and the lobes having numerous and deep denticulations extending on their lateral sides. In the Salt Range, a group of species which were previously assigned to this family do not show any of these characteristics of the suture line. They all have a tabulate or sub-tabulate venter at least in the inner whorls, compressed inner whorls, and the most robust specimens within each species have two or more spiral ridges on the flanks. They also all have a very similar suture line with moderately to very deep lobes and elongated saddles, the lobes having a few rather deep indentations restricted to their base, the third lateral saddle being elongated, and a long auxiliary series with a poorly differentiated auxiliary lobe. These forms can therefore clearly be differentiated from Proptychitidae, as already noticed by Spath (1934) who created the family Paranoritidae to group them. Spath also suggested that these two families were closely related. This opinion was followed by the subsequent authors, who even considered them as synonyms. Spath (1934) based his opinion on the similarity between the suture lines of *Paranorites gigas* Waagen, 1895 and that of one specimen assigned to *Meekoceras markhami* (Diener, 1897) by von Krafft (1909). This last species is of uncertain affinity, so this similarity of suture line is not sufficient to demonstrate a phylogenetic link between the two families. On the other hand, some characteristics of Paranoritidae can also be found in Gyronitidae, such as the tabulate venter, spiral ridges on the flanks, a strigation on the ventro-lateral shoulders, and the compressed inner whorls. Moreover, the youngest species of *Ambites*, *Ambites lilangensis*, has a more differentiated suture line than that of other members of the genus *Ambites*, with less numerous, deeper indentations on the lobes and sometimes an auxiliary lobe. Some specimens of *Ambites lilangensis* lose their bottleneck venter, their venter becoming then sub-tabulate, low arched with sharp ventro-lateral shoulders, as embodied by *Vavilovites*. Paranoritidae are therefore more likely to root within Gyronitidae, as suggested by *Vavilovites* which is an intermediate form between *Ambites* and *Koninckites*.

Genus *Vavilovites* Tozer, 1971

Type species. – *Paranorites sverdrupi* Tozer, 1963.

Discussion. – The type species of this genus was originally placed in the genus *Paranorites*, within the family Paranoritidae. When Tozer (1971) erected *Vavilovites*, he also synonymized Paranoritidae with Proptychitidae. We already justified above that the two families should be kept separate, and that Paranoritidae is a derivative of Gyronitidae (a family which Tozer considered as a synonym of Meekoceratidae). We therefore here re-assign this genus to Paranoritidae and exclude it from Proptychitidae.

Vavilovites cf. *V. sverdrupi* (Tozer, 1963)

Plate 13, figures 27–35; Figure 35

? 1963   *Paranorites sverdrupi* n. sp. Tozer, pp. 12–15, pl. 4, figs 1, 2 (holotype), 3–6.

? 1967   *Paranorites sverdrupi* Tozer, 1963; Tozer, pl. 4, fig. 2.

? 1970   *Paranorites sverdrupi* Tozer, 1963; Tozer, pl. 16, fig. 7 (holotype).

? 1994   *Vavilovites sverdrupi* n. gen. (Tozer, 1963); Tozer, p. 63, fig. 16A, pl. 15, figs 1, 2 (holotype), 3; pl. 20, fig. 2.

? 1996   *Vavilovites tabulatus* n. sp. Waterhouse, pp. 74, 75, text-fig. 10C, pl. 8, figs 13–16, pl. 9, figs 1–3 (holotype), 4–6.

v 2011   *Vavilovites* sp. indet. Ware, Jenks, Hautmann & Bucher, pp. 176, 177, fig. 19.

Material. – 5 specimens.

Description. – Involute (U/D ≈ 15.5%) relatively thick (W/H ≈ 55.7%, W/D ≈ 28.6%) discoidal shell with broad tabulate venter. Flanks nearly flat with maximum whorl width just before the umbilical wall, giving the whorl section a sub-trapezoidal outline. Umbilical wall high, vertical, individualized by a narrowly rounded umbilical shoulder. Flanks nearly smooth, with very weak sigmoid folds. Growth lines not visible. Suture line typical of Paranoritidae, with a clearly individualized auxiliary lobe, and a relatively large third lateral saddle.

Measurements. – See Figure 35 and Table 1.

Discussion. – Only four poorly preserved specimens belonging to this species have been found. The

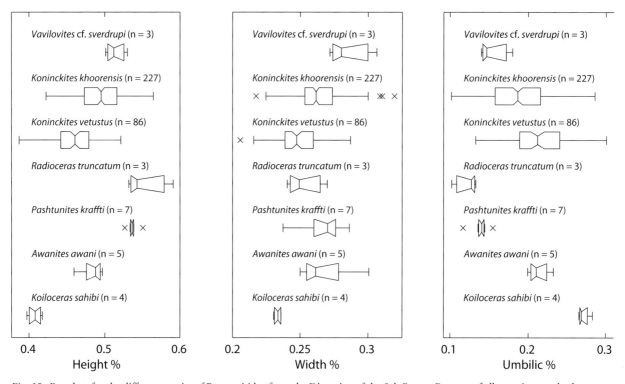

Fig. 35. Boxplots for the different species of Paranoritidae from the Dienerian of the Salt Range. Because of allometric growth, these were calculated only for specimens of 30 to 70 mm in diameter. Abbreviations as in Figure 15.

smallest (Pl. 13, figs 30, 31) is slightly more evolute and has a thicker whorl section than the three other specimens. It has exactly the same morphology as the specimens described by Ware *et al.* (2011) as *Vavilovites* sp. indet. from Nevada. The difference between this specimen and the other three is here interpreted as intraspecific variability. Although Tozer (1994) based his species on a total of 63 specimens, he did not provide measurements for all of them nor any biometrical analyses concerning intraspecific variability and growth allometry. Moreover, the specimens for which he provides measurements are generally much larger (between 8 and 20 cm) than our specimens. It is therefore very difficult to estimate whether our specimens are conspecific with those of Tozer or not. The two smallest specimens measured by Tozer (1963) agree in proportions with our material.

Occurrence. – Nammal Nala, samples Nam316 (*n* = 1), Nam318 (*n* = 1) and Nam396 (*n* = 3).

Genus *Koninckites* Waagen, 1895

Type species. – *Koninckites vetustus* Waagen, 1895.

Composition of the genus. – *Koninckites vetustus* Waagen, 1895; *Proptychites khoorensis* Waagen, 1895.

Emended diagnosis. – Compressed platyconic to discoidal shells whose involution shows an important intraspecific variability, from moderately to very involute shells. Maximum diameter of ca. 15 cm, with a strong allometric growth. The innermost whorls (up to D = 15 mm) are moderately to very evolute, with a moderately compressed whorl section, sometimes sub-serpenticonic. With further growth, the shell becomes more involute and compressed, reaching its maximal whorl compression at a diameter of ca. 15 mm and its maximal involution at a diameter of ca. 30 mm. Venter tabulate to well rounded, the innermost whorls of forms with a rounded venter being sub-tabulate. Umbilical wall varying from very oblique and concave to vertical, always well individualized by a clear angular umbilical shoulder. Ornamentation always weak, consisting of low sigmoid folds parallel to the growth lines and of two low spiral ridges on the flanks, nearly absent on the most involute forms. Sculpture fading away on adult body chamber. The juvenile ornamentation of the more evolute variants is more pronounced (following the first Buckman law of covariation) and variable, consisting usually of deep constrictions, more rarely of ribs and sometimes of a strigation

on the flanks. Suture line typical of Paranoritidae with a poorly to well individualized auxiliary lobe, the auxiliary series being long on involute forms, short on evolute ones.

Occurrence. – Late Dienerian of the Salt Range (Pakistan), Central Himalayas (India, Nepal), Guangxi (China).

Discussion. – This genus has been used by many authors as a wastebasket taxon for every compressed, involute and tabulate species with a simple suture line with one individualized auxiliary lobe. We here redefine this genus on the basis of its type species and give it a more restricted and precise scope.

Koninckites vetustus Waagen, 1895

Plate 14, figures 1–36; Plate 15, figures 1–20;
Plate 16, figures 1–15; Plate 17, figures 1–4;
Figures 35–37

1895   *Koninckites vetustus* n. gen., n. sp. Waagen, pp. 261, 262, pl. 27, figs 4 (lectotype), ?5.

1895   *Prionolobus rotundatus* n. gen., n. sp. Waagen, p. 310, pl. 34, figs 1 (lectotype), 2, 3.

1905   *Prionolobus rotundatus* Waagen, 1895; Noetling, pl. 23, fig. 1.

1909   *Meekoceras waageni* n. sp. Diener, p. 16.

1909   *Meekoceras smithii* n. sp. von Krafft, p. 39, pl. 4, fig. 1 (holotype).

1934   *Koninckites apertus* n. sp. Koken p. 154, pl. 11, fig. 1 (holotype).

? 1996   *Koninckites apertus* Koken, 1934; Waterhouse, pp. 46–48, Text-fig. 4H, pl. 2, figs 9, 11, 13–15, pl. 9, figs 7–12.

v 2010a   *Prionolobus rotundatus* Waagen, 1895; Brühwiler, Ware, Bucher, Krystyn & Goudemand, p. 728, figs 6, 7.

Emended diagnosis. – Moderately to very involute and discoidal *Koninckites* with a tabulate venter persisting until transition to mature body chamber. Flanks strongly convex, umbilical wall oblique, sometimes slightly concave.

Material. – 189 specimens.

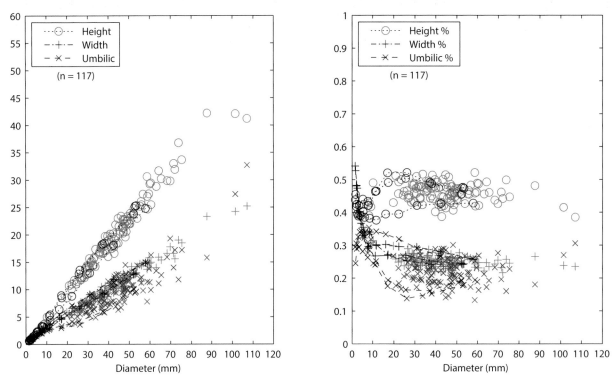

Fig. 36. Koninckites vetustus Waagen, 1895;. Scatter diagrams of H, W and U, and of H/D, W/D, and U/D (abbreviations as in Fig. 15). Ontogenetic trajectories obtained from sectioned specimen illustrated on Plate 14, figures 31–36 are shown in black.

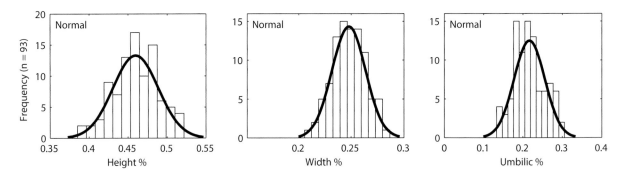

Fig. 37. Koninckites vetustus Waagen, 1895;. Histograms of H/D, W/D, and U/D (abbreviations as in Fig. 15). Because of allometric growth, specimens smaller than 30 mm in diameter have been excluded.

Description. – Moderately to very involute compressed discoidal shells with a marked, although highly variable allometry. Following the evolute neanic stage (where U/D varies between 30 and 35%), the most involute variants show the same allometry as *Koninckites khoorensis* (see below), becoming very involute and compressed up to a diameter of ca. 30 mm, reaching a relative umbilical width of ca. 15%, and then becoming gradually more evolute. Some involute variants show a strong egression of the mature body chamber. The involution of the most evolute variants increases slowly and steadily until maturity, where the relative umbilical width still amounts to as much as 25%. Both extremes have an equivalent whorl width (W/D between 25 and 30%) which they reach very

early, at a diameter of ca. 20 mm. The venter becomes clearly tabulate with almost sharp ventro-lateral shoulders after the neanic stage (at a diameter of ca. 5 mm), and stays so until the beginning of the mature body chamber where the ventro-lateral shoulders become more rounded or even indistinct, the venter becoming then narrowly rounded. The flanks are strongly convex, with maximum width between inner third and mid-flanks. Umbilical wall oblique, sometimes slightly concave. It forms a very obtuse angle with the flanks, the umbilical shoulder being narrowly rounded on the outer shell, absent or nearly absent on the inner mould. It tends to disappear on the body chamber. On involute variants, the umbilical seam follows the umbilical shoulder of the preceding whorl, thus imparting the umbilicus a funnel shape. Inner-most whorls with marked prorsiradiate constrictions which disappear at a diameter of ca. 10 mm. The ornamentation becomes then very weak, generally consisting of very low, thin sigmoid folds following the growth lines and one low spiral ridge at external third of the whorl height, more rarely a second spiral ridge at inner third of the whorl height. These two ridges then bracket the second lateral saddle as observed in the genus *Ambites*. Some very evolute specimens have low broad folds on their flanks. Suture line simple, with moderately deep lobes. Saddles with well rounded tips, sometimes slightly phylloid, the sec-ond lateral saddle being occasionally bent towards the umbilicus. Auxiliary series well developed, although shorter on evolute variants than on involute ones. Auxiliary lobe usually well individualized, but some-times differentiated from the rest of the auxiliary series only by its smaller and denser indentations.

Measurements. – See Figures 35–37.

Discussion. – This species shows a very strong intraspecific variability, especially concerning its umbilical width, a pattern also shared with *Koninck-ites khoorensis*. This strong variability led previous authors to assign them to different species, genera and even families. The type species of *Koninckites*, *Koninckites vetustus*, is mostly based on one, rather poorly preserved specimen from Chiddru, of which a photography is here included (Pl. 15, figs 1, 2). The original drawing of Waagen is thus a reconstitu-tion of this specimen. Waagen originally described it as having a well-rounded venter. However, the pho-tography shows that the venter is actually sub-tabu-late near the beginning of the body chamber, a trait that Waagen apparently did not notice. The phrag-mocone is too damaged for its ventral shape to be clearly seen, but it can be expected to be tabulate. Waagen also did not notice an umbilical shoulder, but the umbilicus of the holotype is poorly pre-served and only visible as an internal mould near the end of the body chamber, where this shoul-der generally fades away. This specimen corre-sponds to an extreme, very involute and thick variant of *Koninckites vetustus*. A nearly identical but well-preserved specimen inclusive of the suture line is figured in Plate 15, figs 7–10. The other specimen which Waagen assigned to this species is too fragmentary to be identified with certainty. The species *Prionolobus rotundatus* cor-responds to the more typical variants of the spe-cies, rather involute and compressed with a funnel shaped umbilicus. This species was renamed *Meekoceras waageni* by Diener (1909).

Koninckites apertus also corresponds to these typical variants. *Meekoceras smithii* corresponds to the most evolute variants. The specimens described by Water-house (1996) as *Koninckites apertus* seem also to cor-respond to this species, but they are too poorly preserved for unambiguous identification. With its strongly oblique umbilical wall, very convex flanks and tabulate venter, *Koninckites vetustus* can easily be distinguished from all other Paranoritidae.

Occurrence. – Amb, sample Amb65 ($n = 9$); Chid-dru, samples Chi51 ($n = 6$), Chi61 ($n = 4$) and Chi151 ($n = 3$); Nammal Nala, samples Nam63 ($n = 15$), Nam70 ($n = 1$), Nam71 ($n = 5$), Nam83 ($n = 5$), Nam101c ($n = 5$), Nam305 ($n = 67$), Nam312 ($n = 12$), Nam347 ($n = 19$) and Nam349 ($n = 11$); Wargal, sample War104 ($n = 27$).

Koninckites khoorensis (Waagen, 1895)

Plate 17, figures 5–37; Plate 18, figures 1–31;
Plate 19, figures 1–12; Plate 20, figures 1–13;
Figures 35, 38–40

> 1895 *Proptychites khoorensis* n. gen., n. sp. Waagen, pp. 176–178, pl. 20, fig. 4 (holotype).

> 1895 *Dinarites minutus* n. sp. Waagen, pp. 31, 32, pl. 7a, figs 1 (lectotype), 2.

> 1895 *Proptychites undatus* n. gen., n. sp. Waagen, pp. 180–182, pl. 24, fig. 4 (holotype).

> 1895 *Meekoceras koninckianum* n. gen., n. sp. Waagen, pp. 245–247, pl. 26, fig. 6 (holotype).

> 1895 *Koninckites ovalis* n. gen., n. sp. Waa-gen, pp. 262, 263, pl. 28, figs 3 (lecto-type), 4.

> 1905 *Aspidites declivis* (Waagen, 1895); Noetling, pl. 22, fig. 8.

> non 1905 *Prionolobus undatus* (Waagen, 1895); Noetling, pl. 25, fig. 2.

> 1909 *Meekoceras tenuistriatum* n. sp. von Krafft, p. 34, pl. 4, fig. 3 (holotype).

> ?p 1909 *Meekoceras varaha* Diener, 1895; von Krafft & Diener, p. 17, pl. 2, fig. 2.

> v 1978 *Proptychites khoorensis* Waagen, 1895; Guex, pl. 2, figs 3, 4, 6.

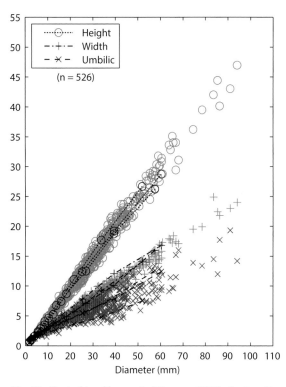
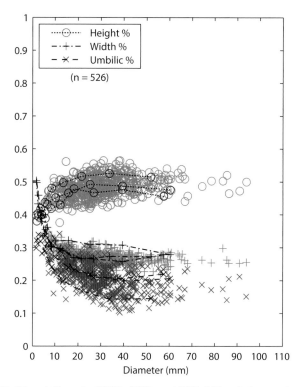

Fig. 38. Koninckites khoorensis (Waagen, 1895). Scatter diagrams of H, W and U, and of H/D, W/D, and U/D (abbreviations as in Fig. 15). Ontogenetic trajectories obtained from sectioned specimen illustrated on Plate 18, figures 10–15 are shown in black.

1996 *Prejuvenites angdawai* n. gen., n. sp. Waterhouse, pp. 55, 56, Text-fig. 7A, pl. 3, figs 5–9, 10 (holotype), 11–13.

1996 *Proptychites chuluensis* n. sp. Waterhouse, pp. 68–70, Text-fig. 11E, pl. 7, figs 7, 8 (holotype).

v 2008 *Proptychites* sp. indet.; Brühwiler, Brayard, Bucher & Guodun, p. 1164, pl. 2, figs 3–5.

Emended diagnosis. – Koninckites whose venter becomes well rounded early in ontogeny (at a diameter of ca. 30 mm). Umbilical wall oblique and concave to vertical. Thin, slightly sinuous to radial lirae appear at a diameter of ca. 20 mm.

Material. – 837 specimens.

Description. – Moderately to very involute, discoidal to platyconic shell with a strong allometry and a broadly rounded venter. At the end of the neanic stage (i.e. at D $\approx$ 5 mm), shells are evolute discoidal (U/D $\approx$ 35%), with a rounded whorl section (W/H $\approx$ 100%) and prorsiradiate deep constrictions (4 or 5 per whorl), which generally cross the venter. Between 5 and ca. 25 mm, the involution

and compression of the shell increase rapidly, the venter becomes sub-tabulate with narrowly rounded ventro-lateral shoulders, the umbilical wall starts to differentiate, the constrictions become more abundant and shallower, and strigation appears. The strength of this allometric growth and change in morphology is very variable, so the intraspecific variability is maximal in these juvenile forms: U/D varies from 15 to 30% and W/H from 50 to 70%. The ornamentation is also very variable and follows Buckman's first law of covariation. The most evolute variants usually have a slightly compressed oval whorl section and numerous deep constrictions crossing the venter. The majority have a moderately compressed (W/H $\approx$ 60%) sub-trapezoidal whorl section, the constrictions becoming shallower and fading on the venter. The most involute variants have a compressed whorl section with sub-parallel flanks and no constriction. The strigation varies independently of the whorl section shape. It can be very weak, visible only on the shell and restricted to the venter, or very strong, visible both on the shell and the inner mould, and present also on the flanks. At a diameter of ca. 20 mm, sigmoid lirae appear, and on specimens with a strong strigation it gives the outer shell a reticulated aspect. For diameters larger than ca. 25 mm, the proportions of the shell stay nearly constant, the umbilicus sometimes

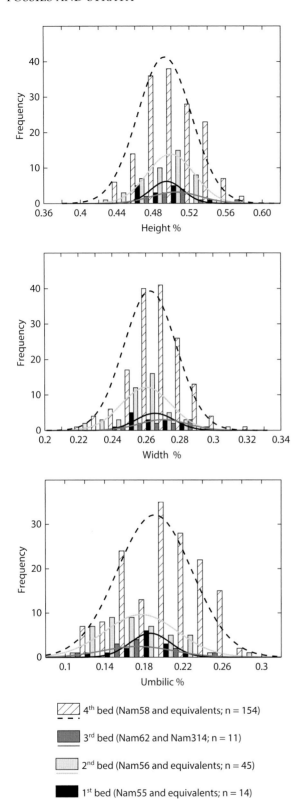

Fig. 39. Koninckites khoorensis (Waagen, 1895). Comparison between the histograms of H/D, W/D, and U/D for four samples occurring in four successive beds at Nammal Nala. Because of allometric growth, only specimens of 30–70 mm in diameter were taken into account.

becomes a little broader, and the ornamentation changes. The broad folds disappear, and a spiral ridge situated at the external third of flank appears, along with an occasional second one emerging at the inner third of flank. These two ridges then bracket the second lateral saddle. The venter usually changes into a rounded outline. However, a few specimens show very weak ventro-lateral shoulders that may persist with further growth. The umbilical wall becomes clearly individualized by a sharp shoulder, and varies from oblique and slightly concave to vertical, independently of the shell proportions. The strigation tends to disappear on the flanks, but 2 to 4 spiral lines remain at a position homologous to that of the ventro-lateral shoulders. On some specimens, the constrictions are replaced by faint sigmoid folds, emphasized by stronger lirae. The lirae become straight on the involute, sub-platyconic variants. At the sub-adult or adult stage, the general morphology of the shells varies from moderately involute (U/D $\approx$ 25%) with convex flanks and a maximum width at the inner third of the flank, to very involute (U/D $\approx$ 15%). On very involute forms, the flanks can be convex with maximal width at inner third, or sub-parallel with maximal width at the umbilical shoulder. The inner third of the flank is generally very flat, sometimes even very slightly concave. Suture line typical of the genus, the auxiliary lobe being clearly individualized even at small size (for D $\approx$ 15 mm), with deep lobes. Lobes deeper and auxiliary series longer on involute specimens than on evolute ones.

Measurements. – See   Figures 35,   38–40   and Tables 2, 3.

Discussion. – More than 500 specimens of this species have been measured for biometrical analyses, the vast majority coming from four consecutive beds in Nammal. The very important intraspecific variability of this species, and especially of its umbilicus size, may actually be increased by the presence of *in vivo* encrusting bivalves in the umbilicus. A vast majority of involute specimens larger than 3 cm in diameter are carrying such epibionts. Hence, measurements of U may be influenced by such noise and absence of unaffected specimens prevents quantifying this effect. The first step of this study is to check if the samples display a normal distribution for the measured characters. To avoid biases linked with allometry, only specimens between 30 and 70 mm of diameter were taken into account. A Lilliefors test was performed. The results of this univariate study

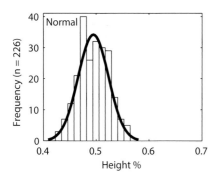 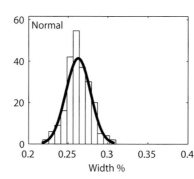 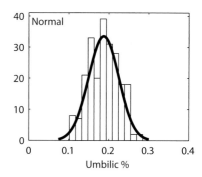

Fig. 40. Koninckites khoorensis (Waagen, 1895). Histograms of H/D, W/D, and U/D (abbreviations as in Fig. 15). Because of allometric growth, only specimens of 30–70 mm in diameter were taken into account.

Table 2. Univariate statistics of H/D, W/D and U/D for each population of *Koninckites khoorensis* (Waagen, 1895) occurring in four successive beds at Nammal Nala.

Bed	N	Mean	Median	Minimum	Maximum	Standard deviation	Lilliefors test (p)
H/D							
1st bed	14	0.495	0.495	0.472	0.543	0.019	0.429
2nd bed	45	0.499	0.506	0.422	0.543	0.026	0.058
3rd bed	11	0.505	0.508	0.465	0.563	0.029	0.860
4th bed	154	0.493	0.491	0.429	0.565	0.030	0.276
W/D							
1st bed	14	0.265	0.266	0.249	0.285	0.012	0.086
2nd bed	45	0.260	0.260	0.227	0.298	0.015	0.156
3rd bed	11	0.270	0.279	0.243	0.291	0.015	0.039
4th bed	154	0.262	0.262	0.217	0.310	0.016	0.675
U/D							
1st bed	14	0.186	0.191	0.139	0.220	0.021	0.238
2nd bed	45	0.177	0.170	0.115	0.286	0.038	0.193
3rd bed	11	0.174	0.163	0.106	0.247	0.038	0.187
4th bed	154	0.191	0.192	0.102	0.278	0.038	0.157

D, Diameter; H, whorl height; W, whorl width; U, umbilical diameter.
Because of allometric growth, only specimens of 30 to 70 mm in diameter were taken into account.

are summarized on Table 2. The corresponding histograms are shown in Figure 39. The null hypothesis (the population is normally distributed) could be rejected only in one case. For the third bed, the relative whorl width does not follow a normal distribution ($p = 0.039$). However, only 11 specimens with a diameter comprised between 30 and 70 mm from this bed could be measured, making this result a not very robust one. It is likely that a larger sample from this bed would modify this result. The second step is to compare the different beds. The results of the One-way ANOVA and Kruskal-Wallis tests (PAST, Hammer *et al.* 2001; see Table 3) show that for all three parameters, the specimens from the four different beds come from a uniform population. Pairwise comparisons show that all beds have the same mean and median, with only one exception, the population from the second bed. These specimens have a median relative umbilicus width which is significantly different from those of the fourth bed (Mann–Whitney

$p = 0.024$). As these two beds are the ones with the maximum number of specimens, this result is *a priori* robust, with a median of U/D of 17% for the second bed, and of 19.2% for the fourth bed. However, the range of variation of the two samples is nearly the same, and a difference of 2.2% in the umbilicus relative width is not very significant from a taxonomical point of view. The interpretation is that both populations represent the same species (i.e. every variant of this species can be found in both beds), but involute forms are more abundant in the second bed than in the fourth bed. A visual examination showed that all the variability of the ornamentation, of the whorl section and of the suture line can be found in the four samples. As a consequence, all these populations are interpreted to represent a single species. Histograms for the pooled samples (between 30 and 70 mm of diameter), including the few specimens from Amb and Chiddru (Fig. 40), show that the resulting distributions are normal.

Table 3. Comparison with statistical tests of H/D, W/D and U/D between the four populations of *Koninckites khoorensis* (Waagen, 1895) occurring in four successive beds at Nammal Nala.

Comparison between beds	One-way ANOVA (p)	Tukey (p)	Kruskal–Wallis (p)	Mann–Whitney (p)
H/D				
1st bed vs. 2nd bed		0.957		0.269
1st bed vs. 3rd bed		0.654		0.366
1st bed vs. 4th bed	0.322	0.994	0.223	0.701
2nd bed vs. 3rd bed		0.916		0.821
2nd bed vs. 4th bed		0.869		0.084
3rd bed vs. 4th bed		0.488		0.200
W/D				
1st bed vs. 2nd bed		0.690		0.350
1st bed vs. 3rd bed		0.720		0.366
1st bed vs. 4th bed	0.224	0.938	0.235	0.537
2nd bed vs. 3rd bed		0.138		0.058
2nd bed vs. 4th bed		0.955		0.423
3rd bed vs. 4th bed		0.360		0.089
U/D				
1st bed vs. 2nd bed		0.851		0.243
1st bed vs. 3rd bed		0.698		0.218
1st bed vs. 4th bed	0.101	0.979	0.079	0.489
2nd bed vs. 3rd bed		0.993		0.984
2nd bed vs. 4th bed		0.623		0.024
3rd bed vs. 4th bed		0.447		0.163

D, Diameter; H, whorl height; W, whorl width; U, umbilical diameter.
Because of allometric growth, only specimens of 30–70 mm in diameter were taken into account.

Inclusion of this species in the genus *Koninckites* may seem odd at first sight. The well rounded venter of adult forms led previous authors to include it in the genus *Proptychites*. However, this species shows all the diagnostic traits listed above as typical for *Koninckites*. The most important similarity between this species and the type species of *Koninckites* (*Koninckites vetustus*) are the ontogenetic trajectories of the two species, which are strictly identical until the adult stage, where *Koninckites vetustus* shows a greater umbilical egression than *Koninckites khoorensis*. Moreover, evolute variants of *Koninckites khoorensis* with an oblique umbilical wall are nearly identical to evolute thick variants of *Koninckites vetustus*. They differ only by their venter (rounded on *Koninckites khoorensis*) and the presence of radial lirae on *Koninckites khoorensis*. Both show a similarly wide range of variation, this range and the median being slightly shifted towards more evolute, more compressed forms in *Koninckites vetustus* than in *Koninckites khoorensis* (see Fig. 35). The innermost whorls (from 5 to 20 mm in diameter) of the two species are nearly identical, the only difference being the venter, which is clearly tabulate in *Koninckites vetustus*, whereas it is sub-tabulate in *Koninckites khoorensis*. The suture lines of the two

species are nearly identical. Considering the similarities of ontogeny, suture line and intraspecific variability, we decided to assign the two species to the same genus. The rounded venter of adults and sub-adults *Koninckites khoorensis* is here interpreted as a convergence towards the genus *Proptychites*, from which they differ by their simpler suture line, their ontogeny, their ornamentation and their clearly differentiated umbilical wall. This species also resembles the co-occurring *Kingites davidsonianus*, from which it differs by its more complex suture line, its ontogeny, its wider umbilicus and, when present, its stronger ornamentation.

The very different morphology of the juveniles compared with that of the adults of this species confused several authors who assigned the juveniles to different families or even superfamilies than the adults. This is the case for *Dinarites minutus* Waagen, 1895 and *Prejuvenites angdawai* Waterhouse, 1996. As a consequence, the genus *Prejuvenites* Waterhouse, 1996, based on these specimens, is here synonymized with *Koninckites*. Waagen (1895) described five species which can be synonymized. As he was the first to describe these forms, we have to choose one of these five names. We here choose to keep the species name *Koninckites khoorensis* because this name corresponds to the first described adult or sub-adult specimen in Waagen's work and because it is the only name which has been used by subsequent authors. *Proptychites khoorensis* Waagen, 1895 (both in the original paper and in Guex, 1978), *Aspidites declivis* (*in* Noetling, 1905) and *Meekoceras tenuistriatum* von Krafft, 1909 correspond to involute variants of this species. *Proptychites undatus* Waagen, 1895; *Meekoceras koninckianum* Waagen, 1895 and *Koninckites ovalis* Waagen, 1895 correspond to evolute variants. *Proptychites chuluensis* Waterhouse, 1996 corresponds to an intermediate variant. *Proptychites* sp. indet. (*in* Brühwiler *et al.* 2008) corresponds to small immature specimens of this species. One specimen identified by Diener (1909, p l. 2, fig. 2) as *Meekoceras varaha*, which is clearly pathological with an irregularly shaped umbilicus and outline, seems to have lirae as in the specimens here described. It may correspond to an involute, pathological variant of *Koninckites khoorensis*, but no lateral view is illustrated, thus preventing any firm assignment of this specimen.

Occurrence. – Amb, samples Amb3 (*n* = 2) and Amb64 (*n* = 3); Chiddru, sample Chi105 (*n* = 4); Nammal Nala, samples Nam55 (*n* = 8), Nam56 (*n* = 2), Nam58 (*n* = 35), Nam59 (*n* = 203), Nam61 (*n* = 70), Nam62 (*n* = 38), Nam67 (*n* =

109), Nam101d (*n* = 39), Nam130 (*n* = 7), Nam304 (*n* = 11), Nam313 (*n* = 22), Nam314 (*n* = 8), Nam315 (*n* = 35), Nam316 (*n* = 3), Nam318 (*n* = 1), Nam319 (*n* = 15), Nam336 (*n* = 9), Nam337 (*n* = 9), Nam346 (*n* = 49), Nam395 (*n* = 4), Nam499 (*n* = 3), Nam502 (*n* = 70), Nam504 (*n* = 27), Nam537 (*n* = 18), Nam721 (*n* = 9), Nam724 (*n* = 22) and Nam727 (*n* = 2).

Genus *Radioceras* Waterhouse, 1996

Type species. – *Meekoceras radiosum* Waagen, 1895.

Composition of the genus. – *Aspidites evolvens* Waagen, 1895 and *Koninckites truncatus* Spath, 1934.

Emended diagnosis. – Compressed and involute shell with tabulate venter, showing a strong allometry during growth. Inner whorls (up to ca. 5 cm) very involute, with nearly flat flanks and angular ventro-lateral and umbilical shoulders imparting the whorl section a compressed trapezoidal outline, with umbilical wall vertical to overhanging. Subsequent whorls more evolute, with more convex flanks and less angular whorl section. Suture line typical of Paranoritidae with one slightly differentiated auxiliary lobe.

Discussion. – Waterhouse (1996) created this genus for several small involute, compressed tabulate forms with a vertical umbilical wall and a suture line with a differentiated auxiliary series. The suture line of this genus is actually diagnostic of Paranoritidae. As already been shown by Brühwiler *et al.* (2010a, 2012), it can actually reach a large size (more than 15 cm) with a strong allometric growth. The diagnosis provided by Waterhouse is therefore incomplete. This genus is closely related to the genus *Koninckites*, from which it differs by its more involute shape in sub-adult specimens, and its more angular whorl section. The species *Meekoceras timorense* Wanner, 1911, which was included in the genus *Koninckites* by Spath (1934) and in the genus *Clypeoceras* by Shigeta & Zakharov (2009), most probably belongs to this genus.

Radioceras truncatum (Spath, 1934)

Plate 21, figures 1–6; Figures 35, 41

1895   *Koninckites davidsonianus* (de Koninck, 1863) n. gen.; Waagen, pp. 272, 273, pl. 33, fig. 4.

? 1897   *Meekoceras (Kingites) varaha* Diener, 1895; Diener, pp. 143, 144, pl. 6, fig. 2, pl. 7, fig. 6.

1934   *Koninckites truncatus* n. sp. Spath, pp. 152, 153, figs. 43c, 44 (holotype) [cop. *Koninckites davidsonianus* in Waagen, 1895].

? 1996   *Radioceras truncatum* (Spath, 1934) n. gen.; Waterhouse, p. 41, Text-fig. 4C, pl. 2, figs 1, 2.

v 2010a   ?*Radioceras* cf. *kraffti* (Spath, 1934); Brühwiler, Ware, Bucher, Krystyn & Goudemand, pp. 728, 729, figs 8, 9.

v 2012   *Radioceras* cf. *kraffti* (Spath, 1934); Brühwiler & Bucher, p. 53, figs 31Q–T, 33A–U, 34A–Z.

Emended Diagnosis. – Involute shell becoming more evolute at maturity. Whorl section sub-trapezoidal. Venter tabulate, with sharp shoulders on inner whorls becoming rounded on outer whorls. High vertical to overhanging umbilical wall with sharp shoulders on inner whorls, becoming less steep and less differentiated on outer whorls.

Material. – 10 specimens.

Description. – Only a few small immature specimens of this species have been found in the late Dienerian. Innermost whorls (up to 2 cm in diameter) are very close to those of *Koninckites vetustus*, slightly more involute (U/D ≈ 24%), with a tabulate venter, weakly convex convergent flanks and weak, slightly sinuous prorsiradiate constrictions. They rapidly become more involute (U/D ≈ 12% for diameter between 3 and 4 cm), and compressed (W/D ≈ 25%, W/H ≈ 46%), and the flanks become nearly flat with maximal width at or slightly above the umbilical shoulder, giving the whorl section a very elongated sub-trapezoidal shape. The umbilical wall is vertical to slightly overhanging, individualized by a sharp shoulder. Some specimens have very weak, slightly sinuous prorsiradiate folds, following the trajectory of the growth lines, which sometimes cross the venter. The only suture line which could be drawn here has slightly deeper lobes than the ones illustrated by Waagen (1895) and by Brühwiler *et al.* (2010a, 2012), which were in both case from larger specimens, but is otherwise very similar with a long auxiliary series and a weakly differentiated auxiliary lobe.

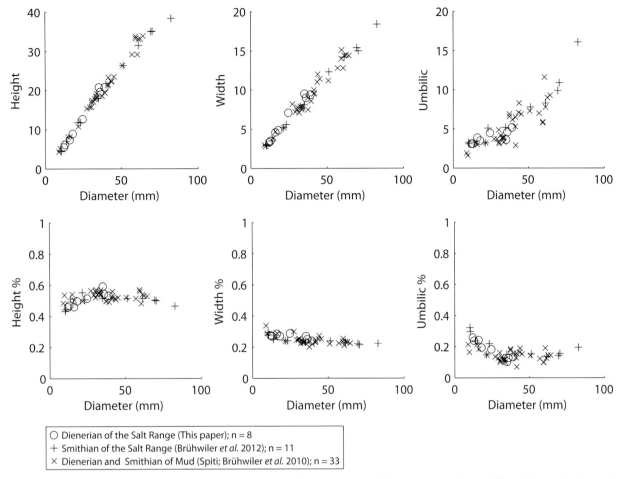

Fig. 41. Radioceras truncatum (Spath, 1934). Scatter diagrams of H, W and U, and of H/D, W/D, and U/D (abbreviations as in Fig. 15), with comparison of the Dienerian of the Salt Range with faunas from the Smithian of the Salt Range (data from Brühwiler *et al.* 2012) and from the Dienerian and Smithian of Mud (Spiti Valley, India; data from Brühwiler *et al.* 2010a).

According to the specimens described in Brühwiler *et al.* (2010a, 2012), adult specimens become more evolute, and the overall whorl section becomes less angular (ventro-lateral and umbilical shoulders become more rounded while the flanks become more convex, with a maximal width around inner third of whorl height), while the umbilical wall changes from vertical to slightly oblique.

Measurements. – See Figures 35, 41.

Discussion. – Spath (1934) created this species for a specimen misidentified by Waagen (1895) as *Koninckites davidsonianus* (de Koninck, 1863). We agree with Waterhouse (1996) who placed this species in a new genus together with *Radioceras radiosum* (Waagen, 1895), a species from which it differs by its less differentiated suture line, its smaller adult size and the more angular whorl section of immature stages. However, the specimen assigned by

Waterhouse to this species is too poorly preserved to be identified with certainty. The two specimens referred to as *Meekoceras (Kingites) varaha* by Diener (1897) agree in their proportions and suture line with the specimens described here. They seem to have a less angular whorl section, but are described by the author as having sharp shoulders, as in the present species. Their assignment to *Radioceras truncatum* cannot be confirmed without examination of Diener's type material. The specimens from the Smithian of the Salt Range and from the Dienerian–Smithian boundary of Mud (Spiti valley, India) were previously ascribed to *Koninckites* cf. *K. kraffti* (Spath, 1934). It is here shown that this latter species differs from *Radioceras truncatum* by a smaller adult size, absence of umbilical egression, less angular whorl section with narrowly rounded ventro-lateral and umbilical shoulders, vertical to slightly oblique umbilical wall and more convex flanks, even at small size.

These involute tabulate Paranoritidae closely resemble some species included below in *Mullericeras*, especially *Mullericeras shigetai* n. sp. and *Mullericeras indusense* n. sp. The main difference between these two species and *Radioceras truncatum* concerns the suture line. Species belonging to *Mullericeras* have a very wide ventral lobe and numerous small indentations at the base of the lobes and in the auxiliary series, generally without well differentiated auxiliary lobe. *Radioceras truncatum* does not have the very wide lateral branches of the ventral lobe characteristic of *Mullericeras*, has fewer and larger indentations at the base of the lobes and in the auxiliary series, and a differentiated auxiliary lobe. *Radioceras truncatum* also tends to have a thinner whorl section than *Mullericeras shigetai* and a smaller umbilicus than *Mullericeras indusense* for specimens smaller than 6 cm in diameter. This close resemblance is here interpreted as another example of convergence of two phylogenetically distant taxa towards the 'meekoceratoid' morphology.

Occurrence. – Chiddru, sample Chi61 ($n = 1$); Nammal Nala, sample Nam350 ($n = 1$); Wargal, sample War104 ($n = 8$).

Genus *Pashtunites* n. gen.

Derivation of name. – Named after the Pashtun people, a major ethnic group in Mianwali District.

Type species. – *Koninckites kraffti* Spath, 1934.

Composition of the genus. – Type species only.

Diagnosis. – Involute, compressed tabulate paranoritid of small size with convex flanks, vertical umbilical wall and narrowly rounded ventro-lateral and umbilical shoulders.

Occurrence. – Late Dienerian of the Salt Range and of Spiti (India).

Discussion. – This genus is very close to *Radioceras*, but no specimens of large size have been found (the largest specimen found has a maximum estimated diameter of ca. 7 cm). Its umbilical wall is slightly oblique to vertical (never overhanging). Between 3 and 7 cm, the shell is slightly more evolute, has a slightly more inflated whorl section, more convex flanks and less angular whorl cross section (i.e. with narrowly rounded ventro-lateral and umbilical shoulders) than *Radioceras*.

Pashtunites kraffti (Spath, 1934)

Plate 21, figures 7–13; Figures 35, 42

p 1909 *Meekoceras varaha* Diener, 1895; von Krafft & Diener, p. 17, pl. 2, fig. 4 (holotype of *Pashtunites kraffti* designated by Spath, 1934).

non p 1909 *Meekoceras varaha* Diener, 1895; von Krafft & Diener, p. 17, pl. 2, figs 2, 3, 5, 6.

1934 *Koninckites kraffti* n. sp. Spath, pp. 155, 156, fig. 43c (holotype).

non v 2010a ?*Radioceras* cf. *kraffti* (Spath, 1934); Brühwiler, Ware, Bucher, Krystyn & Goudemand, pp. 728, 729, figs 8, 9.

non v 2012 *Radioceras* cf. *kraffti* (Spath, 1934); Brühwiler & Bucher, p. 53, figs 31Q–T, 33A–U, 34A–Z.

Emended diagnosis. – As for the genus.

Material. – 11 specimens.

Description. – Involute (U/D ≈ 15%), moderately compressed (W/D ≈ 27%, W/H ≈ 51%), tabulate, small sized shell (maximum diameter estimated at ca. 7 cm) whose flanks are slightly convex with maximum width at inner third. Venter tabulate, becoming slightly convex on the adult body chamber, with narrowly rounded ventro-lateral shoulders. Umbilical wall slightly oblique to vertical with rounded umbilical shoulder. Ornamentation absent except for weak folds on the inner third of the whorl, following the shape of the growth lines. Suture line typical of involute Paranoritidae with deep lobes having few deep indentations at their base, and a long auxiliary series with a poorly differentiated auxiliary lobe.

Measurements. – See Figures 35, 42.

Discussion. – This species was created by Spath (1934) based on a series of specimens considered by Diener (1909) as being conspecific and belonging to *Meekoceras varaha*. However, these specimens most probably belong to different species of different ages. The assignment of our specimens to this species is here only based on their comparison with the specimen figured by von Krafft & Diener (1909, pl. 2, fig.

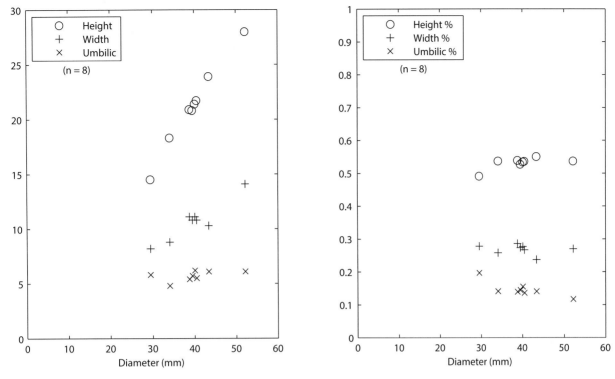

Fig. 42. Pashtunites kraffti (Spath, 1934) n. gen. Scatter diagrams of H, W and U, and of H/D, W/D, and U/D (abbreviations as in Fig. 15).

4) and designated by Spath as the holotype. As Spath (1934) agreed with von Krafft & Diener (1909) that all the specimens they figured were conspecific, he based his diagnosis on all of them, not only on the holotype. His diagnosis is thus not valid and we provide here a more restrictive emended diagnosis.

Occurrence. – Amb, sample Amb65 ($n = 1$); Nammal Nala, samples Nam63 ($n = 2$), Nam83 ($n = 2$), Nam305 ($n = 5$) and Nam349 ($n = 1$).

Genus *Awanites* n. gen.

Derivation of name. – Named after the Awan tribe, a tribe of the Salt Range in the Mianwali District.

Type species. – *Awanites awani* n. sp.

Composition of the genus. – Type species only.

Diagnosis. – Moderately involute (U/D ≈ 21%) and moderately compressed (W/D ≈ 27%, W/H ≈ 55%) paranoritid of small size (maximal diameter of ca. 55 mm) with a very angular sub-trapezoidal whorl cross section (i.e. with very sharp ventro-lateral and umbilical shoulders), thick tabulate venter and

slightly overhanging umbilical wall. Suture line with a wide third lateral saddle and a short auxiliary series with a very small and poorly differentiated auxiliary lobe.

Occurrence. – Latest Dienerian of Nammal Nala, Salt Range.

Discussion. – The umbilical part of the suture line of this new genus is slightly different from that of other paranoritids, so its assignment to this family is arguable. However, it has a poorly differentiated auxiliary lobe (although small), and the lateral and ventral part of the suture line agree with the other genera belonging to this family. Moreover, the suture line could only be drawn from one specimen, so intraspecific variability could not be assessed. As previously discussed, the umbilical portion of the suture line is likely to show more variability than its lateral and ventral portion. The tabulate venter, the sharp umbilical shoulder and the occasional presence of low spiral ridges on the flanks agree with other representatives of Paranoritidae. With its very angular whorl section, it is similar to inner whorls of the genus *Radioceras*, but it is distinguished from these by its more evolute and inflated shape. Some robust variants have radial ribs identical with those of the younger and larger *Paranorites*.

Awanites awani n. gen. et n. sp.

Plate 21, figures 14–23; Figures 35, 43

Derivation of name. – As for the genus.

Holotype. – Specimen PIMUZ30422 (Pl. 21, figs 20–23).

Type locality. – Nammal Nala, Salt Range, Pakistan.

Type horizon. – Sample Nam350 (lower Ceratite Marls, ca. 4.5 m above base), *Awanites awani* Regional Zone, late Dienerian.

Diagnosis. – As for the genus.

Material. – 22 specimens.

Description. – Moderately involute (U/D ≈ 21%) and moderately compressed (W/D ≈ 27%, W/H ≈ 55%) shell with broad tabulate venter bordered by sharp ventro-lateral shoulders. On inner whorls, flanks slightly convex with maximal width at inner third of whorl height. On the single large specimen available (the holotype, PIMUZ30422, Pl. 21, figs 20–23), a very shallow concavity appears on the body chamber between the umbilical shoulder and the point of maximal width. Umbilical wall slightly overhanging, joining the flank with an acute angle, thus forming a sharp umbilical shoulder. Flanks usually nearly smooth, with vague sigmoid folds following the shape of the growth lines and fading on the body chamber. Two weak spiral ridges are sometimes visible on the phragmocone around mid-flanks. One robust variant (specimen PIMUZ30420, Pl. 21, figs 14–16) has relatively strong radial ribs which do not cross the venter. Suture line with deep lateral lobes and few small indentations at their base, first lateral saddle elongated and slightly phylloid, second lateral saddle elongated and dorsally bent, third lateral saddle relatively broad, arched, and somewhat asymmetrical. Auxiliary series with three very small and closely spaced indentations, corresponding to the auxiliary lobe, followed by two larger and more distant indentations.

Measurements. – See Figures 35, 43.

Discussion. – As for the genus.

Occurrence. – Nammal Nala, sample Nam350.

Genus *Koiloceras* Brühwiler & Bucher, 2012

Type species. – *Koiloceras romanoi* Brühwiler & Bucher, 2012.

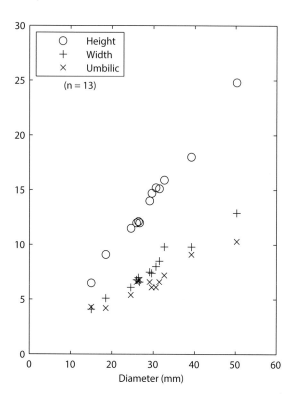
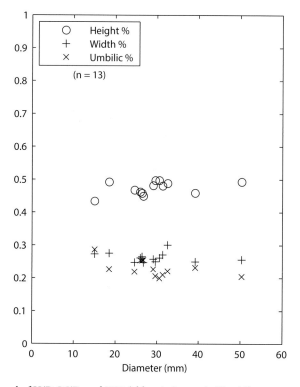

Fig. 43. Awanites awani n. gen. et sp. Scatter diagrams of H, W and U, and of H/D, W/D, and U/D (abbreviations as in Fig. 15).

Discussion. – This genus was originally included in Gyronitidae. With our emended definition of this family, the suture line of this genus does not agree with that of gyronitids. The type species has a slightly differentiated auxiliary lobe, agreeing with our definition of Paranoritidae. Lobes are broader and saddles shallower than those of other members of Paranoritidae. Moreover, the specimens identified by Brühwiler *et al.* (2010a) as Gyronitidae gen. et sp. indet. A are probably conspecific with *Koiloceras romanoi*, and have a suture line which agrees with that of paranoritids. *Koiloceras* also shares with other paranoritids the mature egression of the umbilical suture and the transition from tabulate to rounded venter on adult body chamber. Therefore, assignment of this genus to Paranoritidae is more appropriate.

Koiloceras sahibi n. sp.

Plate 21, figures 24–29; Figures 35, 44

Derivation of name. – Named after Syed Mian Ali Sahib, founder of Mianwali.

Holotype. – Specimen PIMUZ30423 (Pl. 21, figs 24–26).

Type locality. – Chiddru, Salt Range, Pakistan.

Type horizon. – Sample Chi51 (topmost Lower Ceratite Limestone), *Koninckites vetustus* Regional Zone, late Dienerian.

Diagnosis. – *Koiloceras* with a relatively thick whorl section.

Material. – 7 specimens.

Description. – Moderately evolute (U/D ≈ 24%) platyconic shell with a relatively thick whorl section (W/D ≈ 24%, W/H ≈ 57%). Venter tabulate with sharp ventro-lateral shoulders on inner whorls, becoming slightly arched with narrowly rounded shoulders on the last whorl. Maximum width at mid-flanks, the inner half of the flanks being slightly concave while its external half is convex. Umbilical wall not well differentiated, the flanks bending abruptly before the umbilical seam without forming any distinct shoulder. Flanks nearly smooth, with occasional vague broad folds following the shape of the growth lines and reaching their maximal strength at mid-flanks. Suture line too poorly preserved to be drawn or described in detail.

Measurements. – See Figures 35, 44.

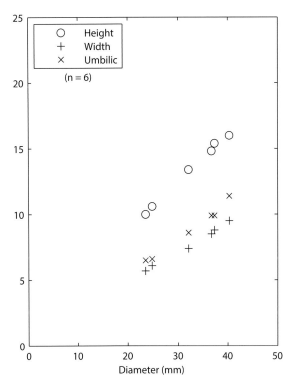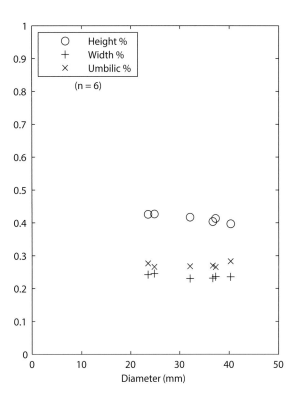

Fig. 44. Koiloceras sahibi n. sp. Scatter diagrams of H, W and U, and of H/D, W/D, and U/D (abbreviations as in Fig. 15).

Discussion. – This new species differs from the type species by its slightly thicker whorl section (in *Koiloceras romanoi*, W/D ≈ 21%, W/H ≈ 54%). However, our specimens are smaller than the ones illustrated by Brühwiler *et al.* (2012), so this difference of only three percent of the whorl width can well be the result of ontogenetic change and/or of intraspecific variability. However, we have too few specimens of this rare genus to test if this difference is significative and do not exclude that this new species may ultimately turn out as a junior synonym of the type species.

Occurrence. – Chiddru, samples Chi51 (n = 1) and Chi151 (n = 3); Wargal, sample War104 (n = 3).

Family Flemingitidae Hyatt, 1900

Genus *Xenodiscoides* Spath, 1930

Type species. – *Xenodiscus perplicatus* Noetling, 1905.

Xenodiscoides? sp. indet.

Plate 21, figures 30–32

Material. – One specimen.

Description. – Moderately evolute (U/D = 32%) and compressed (W/D = 26%, W/H = 65%) sub-platyconic shell. Venter tabulate with sharp ventro-lateral shoulders. Flanks slightly convex with strong and slightly rursiradiate blunt ribs which fade before reaching the venter. Umbilical wall indistinct, the flanks bending abruptly before the umbilical seam without forming any clear shoulder. Suture line too poorly preserved for description.

Measurements. – See Table 1.

Discussion. – This small specimen, with its coarse rursiradiate ribs and its tabulate venter, resembles the genus *Xenodiscoides*. Its rather compressed shape is quite close to *Xenodiscoides variocostatus* Brühwiler, Ware, Bucher, Krystyn & Goudemand, 2010a, but the very small size of this specimen precludes any further comparison with the much larger specimens of the type series. It is therefore placed here in open nomenclature.

Occurrence. – Chiddru, sample Chi51.

Genus *Shamaraites* Shigeta & Zakharov, 2009

Type species. – *Anakashmirites shamarensis* Zakharov, 1968.

Shamaraites? sp. indet.

Plate 21, figures 33–35

Material. – One specimen.

Description. – Evolute (U/D = 40%) sub-serpenticonic shell with a slightly compressed whorl section (W/D = 26%, W/H = 70%). Venter tabulate with sharp ventro-lateral shoulders. Flanks convex, with very vague broad radial folds. Umbilical wall not well differentiated, the flanks bending abruptly before the umbilical seam without forming any clear shoulder. Suture line too poorly preserved for description.

Measurements. – See Table 1.

Discussion. – This very small specimen differs from coeval juvenile of *Koninckites vetustus* by its more evolute coiling and the absence of any constriction. Its general shape is reminiscent of the early Smithian genus *Shamaraites*, but its very small size precludes any direct comparison with the much larger specimens previously illustrated. It is therefore placed in open nomenclature.

Occurrence. – Wargal, sample War104.

Family Proptychitidae Waagen, 1895

Discussion. – According to Tozer (1994), Proptychitidae range from the late Griesbachian to the end of the Spathian. However, this family is in need of revision. Tozer had a very broad definition of this family, considering Paranoritidae as a synonym of Proptychitidae based on the presence of an auxiliary lobe. Proptychitidae have a highly discontinuous record in the Tethys, and many species and genera are based on single specimens whose stratigraphical position is not known precisely. They are more abundant in the Boreal Realm, where Tozer defined two zones with Proptychitidae: the *Bukkenites strigatus* zone in the late Griesbachian and the *Proptychites candidus* zone in the early Dienerian. Tozer (1994) also included the genus *Vavilovites* in this family, but we provide here new evidence that this genus belongs to Paranoritidae (see above). The different species from northern Siberia assigned to

Vavilovites by Dagys & Ermakova (1996) may, at least partially, represent a new genus, the affinity of which needs to be established. To provide a proper emended definition of this family, a broader view is necessary, including the Smithian and Spathian Proptychitidae. This is beyond the scope of the present paper, and although an emended diagnosis of its type genus *Proptychites* is provided, no emended diagnosis is proposed here for this family. Considering that Paranoritidae are not anymore considered as a synonym of this family, and are shown here to be not directly related to Proptychitidae, this family now exclusively comprises forms having a rounded venter. The origin and classification of this family also need to be re-investigated in detail. It is generally considered to be derived from involute Ophiceratidae (e.g. Arkell *et al.* 1957). However, the complex suture line of some Dienerian *Proptychites* is very close to that of *Otoceras*, differing only by the absence of a lobe dividing the third lateral saddle in two parts. These two genera also have inflated juvenile whorls, thus differing from the typically compressed juvenile whorls of Meekocerataceae. These similarities suggest that Otoceratidae and Proptychitidae are closely related, and that Proptychitidae should thus be placed in the superfamily Otocerataceae and not in Meekocerataceae. Such a change of classification would result in a very different picture for the Early Triassic recovery. Otocerataceae would then no longer be short-term survivors, and Triassic ammonoids would then derive from two different Permian groups, not exclusively from Xenodiscidae (e.g. Tozer 1981). Considering the importance of such a change, this hypothesis would need further investigation to be properly tested, especially concerning the latest Otoceratidae and Griesbachian Proptychitidae, which are both absent in our sections. Hence, we provisionally follow the traditional classification, placing Proptychitidae in Meekocerataceae. In the sections studied here, Proptychitidae are present only in the early and middle Dienerian, not in the late Dienerian, unless the genus *Kingites*, whose affinities remain unclear, belongs to Proptychitidae. Proptychitidae 'reappear' in lower Smithian strata (Brühwiler *et al.* 2012). As a consequence, the morphological similarity between Dienerian and Smithian Proptychitidae cannot be excluded to be the result of a convergence, a hypothesis which cannot be tested here owing to the rarity of this family in the Salt Range, especially in the early Smithian.

Genus *Bukkenites* Tozer, 1994

Type species. – *Proptychites strigatus* Tozer, 1961.

Discussion. – Tozer (1994, p. 59) defined this genus mostly on its suture line, differing 'from true *Proptychites* [...] by lacking an auxiliary lobe'. Although most of the species he included in this genus have a subtrigonal whorl section, Tozer did not regard this trait as being of generic significance, an opinion which is here followed. The shape of the ventral lobe is here also regarded as an important additional character. *Bukkenites* has a ventral lobe with narrow lateral branches that exhibit only a few small indentations at their base, whereas *Proptychites* has a ventral lobe with wide lateral branches with deep indentations at their base.

Bukkenites sakesarensis n. sp.

Plate 21, figures 36–39; Plate 22, figures 1–13;
Figures 45, 46

Derivation of name. – Named after the Sakesar Mountain near Amb, the highest point of the Salt Range.

Holotype. – Specimen PIMUZ30429 (Pl. 22, figs 1–3).

Type locality. – Amb, Salt Range, Pakistan.

Type horizon. – Sample Amb104 (lowermost Lower Ceratite Limestone), *Gyronites dubius* Regional Zone, early Dienerian.

Diagnosis. – *Bukkenites* with evolute inner whorls, a sub-rounded whorl section and distant thick radial folds, becoming more involute with growth, with sub-parallel and smooth flanks.

Material. – 21 specimens.

Description. – Involute shell with a broadly rounded venter and a pronounced allometric growth. Juveniles are relatively evolute (for D < 30 mm, U/D ≈ 29%) with a thick whorl section (for D < 30 mm, W/D ≈ 41%, W/H ≈ 96%), most specimens having low broad radial folds on the inner half of the whorl, while some robust variants have a few distant strong blunt ribs fading just above the venter. With growth, the shell becomes smooth,

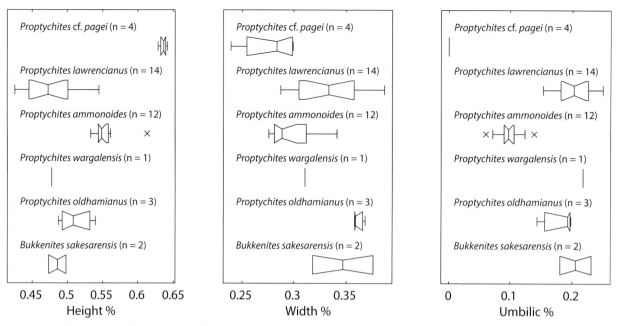

Fig. 45. Boxplots for the different species of Proptychitidae from the Dienerian of the Salt Range. Because of allometric growth, these boxplots were calculated without the specimens smaller than 30 mm in diameter. Abbreviations as in Figure 15.

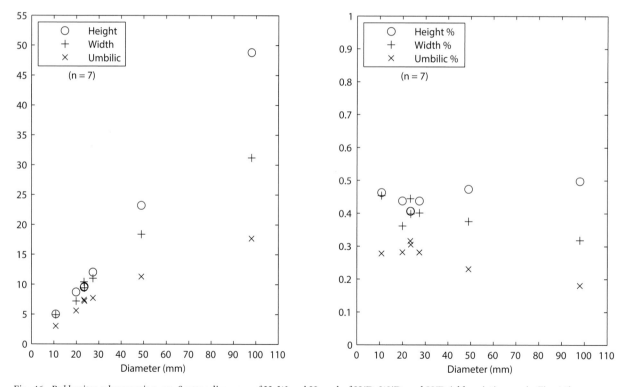

Fig. 46. Bukkenites sakesarensis n. sp. Scatter diagrams of H, W and U, and of H/D, W/D, and U/D (abbreviations as in Fig. 15).

more involute and compressed. For the holotype, which is the largest available specimen, U/D = 18%, W/D = 32% and W/H = 64% at a diameter of 98 mm. Umbilical wall vertical to slightly overhanging, delineated by a very narrowly rounded umbilical shoulder. Flanks with maximal width at the umbilical shoulder in juveniles, at inner third of the whorl section on the holotype. Suture line with a narrow ventral lobe, lateral lobes with a few small indentations at their base, and short auxiliary series with irregular indentations which do not form any clear auxiliary lobe.

Measurements. – See Figures 45, 46.

Discussion. – This species differs from the species assigned to *Bukkenites* by Tozer (1994) mainly by its sub-oval instead of trigonal whorl section, and by the presence of folds or ribs in the inner whorls. However, its suture line is similar to those illustrated by Tozer. Its marked allometry of the shell proportions, ornamentation on inner whorls, sub-oval whorl section and simple suture line allow to distinguish it from all other proptychitids.

Occurrence. – Amb, Sample Amb104.

Genus *Proptychites* Waagen, 1895

Type species. – *Ceratites lawrencianus* de Koninck, 1863.

Composition of the genus. – *Ceratites lawrencianus* de Koninck, 1863; *Proptychites oldhamianus* Waagen, 1895; *Proptychites ammonoides* Waagen, 1895; *Proptychites candidus* Tozer, 1961; *Proptychites subgrandis* Guex, 1978; *Proptychites pagei* Ware, Jenks, Hautmann & Bucher, 2011 and *Proptychites wargalensis* n. sp.

Emended diagnosis. – Involute, moderately compressed to thick discoidal shell with a broadly rounded venter. Suture line complex, with deep lobes. Ventral lobe very broad, divided by a deep ventral saddle into two wide lateral branches with numerous and deep indentations at their base. Lateral lobes very deep, with deep indentations at their base and simpler indentations on their flanks, occasionally reaching up to half of the height of the adjacent saddles in some species. Lateral saddles very elongated, slightly phylloid. Auxiliary series short to long depending on involution. Evolute variants with a short auxiliary series and an auxiliary lobe clearly individualized at least on one side of the shell, involute variants with a very long auxiliary series, a clearly individualized auxiliary lobe and sometimes also an auxiliary saddle. At large sizes, the indentation of the lobes tend to become digitate (e.g. Pl. 23, fig. 4).

Occurrence. – Early to middle Dienerian of the Salt Range (Pakistan), Spiti valley (India), Nepal, Primorye (Russia), Queen Elizabeth Islands and British Columbia (Canada), Nevada (USA).

Discussion. – This genus is here restricted to early and middle Dienerian Proptychitidae, and differs from other members of this family mostly by its

suture line. Griesbachian representatives of Proptychitidae (e.g. *Bukkenites*, *Pachyproptychites* Diener, 1916) differ in having a simpler suture line. Early Smithian representatives of Proptychitidae (such as *Clypeoceras* Smith, 1932; *Paraspidites* Spath, 1934; *Pseudaspidites* Spath, 1934; *Proptychitoides* Spath, 1930; *Eoptychites* Spath, 1930) tend to have more phylloid saddles and more indentations on the lobes. The juvenile suture line of the holotype of *Proptychites candidus* is simpler than that of our specimens, especially the ventral lobe. However, this suture line was drawn from the inner whorls of the holotype and other larger specimens of this species illustrated by Tozer (1994) have a suture line which does agree with that of our definition of *Proptychites*. Hence, in the absence of additional information, *Proptychites candidus* is here kept into this genus. On the other hand, the specimens assigned to *Proptychites candidus* by Brühwiler *et al.* (2008), as well as the previously published specimens from China they synonymized with this species, have a much simpler suture line and cannot be assigned to this genus. Waterhouse (1996) described several new species of *Proptychites*, but they are all based on too poorly preserved material to allow any identification at the species level. The same author also erected the genus *Aspitella* with *Aspidites crassus* von Krafft, 1909 as a type species, which we consider here as being a synonym of *Proptychites ammonoides*. Hence, the genus *Aspitella* is here considered as a junior synonym of *Proptychites*.

Proptychites lawrencianus (de Koninck, 1863) *sensu* Waagen, 1895

Plate 22, figures 14, 15; Plate 23, figures 1–11; Figures 45, 47

? 1863   *Ceratites lawrencianus* n. sp. de Koninck, p. 14, pl. 6, fig. 3 (holotype).

   1895   *Proptychites lawrencianus* (de Koninck, 1863) n. gen.; Waagen, pp. 168–170, pl. 18, fig. 1, pl. 17, fig. 2.

v 1978   *Proptychites ammonoides* Waagen, 1895; Guex, pl. 8, fig. 8.

Material. – 36 specimens.

Description. – Moderately involute, relatively thick shell with a broadly rounded venter and a marked allometric growth. Starting from the neanic stage,

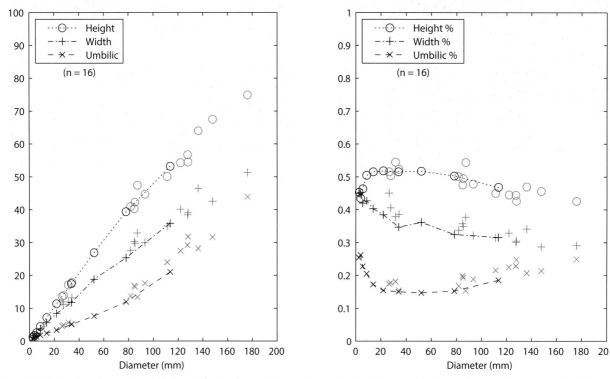

Fig. 47. Proptychites lawrencianus (de Koninck, 1863) *sensu* Waagen, 1895;. Scatter diagrams of H, W and U, and of H/D, W/D, and U/D (abbreviations as in Fig. 15). Ontogenetic trajectories obtained from sectioned specimen illustrated on Plate 22, figures 14, 15 are shown in black.

the shell rapidly becomes very involute, reaching its maximal involution at a diameter of ca. 3 cm, where U/D reaches about 15%. From this point, the shell becomes slowly and progressively more evolute, U/D reaching 25% on the largest specimen measured at a diameter of 17.6 cm. The whorl thickness decreases rather slowly and steadily during growth. As measured on a prepared section, W/D = 45% and W/H = 104% at the end of the neanic stage (D = 4 mm) and decrease to reach 32% and 67% at a diameter of 114 mm, respectively. The maximal whorl thickness is situated at or just above the umbilical shoulder in the inner whorls, and shifts progressively towards the middle part of the flank with growth, the whorl section becoming more and more oval. The umbilical wall is vertical to slightly overhanging on the inner whorls. It becomes progressively slightly oblique with growth, permanently delineated by a broadly rounded umbilical shoulder. The shell is generally smooth, but some specimens have broad radial folds which appear at a minimum diameter of ca. 7 cm. The suture line could only be clearly seen on two specimens, a small and a large one. In both cases, lobes are very deep and saddles are very elongated. Lobes are deeply indented, and the auxiliary series is short. The small specimen, which is very involute, has

phylloid saddles, and a very well individualized auxiliary lobe. On the larger, more evolute specimen, the saddles are not phylloid, and the suture line is strongly asymmetric, especially the auxiliary series, which has a clearly differentiated auxiliary lobe on the right side, whereas it is poorly differentiated on the other (see Pl. 23, fig. 4).

Measurements. – See Figures 45, 47.

Discussion. – The illustration of the holotype of this species given by de Koninck (1863) is not very informative, lacks ventral or apertural view, and is accompanied by a very brief description. Waagen's interpretation of this species is only a guess based on what is visible of the suture line on de Koninck's drawing. However, it is the interpretation of Waagen which has been used by subsequent authors (from Spath, 1934 to Waterhouse, 1996), and which is also adopted here. It should be noted here that the specimen of de Koninck comes from the collections of Dr. A. Fleming, and the author does not indicate where this collection is curated. It is therefore possible that the holotype is lost, so a neotype could be chosen in the collections of Waagen. This, however, would need a careful investigation and the establishment of a neotype requires a formal request to the ICZN. According to the measurements provided by de Koninck, his holotype is much less inflated (W/D = 30%, W/H = 55%) than our specimens and than those of Waagen, all of which are similar to our material. The holotype is thus closer in proportions to the species here referred to as *Proptychites ammonoides*. Until the holotype can be adequately re-described, the definition of Waagen is here kept by default, but no emended diagnosis is provided. Guex (1978) figured a specimen which he assigned to *Proptychites ammonoides*. The thickness of this specimen is actually more in agreement with that of *Proptychites lawrencianus* than with that of *Proptychites ammonoides*.

Occurrence. – Nammal Nala, samples Nam100 ($n = 7$), Nam300 ($n = 1$), Nam301 ($n = 3$), Nam344 ($n = 9$), Nam501 ($n = 12$) and Nam503 ($n = 4$).

Proptychites oldhamianus Waagen, 1895

Plate 24, figures 5–14; Figure 45

1895 *Proptychites oldhamianus* n. gen., n. sp. Waagen, pp. 166, 167, pl. 19, fig. 3 (holotype).

1909 *Proptychites typicus* n. sp. von Krafft, pp. 77–79, pl. 19, figs 4, 5, pl. 20, fig. 6, pl. 21, figs 2, 3 (lectotype), 4.

Emended diagnosis. – Involute, thick *Proptychites* with nearly flat flanks in inner whorls, strongly converging towards the rounded venter, imparting a sub-trigonal shape to the whorl section. Inner whorls bear weak radial folds just above the umbilical wall. The flanks become more convex and the ornamentation disappears with growth. Suture line simple compared to that of other *Proptychites*, with non phylloid saddles and few rather shallow indentations in the lobes.

Material. – 3 specimens.

Description. – Moderately involute (U/D $\approx$ 18%), relatively thick (W/D $\approx$ 36%, W/H $\approx$ 51%) shell with a narrowly rounded venter. Inner whorls have their maximal width at the umbilical shoulder, the flanks being only very slightly convex and strongly converging towards the venter, thus giving the whorl section a sub-trigonal shape. With growth, the flanks become progressively more convex, and the maximal whorl width shifts progressively towards the inner third of the flanks. Umbilical wall vertical at every

stage of growth, delimited by a narrowly rounded umbilical shoulder. On inner whorls, some weak radial folds occur on the dorsal half of the flanks. They disappear completely with growth. Suture line fairly simple for *Proptychites*, but with the characteristic ventral lobe having broad lateral branches and a clearly individualized auxiliary lobe, even at small size. Lateral saddles not phylloid, the third one being strongly bent towards the umbilicus. Lobes with few and rather small indentations.

Measurements. – See Figure 45 and Table 1.

Discussion. – Only three specimens of this species have been documented. They occur in massive limestone beds from the Lower Ceratite Limestone of Nammal Nala, and are very difficult to prepare. The suture line is visible only on the last preserved whorl of the smallest specimen, but despite its small size, it shows already the characteristic ventral and auxiliary lobes of the genus *Proptychites*. Our specimens agree very closely with the holotype of Waagen, except for the low radial folds on the inner whorls. Incidentally, Waagen's specimen is a weathered inner mould, hence this difference is most probably a preservation bias. The strongly bent third lateral saddle of the suture line is somewhat peculiar compared to that of other representatives of *Proptychites*, but considering the large intraspecific variability of this portion of the suture line in other species, this trait is here not considered as being relevant. *Proptychites typicus* as illustrated by von Krafft & Diener (1909) also agrees perfectly with our material. These authors mentioned that 'the periphery of the inner volutions is of indistinctly polygonal outlines', a characteristic which is shared by our smallest specimen. All the specimens of *Proptychites oldhamianus* described so far are of smaller size than the younger representatives of *Proptychites*. However, this small size may well be the result of a taphonomical bias, as they all come from shallower, higher energy environment. Our largest specimen being too recrystallized to see any suture lines, it is not possible to conclude if its body chamber is preserved or not.

Occurrence. – Nammal Nala, samples Nam335a ($n = 2$) and Nam378 ($n = 1$).

Proptychites wargalensis n. sp.

Plate 24, figures 1–4; Figure 45

Derivation of name. – Named after the locality of Wargal.

Holotype. – Specimen PIMUZ30438 (Pl. 24, figs 1–4).

Type locality. – Wargal, Salt Range, Pakistan.

Type horizon. – Sample War5 (middle Lower Ceratite Limestone), *Gyronites plicosus* Regional Zone, early Dienerian.

Diagnosis. – Relatively evolute and thick whorled *Proptychites* with a sub-oval whorl section and a simple suture line with a poorly differentiated auxiliary lobe.

Material. – 5 specimens.

Description. – Relatively evolute (U/D = 22%) and thick (W/D = 31%, W/H = 65%) shell with a broadly rounded venter. Flanks convex with maximal width at the inner third of the whorl. Umbilical wall vertical, grading into a rounded shoulder. No visible ornamentation. Suture line relatively simple, with moderately elongated and non phylloid saddles, a ventral lobe with broad lateral branches, lateral lobes with numerous small indentations at their base, and a very shallow auxiliary lobe individualized by having thinner indentations than the rest of the auxiliary series.

Measurements. – See Figure 45 and Table 1.

Discussion. – This species is based on five poorly preserved specimens, of which only one could be measured. It is very close to *Proptychites oldhamianus*, from which it differs by its more evolute coiling, thinner and more oval whorl section, and its simpler suture line. However, these two species are based on only a few specimens from correlative beds from two localities. Hence, it cannot be excluded that they are conspecific, but sufficient material is not available to demonstrate it. We thus provisionally assign the specimens from Wargal to a different species than those from Nammal Nala. The fairly simple suture line of *Proptychites wargalensis*, with its poorly individualized auxiliary lobe, may possibly lead to exclude this species from the genus *Proptychites*. However, on the other specimens here studied, the suture line is well enough preserved to see that its auxiliary lobe is better individualized than that of the holotype. In the absence of more abundant and better preserved material, we decided to keep this species within *Proptychites*, of which it represents the oldest known representative along with *Proptychites oldhamianus*.

Occurrence. – Wargal, samples War5 (n = 2), War6 (n = 1), War11 (n = 1) and War100 (n = 1).

Proptychites ammonoides Waagen, 1895

Plate 24, figures 15–17; Plate 25, figures 1–14; Plate 26, figures 1, 2; Figures 45, 48

1895 *Proptychites ammonoides* n. gen., n. sp. Waagen, pp. 171–173, pl. 17, fig. 1 (lectotype), pl. 19, fig. 2.

1909 *Aspidites crassus* n. sp. von Krafft, pp. 58, 59, pl. 6, fig. 4, pl. 7, fig. 1, pl. 8, fig. 1.

1909 *Koninckites haydeni* n. sp. von Krafft, pp. 68–70, pl. 17, figs 1, 2, 3 (lectotype), 4–6.

1909 *Koninckites alterammonoides* n. sp. von Krafft, pp. 70–72, pl. 16, figs 1 (lectotype), 2.

? 1996 *Aspitella crassa* (von Krafft, 1909) n. gen.; Waterhouse, p. 51, Text-fig. 41, pl. 3, figs 2–4.

2009 *Proptychites alterammonoides* (von Krafft, 1909); Shigeta & Zakharov, pp. 110–112, figs 98, 99.

v 2011 *Proptychites haydeni* (von Krafft, 1909); Ware, Jenks, Hautmann & Bucher, pp. 173–175, figs 15–17.

Emended diagnosis. – Very involute and compressed *Proptychites* with a highly denticulated suture line.

Material. – 45 specimens.

Description. – Very involute (U/D ≈ 10%) and compressed (W/D ≈ 30%, W/H ≈ 54%) discoidal shell with nearly flat, sub-parallel to slightly convex flanks. Position of maximal whorl width variable, between inner fourth of the flanks and midflanks. The flanks tend to be more convex in juvenile specimens and to become nearly flat with further growth, but convex flanks may persist to larger size in some specimens. Umbilical wall vertical, grading into a rounded shoulder. No clear allometric growth could be detected on our material, but one specimen (PIMUZ30443, Pl. 25, figs 1–5), a complete phragmocone showing the trace of the umbilical seam of the body chamber,

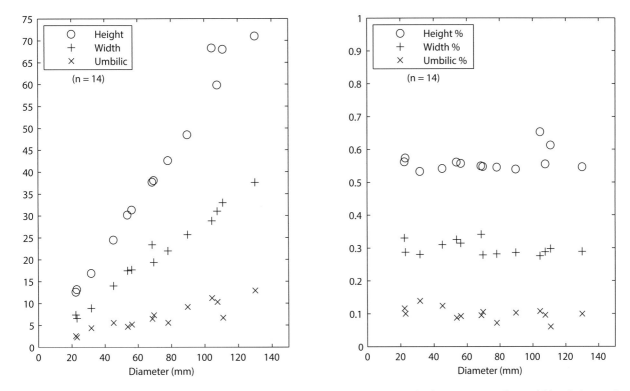

Fig. 48. Proptychites ammonoides Waagen, 1895;. Scatter diagrams of H, W and U, and of H/D, W/D, and U/D (abbreviations as in Fig. 15).

displays a pronounced umbilical egression of the body chamber. This species can reach very large size, the largest specimen being a complete phragmocone of about 17 cm in diameter (PIMUZ30446, Pl. 25, figs 10–14). Adding extra minimum of half a whorl for the body chamber, it may have reached a diameter of at least 26 cm. Some specimens have very vague folds following the shape of the growth lines on the inner half of the flanks. Suture line complex at maturity. Ventral lobe very wide, divided by a deep ventral saddle into two broad branches with numerous very deep indentations at their base. First lateral saddle slightly phylloid. Second lateral saddle bent dorsally and with a subtriangular outline at maturity. Third lateral saddle varying from large and rounded to narrow with a flattened tip. Lateral lobes deep, with many very deep indentations at their base and progressively shallowing on the lateral sides of the lobes. At maturity, lateral denticulations may expand on two-thirds of the sides. On large specimens, the very deep indentations at the bottom of the lobes are further subdivided into smaller indentations. Auxiliary series very long, complex, and variable, with a clearly individualized auxiliary lobe even at small diameter, generally followed by an irregular series of small

lobes or deep indentations, and sometimes with a clearly differentiated large auxiliary saddle.

Measurements. – See Figures 45, 48.

Discussion. – The suture line of the lectotype of this species has a peculiar flattened top of the third lateral saddle with two small incisions, a feature which has not been observed in our material. It is otherwise similar to our specimens. von Krafft & Diener (1909) erected three new species (*Aspidites crassus, Koninckites haydeni* and *Koninckites alterammonoides*) which can be included in the intraspecific variability of *Proptychites ammonoides*. These three species have been assigned to different genera based on a wrong interpretation of *Proptychites*, which von Krafft & Diener (1909, p. 75) defined mainly by the presence of 'globose inner whorls', whereas the two genera in which they included these three species were mainly defined by their suture line. The differences between these three species concern mainly the umbilical portion of the suture line (the third lateral saddle and the auxiliary series), which is shown here to be highly variable and not necessarily diagnostic. Waterhouse (1996) created the new genus *Aspitella* based on von Krafft's species *Aspidites crassus*, and as this species is here considered as a synonym of

Proptychites ammonoides, the genus *Aspitella* becomes a junior synonym of *Proptychites*. The specimens assigned by Waterhouse to this species are too poorly preserved for any identification.

Occurrence. – Amb, samples Amb4 ($n = 1$), Amb5 ($n = 2$), Amb53 ($n = 3$) and Amb54 ($n = 1$); Chiddru, samples Chi 56 ($n = 1$) and CH7A (1); Nammal, samples Nam50 ($n = 7$), Nam52 ($n = 4$), Nam53 ($n = 4$), Nam364 ($n = 2$), Nam370 ($n = 3$), Nam380 ($n = 5$), Nam382 ($n = 2$), Nam384 ($n = 1$), Nam400 ($n = 1$), Nam521 ($n = 1$), Nam522 ($n = 1$), Nam526 ($n = 2$) and Nam527 ($n = 1$); from Wargal, sample War2 ($n = 2$).

Proptychites cf. *P. pagei* Ware, Jenks, Hautmann & Bucher, 2011

Plate 26, figures 3–8; Figure 45

v ? 2011   *Proptychites pagei* n. sp. Ware, Jenks, Hautmann & Bucher, pp. 175, 176, fig. 18.

Material. – 5 specimens.

Description. – Compressed (W/D $\approx$ 28%, W/H $\approx$ 44%) discoidal shell with occluded umbilicus. Flanks slightly convex with maximal width at inner third of whorl height. Shell smooth except for very weak broad folds following the trajectory of the growth lines on the inner half of the flanks. Suture line very complex with elongated and narrow saddles. First lateral saddle occasionally slightly phylloid. Lobes with large, deep indentations at their base and small indentations on their flanks. Auxiliary series very long with a clearly differentiated auxiliary lobe followed by numerous irregular large indentations, occasionally with a poorly differentiated auxiliary saddle.

Measurements. – See Figure 45 and Table 1.

Discussion. – With their characteristic occluded umbilicus, these specimens differ from the type material described by Ware *et al.* (2011) by their larger size only. The smallest specimen found in the Salt Range is 55.6 mm in diameter, the largest is 10 cm in diameter, both being incomplete phragmocones, whereas the largest specimen from the Candelaria Hills (Nevada) is 51.3 mm in diameter and has a part of its body chamber preserved. The specimens from the Candelaria Hills also have a simpler suture line, but this is most likely a consequence of their smaller size. Specimens of intermediate size with preserved body chamber are needed to confirm that the Nevada and the Salt Range material belong to the same species.

Occurrence. – Nammal Nala, samples Nam344 (n = 1), Nam345 (n = 1) and Nam501 (n = 3).

Family Mullericeratidae Ware, Jenks, Hautmann & Bucher, 2011

Type genus. – *Mullericeras* Ware, Jenks, Hautmann & Bucher, 2011.

Composition of the family. – *Mullericeras, Ussuridiscus* Shigeta & Zakharov, 2009.

Emended diagnosis. – Involute, tabulate platyconic shell with a suture line composed of a long auxiliary series and no adventitious element in the ventral lobe.

Discussion. – This family was first erected for a single genus (*Mullericeras*). The more abundant and better preserved specimens presented in this study allow providing a more precise emended diagnosis, and shows that with its involute, tabulate platyconic shell, *Ussuridiscus* is clearly the stem group of *Mullericeras* and should be assigned to Mullericeratidae. Following Tozer's (1994) broad definition of Meekoceratidae, *Ussuridiscus* was originally placed in this family by Shigeta & Zakharov (2009). As discussed previously (see discussion on Gyronitidae), the validity of Meekoceratidae as a monophyletic group is questionable. The superficial resemblance between *Meekoceras*, the type genus of Meekoceratidae, and evolute species of *Mullericeras* is another example of convergence. On the other hand, our new material provides further evidence that Sagecerataceae root into mullericeratids, as demonstrated by the close resemblance between *Mullericeras spitiense* and *Clypites typicus*, which can only be differentiated by their ventral lobe, without adventitious series in *Mullericeras spitiense*.

Genus *Mullericeras* Ware, Jenks, Hautmann & Bucher, 2011

Type species. – *Aspidites spitiensis* von Krafft, 1909.

Discussion. – This genus was originally described as including three species: *Mullericeras spitiense* (von Krafft, 1909), *Meekoceras (Koninckites) vidharba* Diener, 1897 and *Aspidites ensanus* von Krafft,

1909. The more detailed study presented here indicates that *Meekoceras (Koninckites) vidharba* and *Aspidites ensanus* should be assigned to *Ussuridiscus*. Two other new species are here included in the genus *Mullericeras*. It is here defined as including smooth, very involute, compressed and tabulate platyconic to sub-oxyconic shells with a very broad ventral lobe bearing numerous small indentations but without adventive series.

Mullericeras spitiense (von Krafft, 1909)

Plate 27, figures 1–9; Figures 49, 50

1909 *Aspidites spitiensis* n. sp. von Krafft, p. 54, pl. 4, figs 4 (lectotype), 5, pl. 16, figs 3–8.

? 1996 *Clypeoceras spitiense* (von Krafft, 1909); Waterhouse, p. 50, text-fig. 4J, pl. 2, figs 21, 22.

non 2009 *Clypeoceras spitiense* (von Krafft, 1909); Shigeta & Zakharov, pp. 125, 126, figs 113, 114.

v 2010 *Mullericeras spitiense* (von Krafft, 1909) n. gen.; Ware, Jenks, Hautmann & Bucher, pp. 171, 172, figs 11, 12.

Material. – 13 specimens.

Description. – As this species has already been redescribed recently (Ware *et al.* 2011), no detailed description of the general shell morphology is given here as our specimens are strictly identical to the ones already figured. However, additional details of the suture line can be given here. Its lobes and saddles are elongated. The ventral lobe is very broad, with many indentations (9 indentations on the suture line drawn, Pl. 27, fig. 7). First lateral saddle elongated and straight, second and third lateral saddles broad and slightly bent towards the umbilicus, the third one being almost rectangular. Lobes with

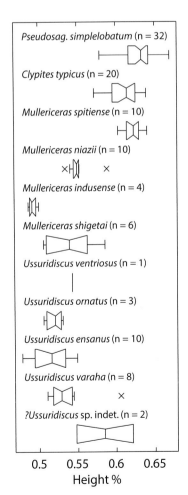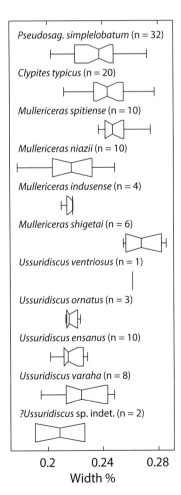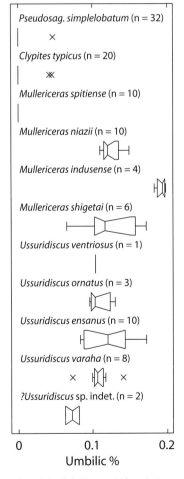

Fig. 49. Boxplots for the different species of Mullericeratidae and Hedenstroemiidae from the Dienerian of the Salt Range. Abbreviations as in Figure 15.

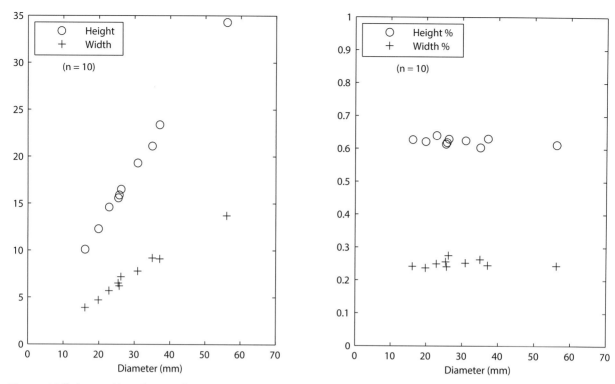

Fig. 50. Mullericeras spitiense (von Krafft, 1909). Scatter diagrams of H and W, and of H/D and W/D (abbreviations as in Fig. 15).

up to six deep indentations. Auxiliary series with one poorly differentiated lobe followed by a series of deep indentations.

Measurements. – See Figures 49, 50.

Discussion. – This species is almost homeomorphic with *Clypites typicus*, from which it differs by the absence of adventive series during the whole ontogeny, its smaller maximum size, broader, less elongated lateral saddles and a slightly simpler auxiliary series. This similarity suggests that *Clypites* and other Hedenstroemiidae originated from this species, but further, more detailed study of these groups, particularly concerning their ontogeny, would be necessary to demonstrate it. The synonym list has already been discussed by Ware *et al.* (2011). However, the specimen illustrated as *Clypeoceras spitiense* by Shigeta & Zakharov (2009) is here assigned to *Clypites typicus* (see below).

Occurrence. – Nammal Nala, samples Nam100 (*n* = 1), Nam344 (*n* = 1) and Nam396 (*n* = 11).

Mullericeras shigetai n. sp.

Plate 27, figures 10–28; Figures 49, 51

?p 1994 *Ambites fuliginatus* n. sp. Tozer, pl. 13, fig. 4.

?p 1994 *Ambites ferruginus* n. sp. Tozer, pl. 14, figs 1, 2, 5, 6.

? 2009 *Ambitoides fuliginatus* (Tozer, 1994) n. gen.; Shigeta & Zakharov, pp. 77–79, figs 63, 64.

?v 2011 *Ussuridiscus* sp. indet.; Ware, Jenks, Hautmann & Bucher, p. 169, fig. 8a.

Derivation of name. – Named after Dr. Yasunari Shigeta (National Museum of Nature and Science, Tokyo, Japan).

Holotype. – Specimen PIMUZ30453 (Pl. 27, figs 10–13).

Type locality. – Nammal Nala, Salt Range, Pakistan.

Type horizon. – Sample Nam381 (lower Ceratite Marls, ca. 1.5 m above base), *Ambites radiatus* Regional Zone, middle Dienerian.

Diagnosis. – Involute, moderately compressed (W/ H ≈ 50%) *Mullericeras* with a very high vertical to

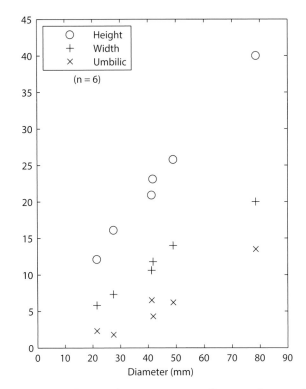
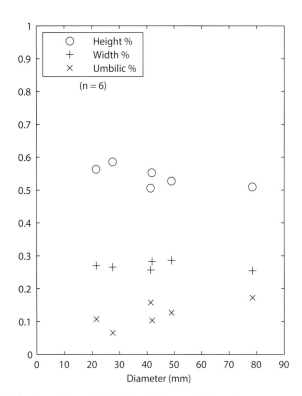

Fig. 51. *Mullericeras shigetai* n. sp. Scatter diagrams of H, W and U, and of H/D, W/D, and U/D (abbreviations as in Fig. 15).

overhanging umbilical wall, a narrow but open umbilicus (U/D ≈ 12%) and a simple suture line without clearly individualized auxiliary lobe or saddle.

Material. – 9 specimens.

Description. – Involute, moderately compressed shell with a broad tabulate venter (on the adult holotype, the width of the venter represents 38% of the whorl width) and sharp ventro-lateral shoulders. Flanks slightly convex with maximum whorl width just below the umbilical wall. Umbilical wall short and sub-vertical in juvenile stages and becoming slightly overhanging at later stages. Umbilical shoulder narrowly rounded. On the holotype, which is the largest and only mature specimen with a partly preserved body chamber, the umbilical wall becomes again vertical on the body chamber, where the umbilical shoulder is underlined by a slight concavity of the upper flanks. Its umbilical shoulders, like its ventro-lateral shoulders, become less angular at larger size. One specimen (PIMUZ30456, Pl. 27, figs 25–28) had remains of an *in vivo* encrusting bivalve on each side of its umbilicus, which explains its strongly overhanging umbilical wall with an extremely sharp umbilical shoulder. Egressive coiling starts from a diameter of ca. 4 cm onward. Flanks smooth, except for occasional very weak folds following the shape of the slightly sigmoid growth lines. Suture line relatively simple, without clearly individualized auxiliary lobe. Lobes and saddles broad, saddles moderately elongated. Ventral lobe divided by a broad saddle into two large rounded branches with numerous and regularly spaced small indentations at their base. Lateral lobes broadly rounded also with numerous and regularly spaced small indentations at their base. The two-first lateral saddles are always moderately elongated, the second one being slightly bent towards the umbilicus. The third lateral saddle is broad, short and sub-rectangular (i.e. about twice broader than high) on the smallest and the largest specimen, as high as broad and sub-rounded on the others. The previously mentioned specimen with a growth disturbance induced by an encrusting bivalve (Pl. 27, figs 25–28) has a slightly different suture line than the others. Its auxiliary series has a slightly individualized auxiliary lobe followed by a series of irregular deep indentations, less numerous than in other normal specimens. Considering the covariation between the shape of the whorl section and the pattern of the suture line, this difference is most probably the consequence of the modified dorsal part of the whorl section.

Measurements. – See Figures 49, 51.

Discussion. – Tozer (1994; pl. 13, fig. 4; pl. 14, figs 1, 2, 5, 6) figured some involute specimens he assigned to *Ambites fuliginatus* and *Ambites ferruginus*. The holotypes of these two species clearly belong to the genus *Ambites*, but these involute specimens are too involute and lack the bottleneck shaped venter characteristic of this genus. Therefore, they most probably belong to the genus *Mullericeras*. They agree in proportions with our specimens, but the exact morphology of their umbilical wall and suture line is not visible. Hence, we refer them only with caution to our species. The specimens assigned to *Ambites fuliginatus* by Shigeta & Zakharov (2009) on which they base their new genus *Ambitoides* are very close to the involute specimens of Tozer (1994) previously discussed and to ours. They differ from our specimens by their more parallel flanks, their non overhanging umbilical wall and a narrower ventral lobe of the suture line. The small number of specimens of *Mullericeras shigetai* does not allow assessing intraspecific variability. Hence, it is not possible to establish whether these small differences can be included into the variability of *Mullericeras shigetai* or if they are characteristic of another new species. Hence, the specimens of Shigeta & Zakharov (2009) are referred with doubts to our species. The specimen described by Ware *et al.* (2011) as *Ussuridiscus* sp. indet. was ascribed to this genus based on its overhanging umbilical wall. This characteristic is also found in the genus *Mullericeras*, and the Nevadan specimen found together with *Ambites* most likely belongs to the genus *Mullericeras*. The morphology of its flanks and umbilicus agrees with that of *Mullericeras shigetai*, but its poor preservation prevents a clear species identification.

Occurrence. – Amb, sample Amb53 (n = 1); Chiddru, sample Chi56 (n = 1); Nammal Nala, samples Nam364 (n = 1), Nam381 (n = 2), Nam382 (n = 1), Nam383 (n = 1), Nam384 (n = 1) and Nam527 (n = 1).

Mullericeras indusense n. sp.

Plate 28, figures 1–9; Figure 49

Derivation of name. – Named after the Indus River.

Holotype. – Specimen PIMUZ30460 (Pl. 28, figs 2–4).

Type locality. – Nammal Nala, Salt Range, Pakistan.

Type horizon. – Sample Nam302 (lower Ceratite Marls, ca. 2.5 m above base), *Ambites superior* Regional Zone, middle Dienerian.

Diagnosis. – *Mullericeras* with a broad umbilicus (U/D ≈ 19%), compressed whorl section (W/H ≈ 44%) and occasional auxiliary lobe on suture line.

Material. – 6 specimens.

Description. – Involute, moderately compressed shell with a broad tabulate venter and sharp ventro-lateral shoulders. Flanks slightly convex with maximum width around mid-flanks. Umbilical wall moderately high, vertical to slightly overhanging with narrowly rounded umbilical shoulder. No ornamentation visible except for very weak folds following the shape of the slightly sigmoid growth lines. Suture line with deep lobes and elongated saddles. Ventral lobe broad, with triangular shaped branches with many denticulations at their base. Auxiliary series long and variable, sometimes without auxiliary lobe and composed of a series of rather regular deep indentations, occasionally with a clearly individualized auxiliary lobe and saddle.

Measurements. – See Figure 49 and Table 1.

Discussion. – This species is very close to *Mullericeras shigetai*, from which it differs by its larger umbilicus, more compressed whorl section and its suture line. More specifically, its ventral lobe branches are intermediate between the large rounded branches of *Mullericeras shigetai* and the triangular branches of *Mullericeras niazii*. With its relatively large umbilicus, it superficially resembles involute variants of *Ambites*, but it differs by the absence of a bottleneck shaped venter, of spiral ridges and its more complex suture line.

Occurrence. – Nammal Nala, samples Nam53 (n = 3) and Nam302 (n = 3).

Mullericeras niazii n. sp.

Plate 28, figures 10–18; Figures 49, 52

Derivation of name. – Named after the Niazi tribe from Mianwali District.

Holotype. – Specimen PIMUZ30464 (Pl. 28, figs 10–13).

Type locality. – Nammal Nala, Salt Range, Pakistan.

Type horizon. – Sample Nam100 (lower Ceratite Marls, ca. 4 m above base), *Ambites lilangensis* Regional Zone, middle Dienerian.

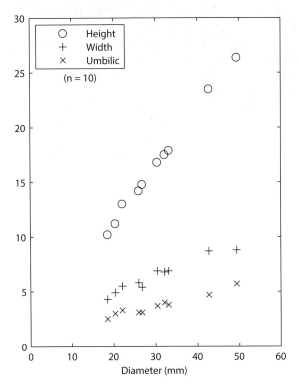
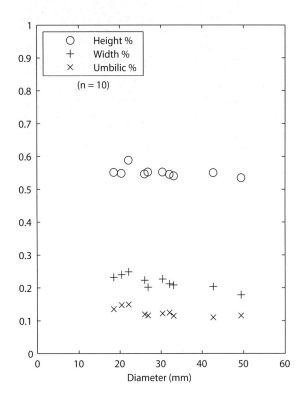

Fig. 52. Mullericeras niazii n. sp. Scatter diagrams of H, W and U, and of H/D, W/D, and U/D (abbreviations as in Fig. 15).

Diagnosis. – Involute, very compressed (W/H ≈ 39%) *Mullericeras* with a very short vertical umbilical wall, open but narrow umbilicus (U/D ≈ 13%) and a clearly individualized auxiliary lobe on the suture line.

Material. – 11 specimens.

Description. – Involute, very compressed shell with a relatively broad, very slightly concave tabulate venter (on the holotype, the venter width represents 36% of the whorl width) and sharp ventro-lateral shoulders. Flanks slightly convex with maximum width around mid-flanks on sub-adult specimens, on inner third of flanks in juveniles. Umbilical wall short and vertical on sub-adults, individualized by narrowly rounded umbilical shoulders. In juveniles, the low umbilical wall is convex, coalescing the preceding whorl with a very acute angle. Flanks smooth except for very weak folds following the shape of the slightly sigmoid growth lines. Ventral lobe broad, divided by the ventral saddle into two large triangular branches with numerous, regularly spaced small indentations at their base and ventral flank. Lateral lobes with nearly straight base, bearing a few deep denticulations. The two-first lateral saddles have a narrowly rounded tip, and the third one has a flattened tip. Auxiliary series very long, with a clearly individualized auxiliary lobe,

sometimes also an auxiliary saddle, followed by a series of irregularly spaced denticules.

Measurements. – See Figures 49, 52.

Discussion. – This species, with its very compressed whorl section, narrow umbilicus and peculiar ventral lobe differs clearly from other species of *Mullericeras*. It differs from the compressed forms of *Ussuridiscus* by its thicker venter, its absence of ornamentation and its wide ventral lobe. At small size (i.e. D < 2 cm.), some *Ambites* have almost the same morphology and can be confused with it, but the juveniles of *Ambites* are usually thicker, have a slightly bottleneck shaped venter and, more importantly, a simpler suture line with thinner ventral lobe branches and without differentiated auxiliary lobe.

Occurrence. – Nammal Nala, samples Nam100 (n = 3), Nam344 (n = 3), Nam501 (n = 2) and Nam503 (n = 1).

Genus *Ussuridiscus* Shigeta & Zakharov, 2009

Type species. – *Meekoceras (Kingites) varaha* Diener, 1895.

Composition of the genus. – *Meekoceras (Kingites) varaha* Diener, 1895, *Aspidites ensanus* von Krafft,

1909, *Ussuridiscus ventriosus* n. sp. and *Ussuridiscus ornatus* n. sp.

Emended diagnosis. – Very involute, compressed tabulate shell. Suture line with a narrow ventral lobe bearing few small indentations.

Discussion. – Shigeta & Zakharov (2009) erected this genus for a single species, and considered its overhanging umbilical wall as being its main characteristic. Our new abundant material shows that beside the type species, three other species from correlative strata in the Salt Range can be grouped within this genus, leading to an emendation of the original diagnosis. One species (*Ussuridiscus ventriosus*) does not have this overhanging umbilical. We however decided to group it with the others as it is otherwise very similar, and this distinction is not significant enough to create a new genus. We therefore consider this genus as including very involute, compressed platyconic to sub-oxyconic shells with a tabulate venter, a vertical to overhanging umbilical wall and a suture line with narrow ventral lobe with only a few little indentations at its base. With its involute, compressed, tabulate shell, it is very close to the genus *Mullericeras*, from which it differs by its simpler suture line with narrower ventral lobe and less numerous indentations in its lateral lobes, and by its more convex flanks. The similarity between these two genera and the

occurrence of *Ussuridiscus* in older strata than *Mullericeras* suggests that the latter derives directly from the former through a complexification of the suture line. The two genera should thus be grouped within the same family (Mullericeratidae). Following the broad definition of Meekoceratidae by Tozer (1994), *Ussuridiscus* was originally placed in this family, an interpretation which is not followed here.

Occurrence. – Early Dienerian of the Salt Range (Pakistan), Central Himalayas (India, Nepal), Dienerian of Guangxi (China), late Griesbachian and early Dienerian of South Primorye (Russia).

Ussuridiscus varaha (Diener, 1895)

Plate 28, figures 19–31; Figures 49, 53

1895   *Meekoceras (Kingites) varaha* n. sp. Diener, p. 52, pl. 1, fig. 2 (holotype).

non 1897   *Kingites varaha* (Diener, 1895); Diener, pp. 143, 144, pl. 6, fig. 2, pl. 7, fig. 6.

1905   *Meekoceras (Kingites) varaha* Diener, 1895; Noetling, pl. 32, fig. 5 [cop. Diener, 1895].

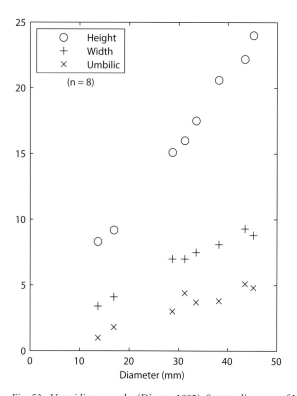
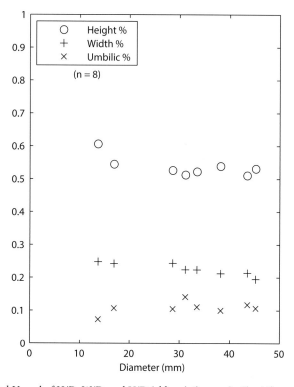

Fig. 53. Ussuridiscus varaha (Diener, 1895). Scatter diagrams of H, W and U, and of H/D, W/D, and U/D (abbreviations as in Fig. 15).

non 1909   *Meekoceras varaha* Diener, 1895; von Krafft & Diener, pp. 17–20, pl. 2, figs 2–6, pl. 14, figs 7, 8.

  1968   *Koninckites varaha* (Diener, 1895); Zakharov, p. 91, text-fig. 20b, pl. 17, figs 4, 5.

  2007   *Hubeitoceras* (?) *wangi* n. sp. Zakharov & Mu, p. 871, figs 13.17–13.19, 15.2–15.5.

v p 2008   '*Koninckites*' cf. *timorense* (Wanner, 1911); Brühwiler, Brayard, Bucher & Guodun, pp. 1165–1166, pl. 3, figs 1–4, pl. 4, fig. 1.

non v p 2008   '*Koninckites*' cf. *timorense* (Wanner, 1911); Brühwiler, Brayard, Bucher & Guodun, pl. 4, fig. 2.

  2009   *Ussuridiscus varaha* (Diener, 1895); Shigeta & Zakharov, pp. 69–73, figs 50.5, 50.6, 55–57.

Material. – 7 specimens.

Measurements. – See Figures 49, 53.

Discussion. – Our material only differs from that described by Shigeta & Zakharov (2009) in the suture line, which shows sub-phylloid saddles and finer crenulated lobes. These small differences may express intraspecific variability, which seems important according to the illustrations of Shigeta & Zakharov (2009). Note that this species is here restricted to a single bed of earliest Dienerian age, whereas in Primorye it occurs in five beds, in the equivalent of what we consider here to be the early and middle Dienerian. The synonymy list was also already discussed by Shigeta & Zakharov (2009). We confirm here that the specimens described by Brühwiler *et al.* (2008) under the name '*Koninckites*' cf. *timorense* correspond to this species, except one specimen which has a strigation on the venter and an unclear bottleneck shaped venter. *Ussuridiscus varaha* superficially resembles *Ussuridiscus ensanus* with which it shares the same shell proportions, but it clearly differs from this species by the shape of its whorl section, which has a thicker venter and nearly flat flanks unlike the typically sub-lanceolate whorl section of *Ussuridiscus ensanus*.

Occurrence. – Amb, sample Amb104 (n = 7).

Ussuridiscus ensanus (von Krafft, 1909)

Plate 3, figure 1; Plate 28, figures 32–41; Plate 29, figures 1–14; Figures 49, 54

  1909   *Aspidites ensanus* n. sp. von Krafft, pp. 56, 57, pl. 5, figs 3, 4, 5(lectotype), 6, 7, pl. 6, fig. 1, pl. 14, fig. 6.

p ? 1909   *Meekoceras varaha* Diener, 1895; von Krafft & Diener, p. 17, pl. 2, fig. 3.

p ? 1909   *Meekoceras hodgsoni* Diener, 1897; von Krafft & Diener, pp. 26–28, pl. 30, fig. 1.

  1934   *Clypeoceras ensanum* (von Krafft, 1909); Spath, pp. 160, 161, pl. 12, fig. 3 [cop. von Krafft 1909].

? 1996   *Clypeoceras ensanum* (von Krafft, 1909); Waterhouse, pp. 49, 50, pl. 2, figs 16–20, pl. 3, fig. 1.

Emended diagnosis. – Compressed (W/D ≈ 22%) *Ussuridiscus* with a narrowly tabulate venter, maximum whorl width at inner third of the whorl, the flanks being strongly convex on the inner half of the whorl and flat, converging towards the venter on the outer half, imparting the whorl section a very compressed sub-lanceolate shape. Suture line with a narrow and simple ventral lobe whose branches are very thin and pointed.

Material. – 20 specimens.

Description. – Involute (U/D ≈ 12%), compressed (W/D ≈ 22%, W/H ≈ 42%) sub-oxyconic shell with a narrowly tabulate venter delimited by sharp ventro-lateral shoulders. Maximum whorl width at inner third of the whorl. The flanks are strongly convex in the inner half of the whorl section, while they become flat and finally converging towards the venter on the outer half, giving the whorl section a very compressed sub-lanceolate shape. Umbilical wall very short, overhanging with narrowly rounded umbilical shoulder. Umbilicus very narrow at small size, becoming slightly egressive from a diameter of ca. 4 cm onward. Sigmoid ribs following the trajectory of growth lines on the flanks, reaching their maximal strength on inner flank, fading towards the venter. Suture line with elongated lateral saddles, the two-first ones being rounded or slightly tapered, the third one being flattened. Ventral lobe divided by a

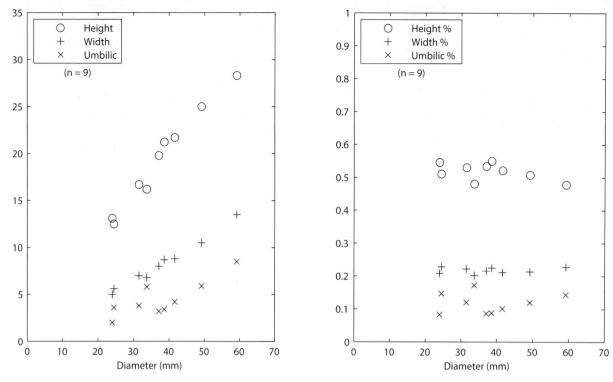

Fig. 54. Ussuridiscus ensanus (Krafft, 1909 *in* Krafft & Diener). Scatter diagrams of H, W and U, and of H/D, W/D, and U/D (abbreviations as in Fig. 15).

low ventral saddle. Its branches are very thin and simple, forming a small spike with occasional few very small indentations on the sides. Auxiliary series long, with a well differentiated auxiliary lobe followed by a series of irregular indentations.

Measurements. – See Figures 49, 54.

Discussion. – The shell shape of our specimens is strictly identical to that of the lectotype and other specimens figured by von Krafft & Diener (1909). The only difference is found in the suture line. Their drawing shows that the lateral branches of the ventral lobe are bifid, whereas they are clearly pointed in our specimens. This small difference does not justify the erection of a new species. One specimen ascribed to *Meekoceras varaha* by Diener (*in* von Krafft & Diener, 1909; pl. 2, fig. 3) may also belong to this species, but no firm assignment can be made as no ventral view is provided. One specimen assigned to *Meekoceras hodgsoni* by Diener (*in* von Krafft & Diener, 1909; pl. 30, fig. 1) is here tentatively included in *Ussuridiscus ensanus* as it shares the same general shell outline, but it lacks the weak ornamentation and seems to have a clearly defined vertical umbilical wall with a sharp umbilical shoulder. The specimens ascribed by

Waterhouse (1996) to this species are too poorly preserved for any tentative identification.

Occurrence. – Amb, samples Amb2 (*n* = 1) and Amb62 (*n* = 1); Nammal Nala, samples Nam339 (*n* = 1), Nam354 (*n* = 1), Nam377 (*n* = 5), Nam378 (*n* = 1) and Nam393 (*n* = 1); Wargal, samples War9 (*n* = 1) and War100 (*n* = 8).

Ussuridiscus ventriosus n. sp.

Plate 29, figures 28–31; Figure 49

Holotype. – Specimen PIMUZ30482 (Pl. 29, figs 28–31).

Type locality. – Nammal Nala, Salt Range, Pakistan.

Type horizon. – Sample Nam377 (middle Lower Ceratite Limestone), *Gyronites plicosus* Regional Zone, early Dienerian.

Derivation of name. – From the Latin *ventriosus*, meaning 'potbellied', in reference to its thick venter.

Diagnosis. – Ussuridiscus with a thick venter and whorl section, without ornamentation.

Material. – One specimen.

Description. – Very involute compressed sub-platyconic shell with broad tabulate venter and sharp ventro-lateral shoulders. Flanks slightly convex, with maximal whorl width at mid-flanks. The tabulate venter amounts to ca. 65% of the whorl width. Umbilicus narrow (U/D = 13%), with a vertical umbilical wall and rounded umbilical shoulders. Shell smooth except for growth lines visible only near the umbilicus of the body chamber. Suture line with broad lateral lobes and saddles. Ventral lobe relatively broad, asymmetrical, with narrow branches having just two deep indentations on the right side and four small indentations on the other. Auxiliary series short, with a poorly individualized auxiliary lobe followed by a few regular indentations.

Measurements. – See Figure 49 and Table 1.

Discussion. – This specimen differs from all other species included here in *Ussuridiscus* by its thicker venter, its vertical umbilical wall, its absence of ornamentation and its suture line with a shorter auxiliary series. Its general outline is close to that of *Mullericeras shigetai*, but it differs clearly from this species by its less sharp ventro-lateral shoulders, its maximum width situated at mid-flanks, its rounded umbilical shoulder and, most importantly, its simpler suture line with less indentations and a ventral lobe with narrow branches. This last characteristic excludes its assignment to the genus *Mullericeras* and allows placing it within the genus *Ussuridiscus* despite the fact that it lacks the overhanging umbilical wall.

Occurrence. – Nammal Nala, sample Nam377.

Ussuridiscus ornatus n. sp.

Plate 29, figures 15–21; Figure 49

p 1897   *Koninckites vidharba* n sp. Diener, pp. 139–141, pl. 7, fig. 8.

non 1897   *Koninckites vidharba* n. sp. Diener, pp. 139–141, pl. 7, fig. 9 (holotype).

1909   *Aspidites vidharba* (Diener, 1897); von Krafft & Diener, pp. 63, 64, pl. 5, figs 1, 2, pl. 14, fig. 14.

Derivation of name. – Species name refers to the ornamentation of this species.

Holotype. – Specimen PIMUZ30481 (Pl. 29, figs 18–21).

Type locality. – Nammal Nala, Salt Range, Pakistan.

Type horizon. – Sample Nam377 (middle Lower Ceratite Limestone), *Gyronites plicosus* Regional Zone; early Dienerian.

Diagnosis. – Ussuridiscus with a thin whorl section and thick, wavy convex ribs.

Material. – 3 specimens.

Description. – Involute (U/D ≈ 12%) compressed (W/D ≈ 22%, W/H ≈ 43%) sub-oxyconic shell with a tabulate venter and sharp ventro-lateral shoulders. Flanks convex with maximum width at or just below mid-flanks. Umbilical wall very short and overhanging, with a broadly rounded umbilical shoulder. Blunt convex ribs on mid-flanks, fading both on the outer half of the flanks and near the umbilical margin. Suture line with an arched outline and shallow lobes and short saddles. Lobes with very few small indentations. Auxiliary series long, with numerous irregular indentations and no auxiliary lobe. The holotype shows low and almost rectangular first two lateral saddles, the third one having a low dome outline. On the other specimen, the lateral saddles are more elongated, the two-first ones being rounded and the third one being rectangular.

Measurements. – See Figure 49 and Table 1.

Discussion. – This species clearly differs from other species of *Ussuridiscus* by its strong ribbing. The specimens ascribed by von Krafft & Diener (1909) to *Aspidites vidharba* are nearly identical with our specimens. They differ slightly by the position of their maximal whorl width, situated lower on the flanks, but this can possibly be interpreted as intraspecific variability. They related their specimens with one of the two specimens figured by Diener (1897, pl. 7, fig. 8) named *Koninckites vidharba*, and therefore kept this species name. However, Diener (1897) explicitly designated the specimen figured on Plate 7, fig. 9 as being the type of this species, and von Krafft & Diener (1909) synonymized it with *Meekoceras hodgsoni*. Therefore, this species needed to be renamed. The illustration given by Diener (1897, pl. 7, fig. 8) does not really agree with our specimens, nor those of von Krafft & Diener (1909) because of their nearly straight instead of convex ribs. However, Diener's (1897) description is perfectly in agreement with ours; therefore, we keep it provisionally in our synonymy list. It should be noted here that the type specimen of *Koninckites vidharba* is a small and immature specimen, and that the illustration given by Diener (1897) is quite different from the illustration of the same specimen shown by von Krafft & Diener (1909). It chiefly differs from our specimens by its larger umbilicus. As the type is such a small specimen, the species name *vidharba* is here considered as a *nomen dubium*.

Occurrence. – Nammal Nala, sample Nam377 (n = 1); Wargal, samples War6 (n = 1) and War9 (n = 1).

Ussuridiscus? sp. indet.

Plate 29, figures 22–27; Figure 49

Material. – Two specimens.

Description. – Very involute (U/D ≈ 7%), very compressed (W/D ≈ 20%, W/H ≈ 36%) platyconic shell with tabulate venter individualized by narrowly rounded ventro-lateral shoulders. Flanks nearly flat and sub-parallel with very weak sigmoid folds. Umbilicus very narrow, with an overhanging umbilical wall individualized by a sharp shoulder. Suture line simple, slightly different on the two specimens. The first one (Pl. 29, figs 22–24) has relatively short and well rounded lateral saddles, while the second one (Pl. 29, figs 25–27) has more elongated saddles, the first one being rounded, the second one slightly bent towards the umbilicus. Its third lateral saddle is very small with a flattened top. Both specimens have a rather shallow ventral lobe, the branches of which are narrow and pointed, and a long auxiliary series with numerous irregularly spaced denticulations.

Measurements. – See Figure 49 and Table 1.

Discussion. – Since both specimens are incomplete, slightly distorted phragmocones, it is not possible to identify them with certainty. Their general shape, very involute with a tabulate venter and a narrow ventral lobe is in agreement with our definition of *Ussuridiscus*, which entails that these specimens are the oldest representatives of this genus in the Salt Range.

Occurrence. – Nammal Nala, sample Nam391.

Family incertae sedis

Genus *Kingites* Waagen, 1895

Type species. – *Kingites lens* Waagen, 1895.

Composition of the genus. – *Kingites davidsonianus* (de Koninck, 1863) and *Kingites korni* Brühwiler, Ware, Bucher, Krystyn & Goudemand, 2010a.

Emended diagnosis. – Very involute, moderately compressed and discoidal shell with arched venter. Suture line with moderately deep lobes, a second saddle somewhat dorsally bent, a broad third lateral saddle with a rectangular outline. Lobes with numerous denticulations restricted to their base, and a rather long auxiliary series with an occasional poorly individualized auxiliary lobe.

Discussion. – This genus, originally ascribed to Meekoceratidae by Waagen (1895), has been placed in Paranoritidae by Spath (1934) and in Proptychitidae by Tozer (1994) who considered the two families as being synonyms. Here, we have brought evidence that Paranoritidae are not related to Proptychitidae but to Gyronitidae (see discussion on Paranoritidae above). There is no direct link between Paranoritidae and Proptychitidae. The genus *Kingites* should therefore belong to one of these two families. It is however very difficult to decide which one. Its general shape, with its broadly rounded venter, its absence of strigation and its ontogeny with very involute juvenile stages suggest affinity with compressed species of *Proptychites* such as *Proptychites ammonoides*. On the other hand, its simple suture line without denticulation on the side of the lobes, its differentiated oblique umbilical wall and the presence of radial lirae suggest a connection with *Koninckites khoorensis*, and thus to Paranoritidae, especially since some variants of the type species have a suture line very similar to Paranoritidae, with an auxiliary lobe (see below). In both case, it would imply a simplification of the suture line, although much stronger if considered as related to Proptychitidae rather than to Paranoritidae. Here, we decided to keep this genus in open classification. The type species of this genus is here synonymized with the species of de Koninck (1863), i.e. *Ceratites davidsonianus*. The species ascribed to this genus by Tozer (1994), *Kingites discoidalis* and *Kingites thulensis*, as well as *Kingites korostolevi* described by Zakharov (1978) and Dagys & Ermakova (1996) have a very narrowly rounded, almost acute venter, and a more complex suture line than the species here considered. Therefore, they must represent a distinct and probably new genus.

Occurrence. – Late Dienerian and early Smithian of the Salt Range (Pakistan), Spiti valley (India) and Nepal.

Kingites davidsonianus (de Koninck, 1863)

Plate 30, figures 1–18; Plate 31, figures 1–4; Figure 55

1863 *Ceratites davidsonianus* n. sp. de Koninck, p. 13, pl. 6, fig. 2 (holotype).

1895 *Kingites lens* n. gen., n. sp. Waagen, pp. 232, 233, pl. 26, fig. 4 (holotype).

1895 *Kingites declivis* n. gen., n. sp. Waagen, pp. 233–235, pl. 26, fig. 2 (holotype).

1934 *Koninckites davidsonianus* (de Koninck, 1863); Spath, pp. 151, 152, text-fig. 43a, pl. 5, fig. 2 (holotype).

1934 *Koninckites occlusus* n. sp. Spath, pp. 154, 155, pl. 5, fig. 1 (holotype).

1934 *Kingites lens* Waagen, 1895; Spath, pp. 157, 158, fig. 45 [cop. Waagen, 1895].

? 1996 *Kingites lens* Waagen, 1895; Waterhouse, p. 43, text-fig. 4E, pl. 2, fig. 3.

1996 *Ceratites davidsonianus* de Koninck, 1863; Waterhouse, p. 46, text-fig. 5 (holotype).

? 1996 *Aspitella crassa* (von Krafft, 1909); Waterhouse, p. 51, text-fig. 4I, pl. 3, figs 2–4.

1996 *Kingites elegans* n. sp. Waterhouse, pp. 73, 74, text-fig. 10D, pl. 8, figs 4–6, 8, 9 (holotype), 10, 11.

Emended diagnosis. – *Kingites* with a narrow, generally funnel-shaped umbilicus and a suture line with occasional differentiated auxiliary lobe.

Material. – 78 specimens.

Description. – Very involute (U/D ≈ 9%), moderately compressed (W/D ≈ 28%, W/H ≈ 50%) discoidal shell with broadly arched venter. Flanks slightly convex with maximal width between inner third and mid-flanks. Umbilical wall generally

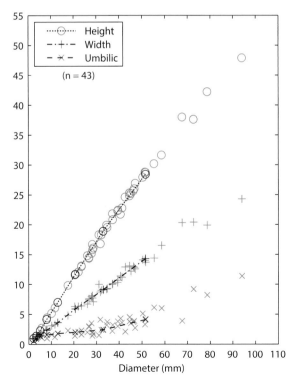
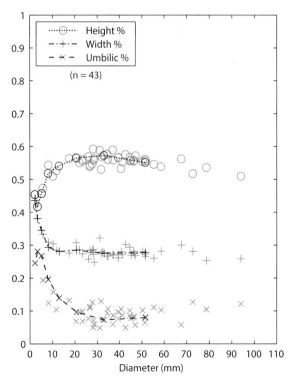

Fig. 55. Kingites davidsonianus (de Koninck, 1863). Scatter diagrams of H, W and U, and of H/D, W/D, and U/D (abbreviations as in Fig. 15). Ontogenetic trajectories obtained from sectioned specimen illustrated on Plate 31, figures 3, 4 are shown in black.

oblique, the umbilical seam being situated just above the umbilical shoulder of the preceding whorl, imparting the umbilicus a funnel shape. Umbilical shoulder sub-angular on the shell, more rounded on the inner mould. Some specimens have a sub-vertical umbilical wall. Just after the neanic stage, the juveniles become rapidly very involute to reach sub-mature proportions at a diameter of ca. 20 mm. Some large specimens have a slight egression of the adult body chamber. Ornamentation nearly absent, with slightly sigmoid very thin radial lirae and sometimes very vague folds following the trajectory of the lirae. Suture line with moderately deep lobes and elongated saddles, lobes having small indentations at their base. Second lateral saddle slightly bent dorsally, third lateral saddle sub-rectangular, as broad as or broader than high. Auxiliary series long, with small irregular indentations. Poorly individualized auxiliary lobe only occasional.

Measurements. – See Figure 55.

Discussion. – The type species of *Kingites*, *Kingites lens*, was originally based on one single, relatively well preserved specimen which is here re-illustrated (Pl. 30, figs 4, 5). Waagen did not have access to the types of de Koninck, and thus could not notice the similarity of his specimen with *Ceratites davidsonianus* as established by de Koninck (1863). The drawing of de Koninck is merely a sketch and the suture line is not reliably drawn, especially the third lateral saddle which has been re-illustrated by Spath (1934) and Waterhouse (1996). This specimen was considered by Spath (1934) as a representative of *Koninckites*, an erroneous assignment first pointed out by Waterhouse (1996). The different species here synonymized with *Kingites davidsonianus* differ merely by small differences in their auxiliary series and their more or less oblique umbilical wall. These differences are here shown to be part of intraspecific variability. The specimens ascribed to *Kingites lens* and *Aspitella crassa* by Waterhouse (1996) are too poorly preserved to be clearly assigned to the present species. *Kingites davidsonianus* looks superficially similar to *Proptychites ammonoides*, but can easily be differentiated by its much simpler suture line and its individualized umbilical wall. It also resembles closely to involute variants of the co-occurring *Koninckites khoorensis*, especially variants with a differentiated auxiliary lobe, but can be distinguished by its stronger involution and the absence of strigation on its venter. One specimen (Pl. 31, figs 1, 2) has its lower jaw preserved almost *in situ* at the beginning of its body chamber.

Occurrence. – Amb, sample Amb64 ($n = 1$); Nammal Nala, samples Nam58 ($n = 8$), Nam59 ($n = 13$), Nam61 ($n = 21$), Nam67 ($n = 1$), Nam101d ($n = 9$), Nam130 ($n = 3$), Nam304 ($n = 3$), Nam313 ($n = 6$), Nam315 ($n = 1$), Nam336 ($n = 1$), Nam346 ($n = 9$) and Nam504 ($n = 2$).

Kingites korni Brühwiler, Ware, Bucher, Krystyn & Goudemand, 2010a

Plate 31, figures 5–9; Figure 56

v 2010a *Kingites korni* n. sp. Brühwiler, Ware, Bucher, Krystyn & Goudemand, p. 734, fig. 15 (holotype).

v 2010a *Kingites parkashi* n. sp. Brühwiler, Ware, Bucher, Krystyn & Goudemand, p. 734, fig. 16 (holotype).

v 2012 ?*Kingites parkashi* Brühwiler, Ware, Bucher, Krystyn & Goudemand, 2010a; Brühwiler & Bucher, p. 53, fig. 36E–I.

Material. – 14 specimens.

Description. – Extremely involute, discoidal to sub-oxyconic shell with occluded umbilicus and rounded venter. Flanks slightly convex, with maximal width at inner third or mid-flank. Surface smooth apart for very vague sigmoid folds following the trajectory of the growth lines. Suture line with relatively shallow lobes and short saddles, the third lateral saddle being low and sub-rectangular. Lateral and ventral lobes with few indentations at their base. Auxiliary series long, with numerous irregular indentations, but no clearly differentiated auxiliary lobe.

Measurements. – See Figure 56.

Discussion. – The two new species originally described by Brühwiler *et al.* (2010a) were based on very few specimens. They were considered as different on the basis of their whorl section, with more convergent flanks and a more narrowly rounded venter in *Kingites parkashi* than in *Kingites korni*. The specimens from the Salt Range show all intermediate shapes between these two species. Therefore, they should be treated as synonyms. The specimen here figured is an inner mould, its umbilicus appears thus to be open, but it is in reality closed by a thickening of the shell on the umbilical wall. *Kingites korni* superficially resembles *Proptychites pagei*, from

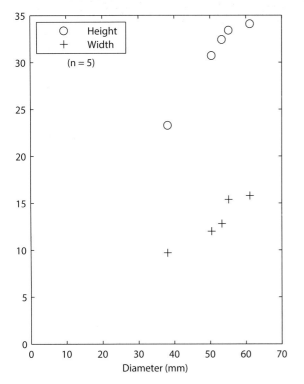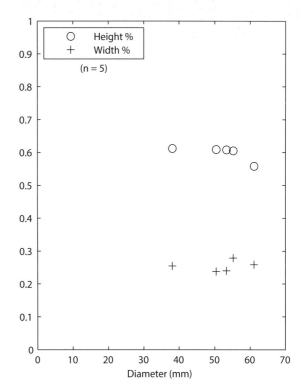

Fig. 56. Kingites korni Brühwiler, Ware, Bucher, Krystyn & Goudemand, 2010a. Scatter diagrams of H, W and U, and of H/D, W/D, and U/D (abbreviations as in Fig. 15).

which it differs by its much simpler suture line with fewer indentations and shorter lobes and saddles.

Occurrence. – Chiddru, sample Chi51 ($n = 4$); Nammal Nala, samples Nam63 ($n = 1$), Nam70 ($n = 1$), Nam305 ($n = 6$), Nam312 ($n = 1$), Nam349 ($n = 1$).

<h1 style="text-align:center">Superfamily Sagecerataceae Hyatt, 1884</h1>

<h1 style="text-align:center">Family Hedenstroemiidae Hyatt, 1884</h1>

Genus *Clypites* Waagen, 1895

Type species. – *Clypites typicus* Waagen, 1895.

Composition of the genus. – Type species only.

Emended diagnosis. – Very involute, moderately compressed, sub-oxyconic shell with occluded umbilicus. Venter tabulate, with sharp ventro-lateral shoulders at small size, becoming rounded on the body chamber of large specimens (i.e. at D ≥ 7 cm). Suture line with a poorly developed adventive series visible only on small sized specimens, disappearing at a diameter of ca. 5 cm and composed of only one adventive lobe and no adventive saddles.

Discussion. – This genus clearly differs from *Pseudosageceras* by its broader venter and simpler suture line. The suture line is close to that of involute forms of Meekocerataceae such as Mullericeratidae but differs by the presence of an adventive lobe in small specimens. This similarity suggests that Hedenstroemiidae root into Mullericeratidae, *Clypites* being the link between the two families. A derivation from Prolecanitida or Otocerataceae as suggested by other authors (e.g. Spath 1934; Arkell *et al.* 1957; Tozer 1981) is consequently rejected.

Occurrence. – Late Dienerian of the Salt Range (Pakistan), Dienerian of South Primorye (Russia) and Guangxi (China).

Clypites typicus Waagen, 1895

Plate 31, figures 10–19; Plate 32, figures 1–4; Figures 49, 57

<blockquote>
1895   *Clypites typicus* n. gen., n. sp. Waagen, pp. 143, 144, pl. 21, fig. 7 (lectotype).

1895   *Clypites kingianus* n. gen., n. sp. Waagen, pp. 144–146, pl. 21, fig. 8 (lectotype), pl. 22, fig. 3.
</blockquote>

1895 *Clypites evolvens* n. gen., n. sp. Waagen, pp. 146–148, pl. 22, fig. 2(lectotype).

non 1905 *Clypites evolvens* Waagen, 1895; Noetling, pl. 23, fig. 3.

1934 *Clypites typicus* Waagen, 1895; Spath, p. 220, fig. 69 [cop. Waagen, 1895].

?2008 *Clypites* sp. indet.; Brayard & Bucher, p. 72, pl. 38, figs 1–4.

2009 *Clypeoceras spitiense* (von Krafft, 1909); Shigeta & Zakharov, pp. 125, 126, figs 113, 114.

non 2010a *Clypites typicus* Waagen, 1895; Brühwiler, Ware, Bucher, Krystyn & Goudemand, pp. 736, 737, fig. 21.

Emended diagnosis. – As for the genus.

Material. – 25 specimens.

Description. – Very involute, moderately compressed (W/D ≈ 25%) sub-oxyconic shell with a broad tabulate venter and slightly convex flanks giving the whorl section an elongated and sub-

trapezoidal shape with maximal width at the inner third of the flanks. Some specimens have a very low, broad keel as in *Mullericeras spitiense*, their venter being then close to tectiform. Ventro-lateral shoulders very sharp in small specimens, becoming rounded on the body chamber of large specimens. Umbilicus occluded, with a strong thickening of the shell similar to that observed in *Pseudosageceras simplelobatum* (see below). Umbilical wall vertical and well differentiated, visible only on inner moulds. Surface nearly smooth, with only low sigmoid folds following the trajectory of the growth lines. Some specimens have one to three very faint spiral folds on the outer half of the flanks (e.g. Pl. 32, figs 1, 2). Suture line with deep lobes and elongated saddles and a poorly developed adventive series with only one adventive lobe which disappears at large size (D ≥ 5 cm, but the diameter at which it disappears is most probably subject to intraspecific variability). One specimen (Pl. 31, fig. 18) with a very asymmetric suture line also has an adventive saddle on its right side. Lobes have three to six deep indentations. Lateral saddles narrow and elongated, the second and third being bent dorsally. Long auxiliary series with one individualized auxiliary lobe followed by a long, rather regular series of indentations.

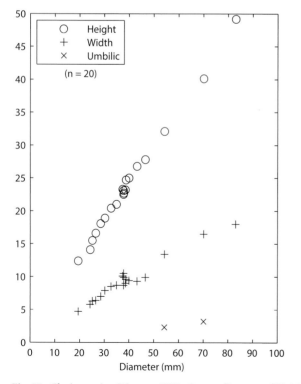
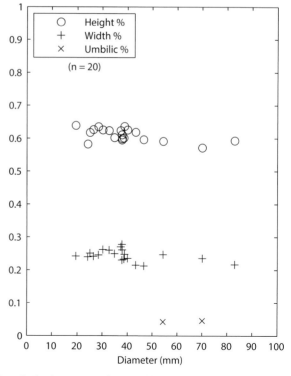

Fig. 57. Clypites typicus Waagen, 1895;. Scatter diagrams of H, W and U, and of H/D, W/D, and U/D (abbreviations as in Fig. 15). The two values for the umbilicus width were measured on internal moulds, the umbilicus being otherwise closed by an umbilical lid.

Measurements. – See Figures 49, 57.

Discussion. – Due to the disappearance of the adventive series, large specimens are nearly identical to *Mullericeras spitiense*, from which they then differ only by their larger size, more elongated saddles and slightly more developed auxiliary series. Waagen (1895) initially erected three different species, which are here considered as synonyms. *Clypites kingianus* is nearly identical to the holotype of *Clypites typicus*, the small differences noted by Waagen being just the result of its larger size. *Clypites evolvens* looks different at first sight, but its open umbilicus is probably a preservation bias, as its outer whorl is an internal mould (unlike the specimens of the two other species he describes), the umbilical lid being not preserved and therefore revealing an open umbilicus. The specimens assigned by Brayard & Bucher (2008) to *Clypites* sp. indet. are close to this species, but as mentioned by the authors, they are too poorly preserved to be identified with confidence. The specimen ascribed by Shigeta & Zakharov (2009) to *Clypeoceras spitiense* is here assigned to *Clypites typicus* as its suture line has a very long and well developed auxiliary series and more importantly, a slightly differentiated adventive lobe.

Occurrence. – Amb, sample Amb65 ($n = 1$); Nammal Nala, samples Nam61 ($n = 8$), Nam62 ($n = 1$), Nam70 ($n = 1$), Nam83 ($n = 2$), Nam130 ($n = 3$), Nam305 ($n = 2$), Nam312 ($n = 1$), Nam319 ($n = 1$), Nam349 ($n = 1$) and Nam504 ($n = 1$); Wargal, sample War104 ($n = 3$).

Genus *Pseudosageceras* Diener, 1895

Type species. – *Pseudosageceras* sp. indet. Diener, 1895.

Pseudosageceras simplelobatum n. sp.

Plate 32, figures 5–16; Plate 33, figures 1–10;
Figures 49, 58–59

1909   *Hedenstroemia lilangense* n. sp. von Krafft, pp. 151, 152, pl. 9, fig. 1.

? 1985   *Pseudosageceras multilobatum* Noetling, 1905; Pakistani-Japanese Research Group, pl. 12, figs 5–7, pl. 14, figs 3, 6.

? 1994   *Tellerites* sp. indet.; Tozer, p. 84, pl. 20, fig. 10.

1996   *Lilastroemia lilangensis* (von Krafft, 1909) n. gen.; Waterhouse, pp. 76, 77.

non 1996   *Lilastroemia lilangensis* (von Krafft, 1909) n. gen.; Waterhouse, Text-fig. 10E, pl. 9, figs 13, 16, 17.

2010a   *Clypites typicus* Waagen, 1895; Brühwiler, Ware, Bucher, Krystyn & Goudemand, pp. 736, 737, fig. 21.

Derivation of name. – From Latin *simplex* and *lobatum*, meaning 'simple' and 'lobe', in reference to its suture line which is simpler than that of other species of *Pseudosageceras*.

Holotype. – Specimen PIMUZ30506 (Pl. 33, figs 1–3).

Type locality. – Nammal Nala, Salt Range, Pakistan.

Type horizon. – Sample Nam346 (lower Ceratite Marls, ca. 6 m above base), top of *Kingites davidsonianus* Regional Zone; late Dienerian.

Diagnosis. – *Pseudosageceras* characterized by a relatively simple suture line, with only two adventitious saddles and one auxiliary lobe at maturity. Lobes and saddles relatively wide and not very elongated.

Material. – 70 specimens.

Description. – Very involute, compressed oxyconic shell with a narrow concave, bicarinate venter and weakly convex flanks conferring the whorl section an elongated sub-trapezoidal shape with maximal width close to the umbilical wall. Umbilicus occluded by an umbilical lid visible on the specimen which has been cut to study its ontogeny (Pl. 33, figs 4, 5). Its shell thickens towards the umbilicus and forms an extension which almost closes the umbilicus. On the internal mould, the umbilical shoulder is rounded and grades into a vertical umbilical wall. The shell reaches its typical oxyconic shape at a diameter of about 1 cm. From there on, its morphology barely changes, the umbilicus relative width (here measured below the umbilical lid) increasing slightly, from 2.9% at a diameter of 15.4 mm to 6% at D = 83.3 mm. Surface almost smooth, with very weak and broad folds following the trajectory of the sigmoid growth lines. One specimen (Pl. 33, figs 8–10) shows a very faint strigation on the inner mould of its body chamber. Suture line typical of the genus, with a trifid first lateral lobe and a well developed

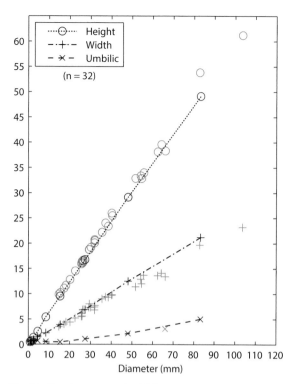
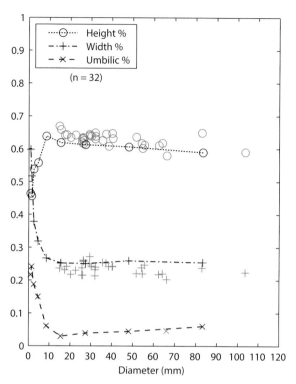

Fig. 58. Pseudosageceras simplelobatum n. sp. Scatter diagrams of H, W and U, and of H/D, W/D, and U/D (abbreviations as in Fig. 15). Ontogenetic trajectories obtained from sectioned specimen illustrated on Plate 33, figures 4, 5 are shown in black. In addition to the ontogenetic trajectory of the sectioned specimen, a single additional value for the umbilicus width was measured on an internal mould, the umbilicus being otherwise occluded.

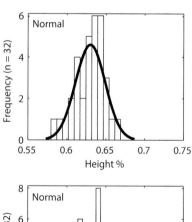
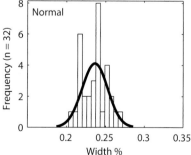

Fig. 59. Pseudosageceras simplelobatum n. sp. Histograms of H/D and W/D (abbreviations as in Fig. 15).

adventive series including two adventive saddles. The second lateral lobe becomes bifid in large specimens, the adventive and auxiliary lobes bear only simple indentations. Lateral lobes and saddles broad, the lateral saddles being elongated and bent towards the umbilicus. Auxiliary series very well developed with only one weakly discernible auxiliary lobe followed by a long, rather regular series of indentations.

Measurements. – See Figures 49, 58, 59.

Discussion. – As the name given by von Krafft (1909) to this species is preoccupied by *Ambites lilangensis* (von Krafft, 1909), we decided here to give it a new name to avoid confusion with this species which can also be found in the Dienerian. This species differs from all other known species of *Pseudosageceras* by its simpler suture line with only two adventive saddles and one auxiliary lobe, the auxiliary lobes and saddles being not bifid, with broader and less elongated lateral saddles. The overall shape of the suture line is closer to that of involute representatives of Meekocerataceae (e.g. Mullericeratidae) than to that of other species of the same lineage (e.g. other species of *Pseudosageceras* or *Cordillerites*). It also differs by its ornamentation, which is very weak, all other species referred to the genus being smooth. This ornamentation is also reminiscent of the weak ornamentation of involute forms of Meekocerataceae, which brings additional support to a derivation of *Pseudosageceras* from Meekocerataceae via Mullericeratidae. The presence of an umbilical lid is here reported for the first time among hedenstroemiids, but as only one specimen has been sectioned, it is not possible to decide whether this is a characteristic trait of the species, genus or even family or if it is present in only some specimens. The specimens described as *Hedenstroemia lilangense* by von Krafft (1909); for which Waterhouse created the new genus *Lilastroemia*) and as *Clypites typicus* by Brühwiler *et al.* (2010a) both show a clear trifid first lateral lobe and therefore must be assigned to *Pseudosageceras*. Their suture lines show some small differences compared to the ones figured here, with a less developed adventive series and an auxiliary series with slightly more differentiated lobes and saddles. These small differences are here interpreted as intraspecific variability. The specimens assigned to *Lilastroemia lilangensis* by Waterhouse (1996) do not show any adventive lobes and saddles, and are therefore closer to Mullericeratidae than to *Hedenstroemia lilangense*. The specimens referred to as *Pseudosageceras multilobatum* by the Pakistani-Japanese Research Group, which come from the Lower Ceratite Limestone of Chiddru and are therefore Dienerian in age, most probably belong to our new species. Because their suture line is not illustrated, a definitive assignment must be left open. The specimen described by Tozer 1994 as *Tellerites* sp. indet., which is also of late Dienerian age, could belong to this species as suggested by its apparently simple suture line. However, it is too poorly preserved for any further comparison. This species is the oldest representative of this genus described so far, with the possible exception of one extremely poorly preserved specimen described by Waterhouse (1994, p. 49, pl. 2, fig. 16). Waterhouse's specimen is too much weathered to be assigned to any taxon. The vague trace of suture line on which this identification was based could well be the imprint of the dorsal part of suture of the missing subsequent whorl.

Occurrence. – Chiddru, samples Chi51 ($n = 6$), Chi61 ($n = 2$) and Chi105 ($n = 1$); Nammal Nala, samples Nam59 ($n = 9$), Nam61 ($n = 3$), Nam62 ($n = 3$), Nam63 ($n = 1$), Nam67 ($n = 18$), Nam70 ($n = 1$), Nam83 ($n = 2$), Nam101d ($n = 3$), Nam130 ($n = 1$), Nam313 ($n = 1$), Nam319 ($n = 2$), Nam336 ($n = 2$), Nam337 ($n = 2$), Nam346 ($n = 3$), Nam349 ($n = 1$), Nam350 ($n = 2$), Nam502 ($n = 1$), Nam504 ($n = 2$) and Nam724 ($n = 1$); Wargal, sample War104 ($n = 3$).

Superfamily incertae sedis
Family incertae sedis

Genus *Subacerites* n. gen.

Derivation of name. – Latin prefix *sub-*, meaning 'almost', and the Latin word *acer*, meaning 'sharp', in reference to its almost acute venter.

Type species. – *Subacerites friski* n. sp.

Composition of the genus. – Type species only.

Diagnosis. – Very involute compressed oxyconic shell with occluded umbilicus, characterized by its extremely narrowly tabulate, almost acute venter. Suture line without adventive series and with a long auxiliary series comprising one auxiliary lobe.

Discussion. – This new genus is based on only a single specimen found in Amb. Its morphology is very close to the genus *Parahedenstroemia* Spath, 1934, but it differs from the latter by its narrowly tabulate venter and, more importantly, by the absence of adventive lobes and saddles. This last trait excludes it from Hedenstroemiidae. It differs from Mullericeratidae also by its peculiar ventral lobe, by its very narrowly elongated ventral saddle, and by the absence of differentiated umbilical wall. As this genus is defined on a single specimen, we prefer here to keep its systematic position open. More material would be necessary to assess its intraspecific variability and eventually unravel its affinity.

Occurrence. – Late Dienerian of Amb, *Kingites davidsonianus* Regional Zone.

Subacerites friski n. sp.

Plate 33, figures 11–15

Holotype. – Specimen PIMUZ30510 (Pl. 33, figs 11–15).

Type locality. – Amb, Salt Range, Pakistan.

Type horizon. – Sample Amb3 (lower Ceratite Marls, ca. 4 m above base), *Kingites davidsonianus* Regional Zone, late Dienerian.

Derivation of name. – Named after Dr. Åsa Marianne Frisk.

Diagnosis. – As for the genus.

Material. – One single specimen.

Description. – Very involute compressed oxyconic shell with occluded umbilicus. Venter extremely narrow and tabulate, almost acute, delimited by sharp ventro-lateral shoulders which become narrowly rounded on the body chamber. Flanks convex, with maximum width at mid-flanks. Umbilicus occluded by a thickening of the shell. On the internal mould, flanks slope gently towards the umbilicus, without forming a distinct umbilical wall. Very weak ornamentation on the flanks, consisting of very low sinuous radial folds parallel to growth lines. An extremely faint strigation is visible only on the inner mould of the body chamber, on the outer and inner third of the flanks. Growth lines prorsiradiate and sinuous, being slightly offset on the outer flanks. Suture line with a peculiar ventral lobe, divided by a very narrow and elongated ventral saddle, and with broad and shallow branches having numerous, regular and small indentations. These indentations are aligned almost perpendicularly to the ventral edge. Lateral saddles short with rounded tips, lateral lobes with a small number of large indentations. Auxiliary series long, with one poorly individualized auxiliary lobe followed by a long series of irregular indentations.

Measurements. – See Table 1.

Discussion. – As for the genus.

Occurrence. – Amb, sample Amb3.

Gen. et sp. indet.

Plate 33, figures 16–19

Material. – One single specimen.

Description. – Very involute, sub-oxyconic internal mould with tabulate venter. Its umbilicus is very narrow, most likely occluded by a thickened outer shell. Flanks flat, converging towards the venter, imparting the whorl section a trapezoidal shape. No ornamentation visible on the internal mould. Its suture line is peculiar, very complex for such a small sized individual. Its ventral lobe is divided by a wide ventral saddle and a deep indentation occurs between the ventral saddle and the long, narrow and bifid branch of the ventral lobe, thus individualizing an adventive saddle. The lateral saddles are very narrow and elongated. The first one is straight, subphylloid and asymmetric, almost pointed. The second one is very long and strongly bent dorsally. The third one is also elongated, bent towards the umbilicus with a flattened tip. The first lateral lobe is thin, deep and bifid. The second one has a broad base, and lateral indentations on both sides individualize an elongated deep and bifid indentation. The auxiliary series is also very unusual, with a poorly individualized auxiliary lobe. Instead of the usual pointed indentations, a series of smooth undulations confers the auxiliary series a goniatitic pattern.

Measurements. – See Table 1.

Discussion. – The outline of this shell is very close to that of *Pseudosageceras*, but its suture line prevents assignment to this genus, as it lacks the diagnostic trifid second lateral lobe of *Pseudosageceras*. The presence of an adventive saddle may indicate that it belongs to Hedenstroemiidae, but its peculiar second lateral lobe and goniatitic auxiliary series is very different from any species belonging to this group or from any other Dienerian ammonoid group. Considering that we have only one, small and fully septate specimen, we prefer to keep it here in open nomenclature. Additional larger specimens would be necessary to know if it is really a Hedenstroemiidae, if it is a new group converging towards Hedenstroemiidae, or if its peculiar suture line is induced by a pathology of some sort.

Occurrence. – Amb, sample Amb104.

Acknowledgements. – James M. Neenan (Oxford) improved the English text of an earlier version. Claude Monnet (Lille) is thanked for providing his statistical analyses software. Nicolas Goudemand and Séverine Urdy (Lyon) are thanked for their help with Matlab. Mike Orchard (Geological Survey of Canada, Vancouver) is thanked for providing access to the collections and archives of E.T. Tozer. Technical support for preparation and photography was provided by Markus Hebeisen and Rosemarie Roth (Zürich). This work is a contribution to the Swiss National Science Foundation project 200021-135446 to H.B.

References

Agassiz, L. 1847: Lettres sur quelques points d'organisation des animaux rayonnés. *Comptes Rendus de L'Académie des Sciences 25*, 677–682.

Arkell, W.J., Kummel, B. & Wright, C.W. 1957: Mesozoic Ammonoidea. *In* Arkell, W.J., Furnish, W.M., Kummel, B.,

Miller, A.K., Moore, R.C., Schindewolf, O.H., Sylvester-Bradley, P.C., Wright, C.W. (eds): *Treatise on Invertebrate Paleontology, Part L, Mollusca 4: Cephalopoda, Ammonoidea*, 80–436. Geological Society of America and the University of Kansas Press, Lawrence.

Arthaber, G.V. 1911: Die Trias von Albanien. *Beiträge zur Paläontologie und Geologie Österreich-Ungarns und des Orients 24*, 169–276.

Bando, Y. 1981: Lower Triassic Ammonoids from Guryul Ravine and the Spur three kilometres north of Barus. *In* Nakazawa, K., Kapoor, H.M. (eds): *The Upper Permian and Lower Triassic Faunas of Kashmir*. Palaeontologia Indica, volume *46*, 135–178.

Bengtson, P. 1988: Open nomenclature. *Palaeontology 31*, 223–227.

Brayard, A. & Bucher, H. 2008: Smithian (Early Triassic) ammonoid faunas from northwestern Guangxi (South China): taxonomy and biochronology. *Fossils and Strata 55*, 179.

Brayard, A., Bucher, H., Escarguel, G., Fluteau, F., Bourquin, S. & Galfetti, T. 2006: The Early Triassic ammonoid recovery: paleoclimatic significance of diversity gradients. *Palaeogeography, Palaeoclimatology, Palaeoecology 239*, 374–395.

Brayard, A., Escarguel, G., Bucher, H., Monnet, C., Brühwiler, T., Goudemand, N., Galfetti, T. & Guex, J. 2009: Good Genes and Good Luck: ammonoid Diversity and the End-Permian Mass Extinction. *Science 325*, 1118–1121.

Brosse, M., Brayard, A., Fara, E. & Neige, P. 2013: Ammonoid recovery after the Permian-Triassic mass extinction: a re-exploration of morphological and phylogenetical diversity patterns. *Journal of the Geological Society, London 170*, 225–236.

Brühwiler, T. & Bucher, H. 2012: Systematic Palaeontology. *In* Brühwiler, T., Bucher, H., Ware, D., Schneebeli-Hermann, E., Hochuli, P.A., Roohi, G., Rehman, K., Yaseen, A. (eds): *Smithian (Early Triassic) Ammonoids From the Salt Range, Pakistan*. Special Papers in Palaeontology *88*, 1–114.

Brühwiler, T., Brayard, A., Bucher, H. & Guodun, K. 2008: Griesbachian and Dienerian (Early Triassic) Ammonoid Faunas from Northwestern Guangxi and Southern Guizhou (South China). *Palaeontology 51*, 1151–1180.

Brühwiler, T., Ware, D., Bucher, H., Krystyn, L. & Goudemand, N. 2010a: New Early Triassic ammonoid faunas from the Dienerian/Smithian boundary beds at the Induan/Olenekian GSSP candidate at Mud (Spiti, Northern India). *Journal of Asian Earth Sciences 39*, 724–739.

Brühwiler, T., Bucher, H., Brayard, A. & Goudemand, N. 2010b: High-resolution biochronology and diversity dynamics of the Early Triassic ammonoid recovery: the Smithian faunas of the Northern Indian Margin. *Palaeogeography, Palaeoclimatology, Palaeoecology 297*, 491–501.

Brühwiler, T., Bucher, H., Ware, D., Schneebeli-Hermann, E., Hochuli, P.A., Roohi, G., Rehman, K. & Yaseen, A. 2012: Smithian (Early Triassic) ammonoids from the Salt Range, Pakistan. *Special Papers in Palaeontology 88*, 1–114.

Checa, A.G. & Garcia-Ruiz, J.M. 1996: Morphogenesis of the Septum in Ammonoids. *In* Landman, N.H., Tanabe, K. & Davis, R.A. (eds): *Ammonoid Paleobiology*, 253–296. Plenum Press, New York and London.

Cuvier, G.L.C.F.D. An 6 1797: *Tableau Élémentaire de L'histoire Naturelle des Animaux 14*. Baudouin, Paris, 710 pp.

Dagys, A.S. & Ermakova, S. 1996: Induan (Triassic) ammonoids from North-Eastern Asia. *Revue de Paléobiologie 15*, 401–447.

Diener, C. 1895: Triadische Cephalopodenfaunen der Ostsibirischen Küstenprovinz. *Mémoires du Comité Géologique St. Pétersbourg 14*, 1–59.

Diener, C. 1897: Part I: the Cephalopoda of the Lower Trias. *Palaeontologia Indica, Series 15. Himalayan Fossils 2*, 1–181.

Diener, C. 1909: see Krafft, A. von & Diener, C. 1909

Diener, C. 1916: Einige Bemerkungen zur Nomenklatur der Triascephalopoden. *Centralblatt für Mineralogie, Geologie und Palaeontologie, Stuttgart*, 97–105.

Erlich, A., Moulton, D.E., Goriely, A. & Chirat, R. 2016: Morphomechanics and developmental constraints in the evolution of ammonites shell form. *Journal of Experimental Zoology, Part B, Molecular and Developmental Evolution 00B*, 1–14.

Frech, F. 1902: Die Dyas: Lethaea geognostica. Theil 1. *Lethaea Palaeozoica 2*, 579–788.

Gee, E.R. 1980–1981: *Pakistan Geological Maps, Salt Range Series, Sheets 1–6*. Government of Pakistan, s.l.

Griesbach, C.L. 1880: Palaeontological notes on the Lower Trias of the Himalayas. *Records of the Geological Survey of India 13*, 94–113.

Guex, J. 1978: Le Trias inférieur des Salt Ranges (Pakistan): problèmes biochronologiques. *Eclogae Geologia Helvetica 71*, 105–141.

Guex, J. 1991: *Biochronological Correlations*, 252. Springer, Berlin.

Hammer, Ø., Harper, D.A.T. & Ryan, P.D. 2001: PAST: Paleontological statistics software package for education and data analysis. *Palaeontologia Electronica 4*, 9.

Hautmann, M., Ware, D. & Bucher, H. 2017: Geologically oldest oysters were epizoans on Early Triassic ammonoids. *Journal of Molluscan Studies 2017*, 1–8.

Hermann, E., Hochuli, P.A., Bucher, H., Brühwiler, T., Hautmann, M., Ware, D. & Roohi, G. 2011a: Terrestrial ecosystems on North Gondwana following the end-Permian mass extinction. *Gondwana Research 20*, 630–637.

Hermann, E., Hochuli, P.A., Méhay, S., Bucher, H., Brühwiler, T., Ware, D., Hautmann, M., Roohi, G., ur-Rehman, K. & Yaseen, A. 2011b: Organic matter and palaeoenvironmental signals during the Early Triassic biotic recovery: the Salt Range and Surghar Range records. *Sedimentary Geology 234*, 19–41.

Hermann, E., Hochuli, P.A., Bucher, H., Brühwiler, T., Hautmann, M., Ware, D., Weissert, H., Roohi, G., Yaseen, A. & ur-Rehman, K. 2012a: Climatic oscillations at the onset of the Mesozoic inferred from palynological records from the North Indian Margin. *Journal of the Geological Society, London 169*, 227–237.

Hermann, E., Hochuli, P.A., Bucher, H. & Roohi, G. 2012b: Uppermost Permian to Middle Triassic palynology of the Salt Range and Surghar Range, Pakistan. *Review of Palaeobotany and Palynology 169*, 61–95.

Hyatt, A. 1884: Genera of fossil cephalopods. *Proceedings of the Boston Society of Natural History 22*, 253–338.

Hyatt, A. 1900: Cephalopoda. *In* Zittle, K.A.V. (ed.): *Textbook of Palaeontology* vol. *1*, 502–604. C.R. Eastman, London.

Jenks, J.F., Monnet, C., Balini, M., Brayard, A. & Meier, M. 2015: Biostratigraphy of Triassic Ammonoids. *In* Klug, C., Korn, D., De Baets, K., Kruta, I., Mapes, R.H. (eds): *Ammonoid Paleobiology: From Macroevolution to Paleogeography*, 277–298. Topics in Geobiology 44, Springer Verlag, Berlin and Heidelberg, 628 pp.

Kiparisova, L.D. 1961: Paleontological fundamentals for the stratigraphy of Triassic deposits of Primorye region. 1. Cephalopod Mollusca. *Trudy Vsyesoyuzhogo Nauchno-Isslyedovatyel'skogo Geologichyeskogo Instituta (VSEGEI). Novaya Seriya 48*, 1–278 [In Russian].

Koken, E.H.F. 1905: see Noetling, F. 1905.

Koken, E.H.F. 1934: see Spath, L.F. 1934.

de Koninck, L.G. 1863: Description of some fossils from India, discovered by Dr. A. Fleming, of Edinburgh. *The Quarterly Journal of the Geological Society of London 19*, 1–19.

von Krafft, A. & Diener, C. 1909: Lower Triassic Cephalopoda from Spiti, Malla, Johar, and Byans. *Palaeontologia Indica 6*, 1–186.

von Krafft, A. 1909: see von Krafft, A. & Diener, C. 1909.

Krystyn, L., Balini, M. & Nicora, A. 2004: Lower and Middle Triassic stage and substage boundaries in Spiti. *Albertiana 30*, 40–53.

Kummel, B. 1966: The Lower Triassic Formations of the Salt Range and Trans-Indus Ranges, West Pakistan. *Bulletin of the Museum of Comparative Zoology 134*, 361–429.

Kummel, B. 1970: Ammonoids from the Kathwai Member, Mianwali Formation, Salt Range, West Pakistan. *In* Kummel,

B., Teichert, C. (eds): *Stratigraphic Boundary Problems: Permian and Triassic of West Pakistan*, 177–192. Special Publication of the Department of Geology, Vol. 4, University of Kansas.

Kummel, B. & Teichert, C. 1966: Relations between the Permian and Triassic formations in the Salt Range and Trans-Indus ranges, West Pakistan. *Neues Jahrbuch für Geologie Paläontologie. Abhandlungen 125*, 297–333.

Kummel, B. & Teichert, C. 1970: Stratigraphy and Paleontology of the Permian–Triassic Boundary Beds, Salt Range and Trans-Indus Ranges, West Pakistan. *In* Kummel, B. & Teichert, C. (eds): *Stratigraphic Boundary Problems: Permian and Triassic of West Pakistan*, 1–110. Special Publication of the Department of Geology, Vol. 4, University of Kansas.

Lilliefors, H.W. 1967: On the Kolmogorov-Smirnov test for normality with mean and variance unknown. *American Statistical Association Journal 62*, 399–402.

Matthews, S.C. 1973: Notes on open nomenclature and synonymy lists. *Palaeontology 16*, 713–719.

Mojsisovics, E.V., Waagen, W. & Diener, C. 1895: Entwurf einer Gliederung der pelagischen Sedimente des Trias-Systems. *Sitzungberichte der Akademie der Wissenschaften in Wien (I) 104*, 1271–1302.

Monnet, C. & Bucher, H. 2005: New Middle and Late Anisian (Middle Triassic) ammonoid faunas from northwestern Nevada (USA): taxonomy and biochronology. *Fossil and Strata 52*, 121.

Monnet, C., Bucher, H., Wasmer, M. & Guex, J. 2010: Revision of the genus *Acrochordiceras* Hyatt, 1877 (Ammonoidea, Middle Triassic): morphology, biometry, biostratigraphy and intra-specific variability. *Palaeontology 53*, 961–996.

Monnet, C., Bucher, H., Guex, J. & Wasmer, M. 2012: Large scale evolutionary trends of Acrochordiceratidae Arthaber, 1911 (Ammonoidea, Middle Triassic) and Cope's rule. *Palaeontology 55*, 87–108.

Mu, L., Zakharov, Y.D., Li, W.-Z. & Shen, S.-Z. 2007: Early Induan (Early Triassic) cephalopods from the Daye Formation at Guiding, Guizhou Province, South China. *Journal of Paleontology 81*, 858–872.

Noetling, F. 1901: Beiträge zur Geologie der Salt Range, insbesondere der permischen und Triassischen Ablagerungen. *Neues Jahrbuch für Mineralogie, Geologie und Paläontologie, Beilage-Band 14*, 369–471.

Noetling, F. 1905: Die asiatische Trias. *In* Frech, F. (ed.): *Lethaea Geognostica, Das Mesozoicum*, 107–221. Verlag der E. Schweizerbart'schen Verlagsbuchhandlung (E. Nägele), Stuttgart, Germany.

Orchard, M.J. 2007: Conodont diversity and evolution through the latest Permian and Early Triassic upheavals. *Palaeogeography, Palaeoclimatology, Palaeoecology 252*, 93–117.

Orchard, M.J. 2008: Lower Triassic conodonts from the Canadian Arctic, their intercalibration with ammonoid-based stages and a comparison with other North American Olenekian faunas. *Polar Research 27*, 393–412.

Pakistani-Japanese Research Group 1985: Permian and Triassic systems in the Salt Range and Surghar Range, Pakistan. *In* Nakazawa, K. & Dickins, J.M. (eds): *The Tethys, her Paleogeography and Paleobiogeography From Paleozoic to Mesozoic*, 221–312. Tokai University Press, Tokyo.

Romano, C., Goudemand, N., Vennemann, T.W., Ware, D., Schneebeli-Hermann, E., Hochuli, P.A., Brühwiler, T., Brinkmann, W. & Bucher, H. 2013: Climatic and biotic upheavals following the end-Permian mass extinction. *Nature Geoscience 6*, 57–60.

Schindewolf, O.H. 1954: Über die Faunenwende vom Paläozoikum zum Mesozoikum. *Zeitschrift der Deutschen Geologischen Gesellschaft 105*, 153–182.

Schneebeli-Hermann, E., Kürschner, W.M., Hochuli, P.A., Bucher, H., Ware, D., Goudemand, N. & Roohi, G. 2012: Palynofacies analysis of the Permian-Triassic transition in the Amb section (Salt Range, Pakistan): implications for the anoxia on the South Tethyan Margin. *Journal of Asian Earth Sciences 60*, 225–234.

Shigeta, S. & Zakharov, Y.D. 2009: Cephalopods. *In* Shigeta, Y., Zakharov, Y.D., Maeda, H., Popov, A.M. (eds). *The Lower Triassic System in the Abrek Bay Area, South Primorye, Russia*, 44–140. National Museum of Nature and Science Monographs 38, Tokyo.

Smith, J.P. 1932: Lower Triassic ammonoids of North America. *United States Geological Survey, Professional Paper 167*, 1–199.

Spath, L.F. 1919a: V. Notes on Ammonites. *Geological Magazine 6*, 27–35.

Spath, L.F. 1919b: IV. Notes on Ammonites. *Geological Magazine 6*, 115–122.

Spath, L.F. 1930: The Eotriassic Invertebrate Fauna of East Greenland. *Meddelelser om Grönland 83*, 1–90.

Spath, L.F. 1934: *Catalogue of the Fossil Cephalopoda in the British Museum (Natural History), Part IV: The Ammonoidea of the Trias*, 521 pp. The Trustees of the British Museum, London.

Spath, L.F. 1935: Additions to the Eotriassic Invertebrate Fauna of East Greenland. *Meddelelser om Grönland 98*, 1–115.

Tozer, E.T. 1961: Triassic Stratigraphy and faunas, Queen Elizabeth Islands, Arctic Archipelago. *Memoir of the Geological Survey of Canada 316*, 1–116.

Tozer, E.T. 1963: Lower Triassic ammonoids from Tuchodi Lakes and Halfway River areas, northeastern British Columbia. *Bulletin of the Geological Survey of Canada 96*, 1–28.

Tozer, E.T. 1965: Lower Triassic stages and Ammonoid zones of Arctic Canada. *Paper of the Geological Survey of Canada 65–12*, 1–14.

Tozer, E.T. 1967: A Standard for Triassic time. *Bulletin of the Geological Survey of Canada 156*, 1–103.

Tozer, E.T. 1970: Marine Triassic Faunas. *In* Douglas, R.J.W. (ed): *Geology and Economic Minerals of Canada*, 5 edn, 633–640. Geological Survey of Canada, Vancouver, BC.

Tozer, E.T. 1971: Triassic Time and Ammonoids: problems and Proposals. *Canadian Journal of Earth Sciences 8*, 989–1031.

Tozer, E.T. 1981: Triassic Ammonoidea: classification, evolution and relationship with Permian and Jurassic forms. *In* House, M.R., Senior, J.R. (eds): *The Ammonoidea: The Evolution, Classification, Mode of Life and Geological Usefulness of a Major Fossil Group*, 65–100. The Systematics Association, Special Volume 18, Academic Press, London, UK.

Tozer, E.T. 1994: Canadian Triassic Ammonoid Faunas. *Bulletin of the Geological Survey of Canada 467*, 1–663.

Trümpy, R. 1969: Lower Triassic Ammonites from Jameson Land (East Greenland). *Meddelelser om Grönland 168*, 79–121.

Urdy, S., Goudemand, N., Bucher, H. & Chirat, R. 2010a: Allometries and the morphogenesis of the molluscan shell: a quantitative and theoretical model. *Journal of Experimental Zoology 314B*, 280–302.

Urdy, S., Goudemand, N., Bucher, H. & Chirat, R. 2010b: Growth dependent phenotypic variation of molluscan shell shape: a theoretical and empirical comparison using gastropods. *Journal of Experimental Zoology 314B*, 303–326.

Waagen, W. 1895: Salt Ranges Fossils. vol. 2: fossils from the Ceratites formation - Part I – Pisces, Ammonoidea. *Palaeontologia Indica 13*, 1–323.

Wang, Y.G. & He, G.X. 1976: Triassic ammonoids from the Mount Jolmo Lungma region. In: *A Report of Scientific Expedition From the Mount Jolmo Lungma Region (1966–1968)*, 223–502. Palaeontology, fascicule 3. Science Press, Beijing [In Chinese].

Wanner, J. 1911: Triascephalopoden von Timor und Rotti. *Neues Jahrbuch für Mineralogie, Geologie und Paläontologie 32*, 177–196.

Ware, D. & Bucher, H. 2018: Foreword. *Fossil & Strata 63*, 3–9.

Ware, D., Jenks, J.F., Hautmann, M. & Bucher, H. 2011: Dienerian (Early Triassic) ammonoids from the Candelaria Hills (Nevada, USA) and their significance for palaeobiogeography and palaeoceanography. *Swiss Journal of Geoscience 104*, 161–181.

Ware, D., Bucher, H., Brayard, A., Schneebeli-Hermann, E. & Brühwiler, T. 2015: High-resolution biochronology and diversity dynamics of the Early Triassic ammonoid recovery: the Dienerian faunas of the Northern Indian Margin. *Palaeogeography, Palaeoclimatology, Palaeoecology 440*, 363–373.

Ware, D., Bucher, H., Brühwiler, T. & Krystyn, L. 2018: Dienerian (Early Triassic) ammonoids from Spiti, Himachal Pradesh, India. *Fossil & Strata 63*, 179–241.

Wasmer, M., Hautmann, M., Hermann, E., Ware, D., Roohi, G., ur-Rehman, K., Yaseen, A. & Bucher, H. 2012: Olenekian (Early Triassic) Bivalves From The Salt Range And Surghar Range, Pakistan. *Palaeontology 55*, 1043–1073.

Waterhouse, J.B. 1994: The Early and Middle Triassic ammonoid succession of the Himalayas in western and central Nepal. Part 1. Stratigraphy, classification and Early Scythian ammonoid systematics. *Palaeontographica A232*, 1–83.

Waterhouse, J.B. 1996: The Early and Middle Triassic ammonoid succession of the Himalayas in western and central Nepal. Part 2. Systematic studies of the Early Middle Scythian. *Palaeontographica A241*, 27–100.

Westerman, G.E.G. 1966: Covariation and taxonomy of the Jurassic ammonite *Sonninia adicra* (Waagen). *Neues Jahrbuch für Mineralogie, Geologie und Paläontologie, Abhandlungen 124*, 289–312.

Yacobucci, M.M. & Manship, L.L. 2011: Ammonoid septal formation and suture asymmetry explored with a geographic information systems approach. *Palaeontologia Electronica 14* (1), 17.

Zakharov, Y.D. 1968: *Biostratigraphiya i Amonoidei Nizhnego Triasa Yuzhnogo Primorya (Lower Triassic Biostratigraphy and Ammonoids of South Primorye)*. Nauka, Moskva, 175 pp. [In Russian]

Zakharov, Y.D. 1978: *Lower Triassic Ammonoids of East USSR*. Nauka, Moskva, 224 pp. [In Russian]

Zakharov, Y.D. & Mu, L. 2007: *Systematic Paleontology*. In Mu L., Zakharov Y.D., Li W.-Z. & Shen S.-Z. 2007: Early Induan (Early Triassic) cephalopods from the Daye Formation at Guiding, Guizhou Province, South China. Journal of Paleontology 81, 860–872.

Plate 1   All figures natural size unless otherwise indicated; asterisks indicate the position of the last septum.

1–3:   ***Hypophiceras* aff. *H. gracile* (Spath, 1930)**
Lateral, apertural and ventral views. PIMUZ30236.
Loc. Nam361, dolomitic unit of the Kathwai Member, Nammal Nala. ?early Griesbachian.

4–13:   ***Ophiceras connectens* Schindewolf, 1954.**
4–6:   Lateral, apertural and ventral views. PIMUZ30239.
Loc. Nam376, base Lower Ceratite Limestone, Nammal Nala, *Ophiceras connectens* Regional Zone; late Griesbachian.
7–10:   (7–9) Lateral, apertural and ventral views. (10) Suture line at H = 16.6 mm, × 1.5. PIMUZ30237.
Loc. Chi54, limestone unit of the Kathwai Member, Chiddru, *Ophiceras connectens* Regional Zone; late Griesbachian.
11–13:   Lateral, apertural and ventral views. PIMUZ30238.
Loc. Nam390, base Lower Ceratite Limestone, Nammal Nala, *Ophiceras connectens* Regional Zone; late Griesbachian.

14–18:   ***Ophiceras sakuntala* Diener, 1897.**
14, 15:   Lateral and ventral views. PIMUZ30240.
Loc. Nam391, base Lower Ceratite Limestone, Nammal Nala. ?latest Griesbachian – earliest Dienerian.
16–18:   (16, 17) Lateral and ventral views. (18) Suture line at H = 15.5 mm, × 1.5.
PIMUZ30241.
Loc. Nam391, base Lower Ceratite Limestone, Nammal Nala. ?latest Griesbachian – earliest Dienerian.

19–21:   **?Ophiceratidae gen. et sp. indet.**
Lateral, apertural and ventral views. PIMUZ30242.
Loc. Nam391, base Lower Ceratite Limestone, Nammal Nala; ?latest Griesbachian – earliest Dienerian.

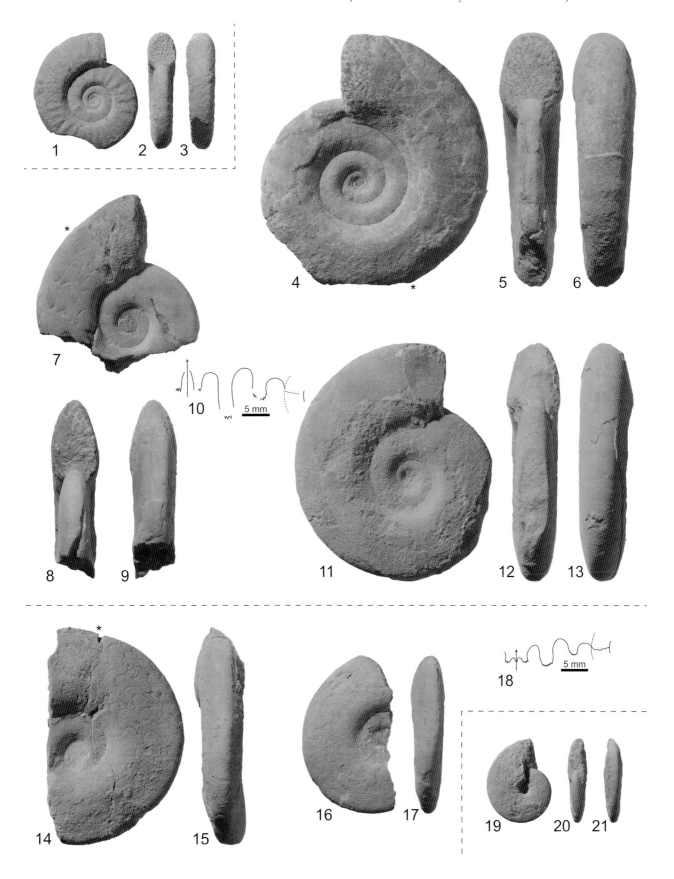

Plate 2   All figures natural size unless otherwise indicated; asterisks indicate the position of the last septum.

1–4:   ***Kyoktites hebeiseni* n. gen. et n. sp.**
      (1–3) Lateral, apertural and ventral views. (4) Suture line at H = 17.5 mm, × 1.5.
      PIMUZ30243. Holotype.
      Loc. Nam391, base Lower Ceratite Limestone, Nammal Nala. ?latest Griesbachian – earliest Dienerian.

5–8:   ***Kyoktites* cf. *K. hebeiseni* n. gen. et n. sp.**
      (5–7) Lateral, apertural and ventral views. (8) Suture line at H = 14.8 mm, × 1.5. PIMUZ30245.
      Loc. Amb104, base Lower Ceratite Limestone, Amb, *Gyronites dubius* Regional Zone; early Dienerian.

9–11:   **?Ophiceratidae n. gen. A n. sp. A**
      (9, 10) Lateral and ventral views. (11) Suture line at H = 23.4 mm. PIMUZ30246.
      Loc. Nam391, base Lower Ceratite Limestone, Nammal Nala; late Griesbachian.

12–24: *Ghazalaites roohii* n. gen. et n. sp.
12–14:   Lateral, apertural and ventral views. PIMUZ30247. Paratype.
      Loc. War8, middle Lower Ceratite Limestone, Wargal, *Gyronites plicosus* Regional Zone; early
      Dienerian.
15–17:   Lateral, apertural and ventral views. PIMUZ30248. Holotype.
      Loc. Nam332, middle Lower Ceratite Limestone, Nammal Nala, *Gyronites plicosus* Regional Zone; early
      Dienerian.
18–20:   Lateral, apertural and ventral views. PIMUZ30249. Paratype.
      Loc. Nam377, middle Lower Ceratite Limestone, Nammal Nala, *Gyronites plicosus* Regional Zone; early
      Dienerian.
21:        Suture line at H = 18.2 mm, × 1.5. PIMUZ30250. Paratype.
      Loc. Nam377, middle Lower Ceratite Limestone, Nammal Nala, *Gyronites plicosus* Regional Zone; early
      Dienerian.
22–24:   Lateral (22, 23) and ventral (24) views. PIMUZ30251. Paratype.
      Loc. Amb104, base Lower Ceratite Limestone, Amb, *Gyronites dubius* Regional Zone; early Dienerian.

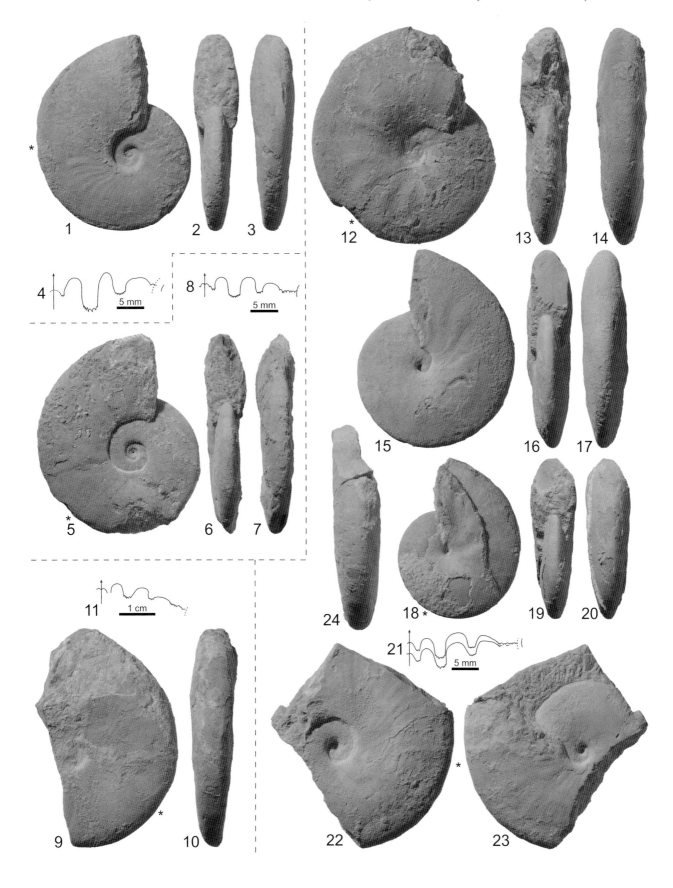

Plate 3 All figures natural size unless otherwise indicated; asterisks indicate the position of the last septum.

1:      Bedding plane with *Gyronites frequens* Waagen, 1895 and one specimen of *Ussuridiscus ensanus* (von Krafft, 1909) (marked 'U'). PIMUZ30252. × 0.5.
        Loc. Nam339, top Lower Ceratite Limestone, Nammal Nala, *Gyronites frequens* Regional Zone; early Dienerian.

2–5: *Gyronites frequens* Waagen, 1895
2:      Upper side of a specimen encrusted by worm tubes. PIMUZ30253.
        Loc. Nam354, top Lower Ceratite Limestone, Nammal Nala, *Gyronites frequens* Regional Zone; early Dienerian.
3–5:   Specimen encrusted by worm tubes. PIMUZ30255.
        Loc. Nam354, top Lower Ceratite Limestone, Nammal Nala, *Gyronites frequens* Regional Zone; early Dienerian.

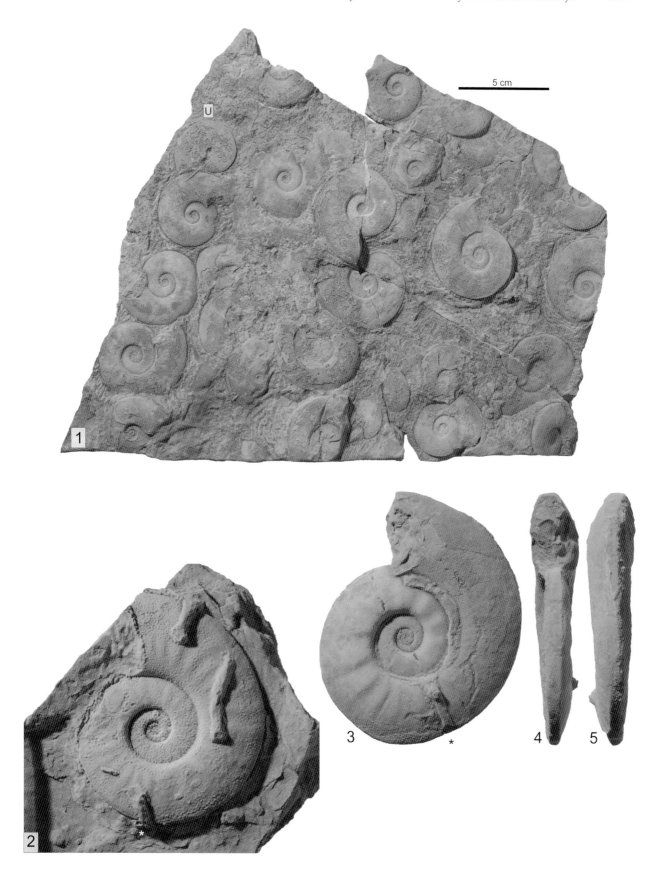

Plate 4 All figures natural size unless otherwise indicated asterisks indicate the position of the last septum.

1–17: *Gyronites frequens* Waagen, 1895

1–3:   Lateral, apertural and ventral views. PIMUZ30254.
        Loc. Nam354, top Lower Ceratite Limestone, Nammal Nala, *Gyronites frequens* Regional Zone; early Dienerian.

4–6:   Lateral, apertural and ventral views. PIMUZ30256.
        Loc. Nam76, top Lower Ceratite Limestone, Nammal Nala, *Gyronites frequens* Regional Zone; early Dienerian.

7–9:   Lateral, apertural and ventral views. PIMUZ30257.
        Loc. War7, top Lower Ceratite Limestone, Wargal, *Gyronites frequens* Regional Zone; early Dienerian.

10, 11: Lateral and ventral views. PIMUZ30258.
        Loc. Nam76, top Lower Ceratite Limestone, Nammal Nala, *Gyronites frequens* Regional Zone; early Dienerian.

12, 13: Lateral and ventral views. PIMUZ30259.
        Loc. Nam76, top Lower Ceratite Limestone, Nammal Nala, *Gyronites frequens* Regional Zone; early Dienerian.

14–17: (14–16) Lateral, apertural and ventral views. (17) Suture line at H = 15.4 mm, × 1.5 PIMUZ30260.
        Loc. War7, top Lower Ceratite Limestone, Wargal, *Gyronites frequens* Regional Zone; early Dienerian.

18–35: *Gyronites dubius* (von Krafft, 1909)

18–21: (18–20) Lateral, apertural and ventral views. (21) Suture line at H = 12.7 mm, × 2 (mirrored image). PIMUZ30261.
        Loc. Amb104, base Lower Ceratite Limestone, Amb, *Gyronites dubius* Regional Zone; early Dienerian.

22–24: Lateral, apertural and ventral views. PIMUZ30262.
        Loc. Amb104, base Lower Ceratite Limestone, Amb, *Gyronites dubius* Regional Zone; early Dienerian.

25–27: Lateral, apertural and ventral views. PIMUZ30263.
        Loc. Amb104, base Lower Ceratite Limestone, Amb, *Gyronites dubius* Regional Zone; early Dienerian.

28–30: Lateral, apertural and ventral views. PIMUZ30264.
        Loc. Amb104, base Lower Ceratite Limestone, Amb, *Gyronites dubius* Regional Zone; early Dienerian.

31, 32: Lateral and apertural views. PIMUZ30265.
        Loc. Nam391, base Lower Ceratite Limestone, Nammal Nala; ?latest Griesbachian – earliest Dienerian.

33–35: Lateral, apertural and ventral views. PIMUZ30266.
        Loc. Nam391, base Lower Ceratite Limestone, Nammal Nala; ?latest Griesbachian – earliest Dienerian.

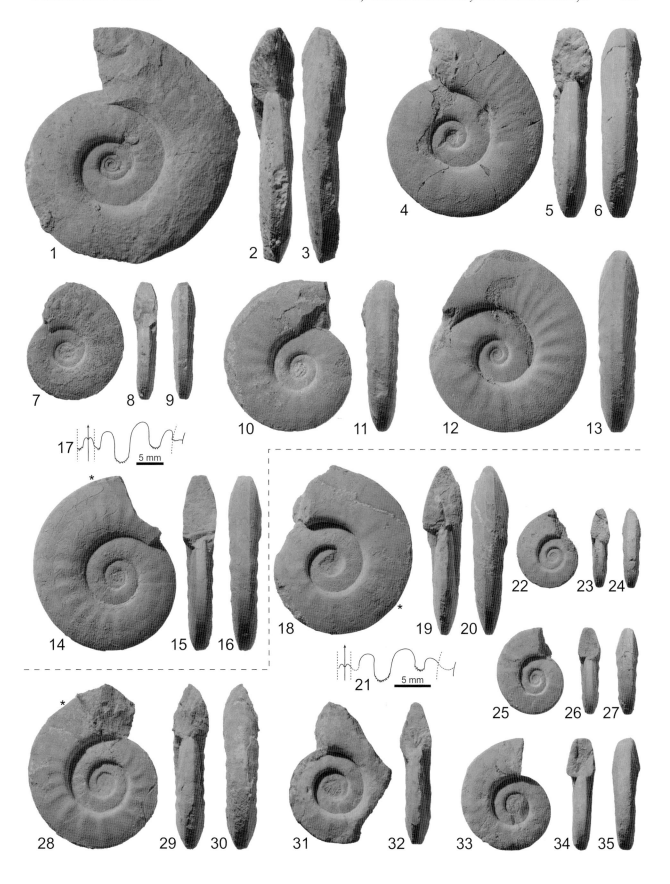

Plate 5   All figures natural size unless otherwise indicated; asterisks indicate the position of the last septum.

1–3:   ***Gyronites rigidus* (Diener, 1897)**
           Lateral, apertural and ventral views. PIMUZ30267.
           Loc. Amb104, base Lower Ceratite Limestone, Amb, *Gyronites dubius* Regional Zone; early Dienerian.

4–28:   ***Gyronites plicosus* Waagen, 1895**
4–6:   Lateral, apertural and ventral views. PIMUZ30268.
           Loc. Nam377, middle Lower Ceratite Limestone, Nammal Nala, *Gyronites plicosus* Regional Zone; early Dienerian.
7–10:   (7–9) Lateral, apertural and ventral views. (10) Suture line at H = 9.3 mm, × 2 (mirrored image). PIMUZ30269.
           Loc. Nam377, middle Lower Ceratite Limestone, Nammal Nala, *Gyronites plicosus* Regional Zone; early Dienerian.
11–13:   Lateral, apertural and ventral views PIMUZ30270.
           Loc. Nam331, middle Lower Ceratite Limestone, Nammal Nala, *Gyronites plicosus* Regional Zone; early Dienerian.
14, 15:   Lateral and ventral views. PIMUZ30271.
           Loc. Nam335, middle Lower Ceratite Limestone, Nammal Nala, *Gyronites plicosus* Regional Zone; early Dienerian.
16–18:   Lateral, apertural and ventral views. PIMUZ30272.
           Loc. Amb2, middle Lower Ceratite Limestone, Amb, *Gyronites plicosus* regional Zone; early Dienerian.
19–21:   Lateral, apertural and ventral views. PIMUZ30273.
           Loc Nam377, middle Lower Ceratite Limestone, Nammal Nala, *Gyronites plicosus* Regional Zone; early Dienerian.
22, 23:   Lateral and ventral views. PIMUZ30274.
           Loc. Amb2, middle Lower Ceratite Limestone, Amb, *Gyronites plicosus* Regional Zone; early Dienerian.
24–26:   Lateral, apertural and ventral views. PIMUZ30275.
           Loc. Nam377, middle Lower Ceratite Limestone, Nammal Nala, *Gyronites plicosus* Regional Zone; early Dienerian.
27, 28:   Lateral and ventral views. PIMUZ30276.
           Loc. Nam377, middle Lower Ceratite Limestone, Nammal Nala, *Gyronites plicosus* Regional Zone; early Dienerian.

29–33:   ***Gyronites sitala* (Diener, 1897)**
29–31:   Lateral, apertural and ventral views. PIMUZ30277.
           Loc. Nam331, middle Lower Ceratite Limestone, Nammal Nala, *Gyronites plicosus* Regional Zone; early Dienerian.
32, 33:   Lateral and ventral views. PIMUZ30278.
           Loc. Nam377, middle Lower Ceratite Limestone, Nammal Nala, *Gyronites plicosus* Regional Zone; early Dienerian.

34–38:   ***Gyronites schwanderi* n. sp.**
34, 35:   Lateral and ventral views. PIMUZ30279. Paratype.
           Loc. War100, from dipslope near top Lower Ceratite Limestone, Wargal, precise age unknown: *Gyronites plicosus* or *Gyronites frequens* Regional zones; early Dienerian.
36–38:   Lateral, apertural and ventral views. PIMUZ30281. Holotype.
           Loc. Nam378, top Lower Ceratite Limestone, Nammal Nala, *Gyronites frequens* Regional Zone; early Dienerian.

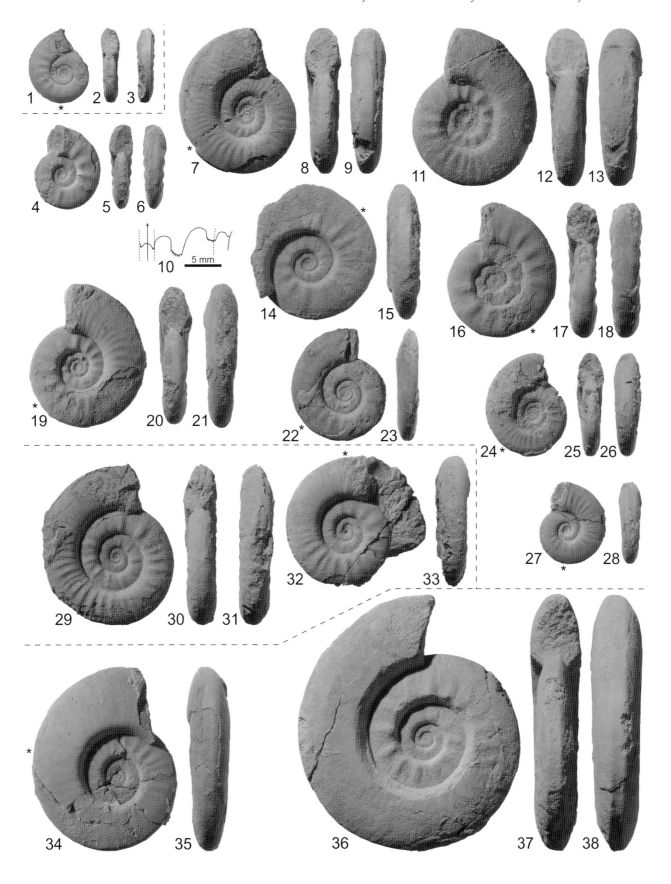

Plate 6 All figures natural size unless otherwise indicated; asterisks indicate the position of the last septum.

1–31: *Ambites discus* **Waagen, 1895**

1, 2:   Lateral and ventral views. 1GSI7138 (photo by B. Kummel). Lectotype.
         Base Ceratite Marls, Amb, bed and locality unknown.

3, 4:   Lateral and ventral views. PIMUZ30283.
         Loc. War2, base Ceratite Marls, Wargal, *Ambites discus* Regional Zone; middle Dienerian.

5–7:   Lateral, apertural and ventral views. PIMUZ30284.
         Loc. Nam527, base Ceratite Marls, Nammal Nala, *Ambites discus* Regional Zone; middle Dienerian.

8–11:   (8–10) Lateral, apertural and ventral views. (11) Suture line at H = 16.6 mm, × 1.5 (mirrored image).
         PIMUZ30285.11.
         Loc. Nam50, base Ceratite Marls, Nammal Nala, *Ambites discus* Regional Zone; middle Dienerian.

12–14:   Lateral, apertural and ventral views. PIMUZ30286.
         Loc. Nam52, base Ceratite Marls, Nammal Nala, *Ambites discus* Regional Zone; middle Dienerian.

15, 16: Lateral and ventral views. PIMUZ30287.
         Loc. Nam50, base Ceratite Marls, Nammal Nala, *Ambites discus* Regional Zone; middle Dienerian.

17, 18: Polished cross section (17) and cross section (18). PIMUZ30288.
         Loc. Nam50, base Ceratite Marls, Nammal Nala, *Ambites discus* Regional Zone; middle Dienerian.

19–21:   Lateral, apertural and ventral views. PIMUZ30289.
         Loc. Nam364, base Ceratite Marls, Nammal Nala, *Ambites discus* Regional Zone; middle Dienerian.

22–24:   Lateral, apertural and ventral views. PIMUZ30290.
         Loc. Nam50, base Ceratite Marls, Nammal Nala, *Ambites discus* Regional Zone; middle Dienerian.

25–28:   (25–27) Lateral, apertural and ventral views. (28) Suture line at H = 29.6 mm.
         PIMUZ30291.
         Loc. Nam50, base Ceratite Marls, Nammal Nala, *Ambites discus* Regional Zone; middle Dienerian.

29:      Suture line at H = 31 mm. PIMUZ30292.
         Loc. Nam52, base Ceratite Marls, Nammal Nala, *Ambites discus* Regional Zone; middle Dienerian.

30:      Suture line at H = 26.2 mm (see also Pl. 7, figs 15–17). PIMUZ30293.
         Loc. Nam50, base Ceratite Marls, Nammal Nala, *Ambites discus* Regional Zone; middle Dienerian.

31:      Suture line at H = 10.9 mm, × 2. PIMUZ30294.
         Loc. Nam364, base Ceratite Marls, Nammal Nala, *Ambites discus* Regional Zone; middle Dienerian.

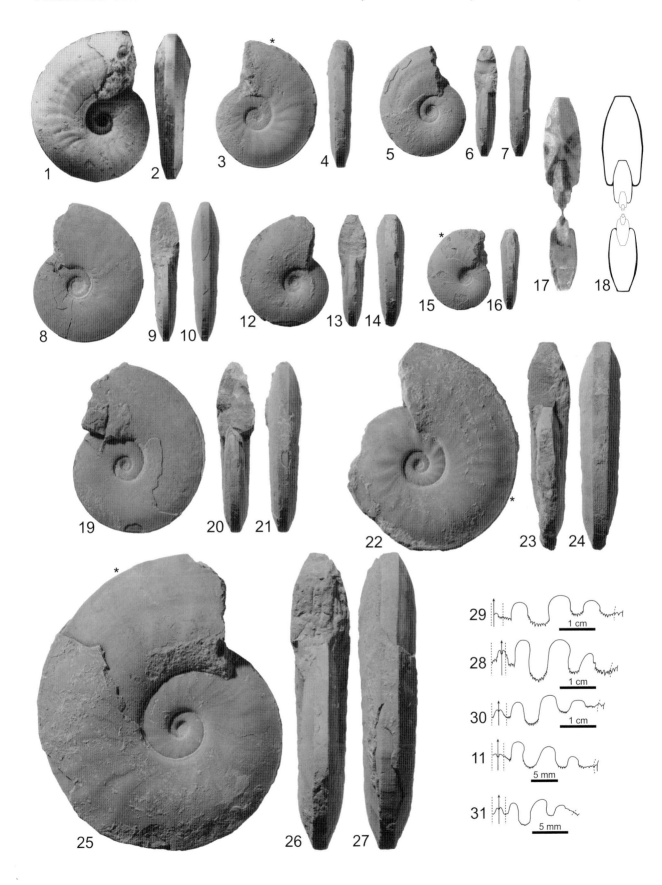

Plate 7 All figures natural size unless otherwise indicated; asterisks indicate the position of the last septum.

1–17:   *Ambites discus* **Waagen, 1895**
1–3:   Lateral, apertural and ventral views. PIMUZ30295.
           Loc. Nam50, base Ceratite Marls, Nammal Nala, *Ambites discus* Regional Zone; middle Dienerian.
4–6:   Lateral, apertural and ventral views. PIMUZ30296.
           Loc. Nam52, base Ceratite Marls, Nammal Nala, *Ambites discus* Regional Zone; middle Dienerian.
7–9:   Lateral, apertural and ventral views. PIMUZ30297.
           Loc. Nam50, base Ceratite Marls, Nammal Nala, *Ambites discus* Regional Zone; middle Dienerian.
10, 11: Lateral and ventral views. PIMUZ30298.
           Loc. Amb54, base Ceratite Marls, Amb, *Ambites discus* Regional Zone; middle Dienerian.
12–14: Lateral, apertural and ventral views. PIMUZ30299.
           Loc. Ch7A, middle Lower Ceratite Limestone, Chiddru, *Ambites discus* Regional Zone; middle
           Dienerian.
15–17: Lateral, apertural and ventral views. PIMUZ30293 (see also Pl. 6, fig. 30).
           Loc. Nam50, base Ceratite Marls, Nammal Nala, *Ambites discus* Regional Zone; middle Dienerian.

Plate 8   All figures natural size unless otherwise indicated; asterisks indicate the position of the last septum.

1– 26:   ***Ambites atavus* (Waagen, 1895)**
1:      Lateral view. GSI7187. Holotype. (Photo by B. Kummel).
        Lower Ceratite Limestone, Wargal, bed and locality unknown.
2–5:   (2–4 Lateral, apertural and ventral views. (5) Suture line at H = 20 mm, × 1.5.
        PIMUZ30300.
        Loc. Amb52, top Lower Ceratite Limestone, Amb, *Ambites atavus* Regional Zone; middle Dienerian.
6–9:   (6–8) Lateral, apertural and ventral views. (9) Suture line at H = 18.6 mm, × 1.5.
        PIMUZ30301.
        Loc. Amb52, top Lower Ceratite Limestone, Amb, *Ambites atavus* Regional Zone; middle Dienerian.
10–12: Lateral, apertural and ventral views. PIMUZ30302.
        Loc. Amb63, top Lower Ceratite Limestone, Amb, *Ambites atavus* Regional Zone; middle Dienerian.
13–16: (13–15) Lateral, apertural and ventral views. (16) Suture line at H = 16.7 mm, × 1.5. PIMUZ30303.
        Loc. Amb11, top Lower Ceratite Limestone, Amb, *Ambites atavus* Regional Zone; middle Dienerian.
17–19: Lateral, apertural and ventral views. PIMUZ30304.
        Loc. Amb63, top Lower Ceratite Limestone, Amb, *Ambites atavus* Regional Zone; middle Dienerian.
20–23: (20–22) Lateral, apertural and ventral views. (23) Suture line at H = 22.9 mm, × 1.5. PIMUZ30305.
        Loc. Nam60, base Ceratite Marls, Nammal Nala, *Ambites atavus* Regional Zone; middle Dienerian.
24–26: Lateral, apertural and ventral views. PIMUZ30306.
        Loc. Amb52, top Lower Ceratite Limestone, Amb, *Ambites atavus* Regional Zone; middle Dienerian.

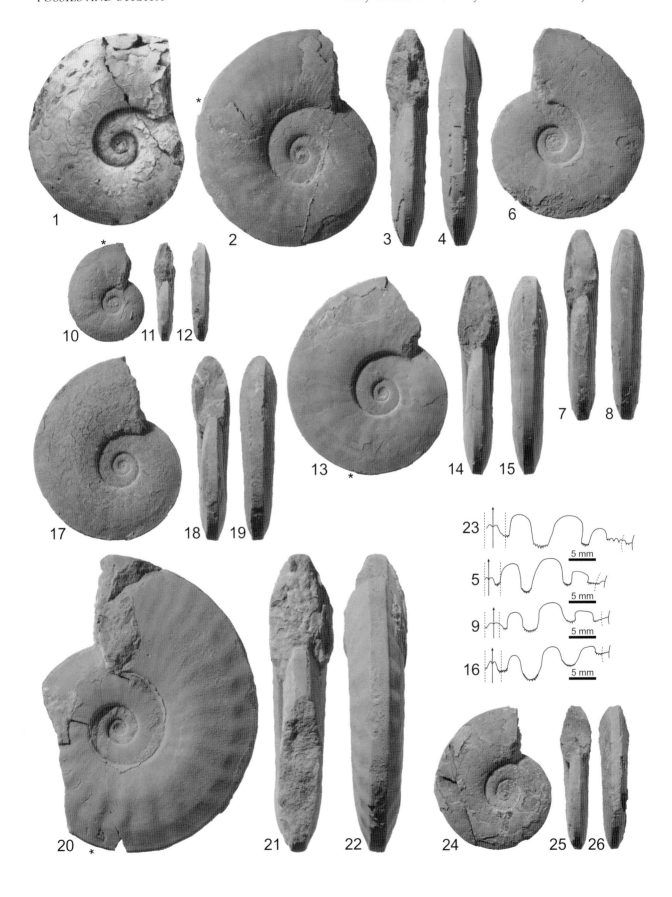

Plate 9   All figures natural size unless otherwise indicated; asterisks indicate the position of the last septum.

1–2:   ***Ambites atavus* (Waagen, 1895)**
Lateral, apertural and ventral views. PIMUZ30307.
Loc. Amb63, top Lower Ceratite Limestone, Amb, *Ambites atavus* Regional Zone; middle Dienerian.

3–10:   ***Ambites tenuis* n. sp.**
3–6:   (3, 4) Lateral, (5) apertural and (6) ventral views. (7) Suture line at H = 12.8 mm, × 2. PIMUZ30308.
Holotype.
Loc. Amb63, top Lower Ceratite Limestone, Amb, *Ambites atavus* Regional Zone; middle Dienerian.
8–10:   Lateral, apertural and ventral views. PIMUZ30309. Paratype.
Loc. Amb63, top Lower Ceratite Limestone, Amb, *Ambites atavus* Regional Zone; middle Dienerian.

11–28:   ***Ambites radiatus* (Brühwiler *et al.*, 2008)**
11–14:   (11, 12) Lateral (13) apertural and (14) ventral views. PIMUZ30310.
Loc. Nam381, base Ceratite Marls, Nammal Nala, *Ambites radiatus* Regional Zone; middle Dienerian.
15, 16:   Lateral and ventral views. PIMUZ30311.
Loc. Nam520, base Ceratite Marls, Nammal Nala, *Ambites radiatus* Regional Zone; middle Dienerian.
17–19:   Lateral, apertural and ventral views. PIMUZ30312.
Loc. Nam72, base Ceratite Marls, Nammal Nala, *Ambites radiatus* Regional Zone; middle Dienerian.
20–21:   Lateral and ventral views. Specimen with partially preserved body chamber, but position of the last
septum unknown. PIMUZ30313.
Loc. Nam381, base Ceratite Marls, Nammal Nala, *Ambites radiatus* Regional Zone; middle Dienerian.
22:   Suture line at H = 10 mm, × 2. PIMUZ30314.
Loc. Nam381, base Ceratite Marls, Nammal Nala, *Ambites radiatus* Regional Zone; middle Dienerian.
23–25:   Lateral, apertural and ventral views. PIMUZ30315.
Loc. Nam381, base Ceratite Marls, Nammal Nala, *Ambites radiatus* Regional Zone; middle Dienerian.
26–28:   Lateral, apertural and ventral views. PIMUZ30316.
Loc. Nam72, base Ceratite Marls, Nammal Nala, *Ambites radiatus* Regional Zone; middle Dienerian.

29–31:   ***Ambites bojeseni* n. sp.**
Lateral, apertural and ventral views. PIMUZ30317. Paratype.
Loc. Nam521, base Ceratite Marls, Nammal Nala, *Ambites radiatus* Regional Zone; middle Dienerian.

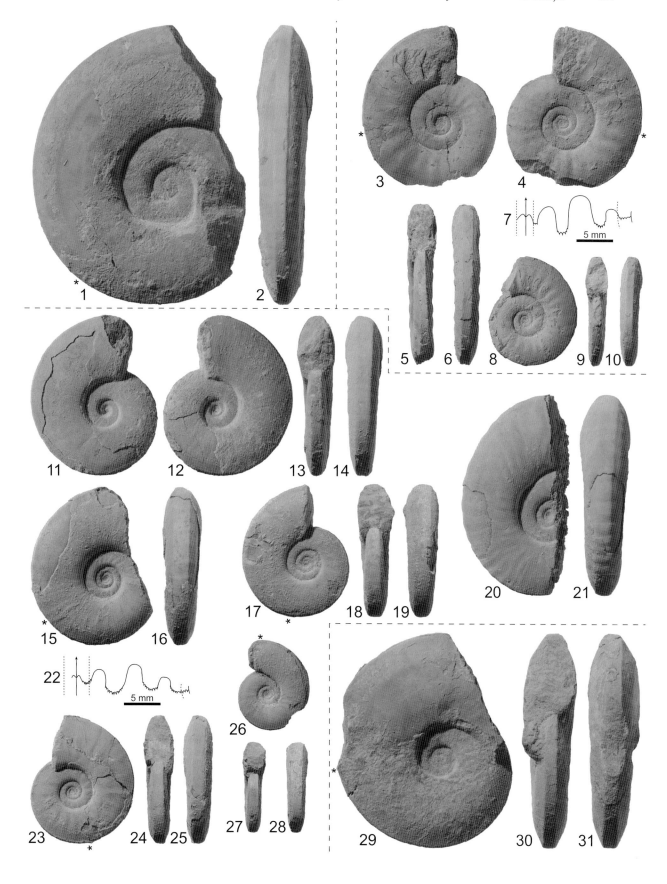

Plate 10 All figures natural size unless otherwise indicated; asterisks indicate the position of the last septum.

1–11: *Ambites bojeseni* n. sp.
1–3:   Lateral, apertural and ventral views. PIMUZ30318. Holotype.
          Loc. Nam384, base Ceratite Marls, Nammal Nala, *Ambites radiatus* Regional Zone; middle Dienerian.
4–7:   (4–6) Lateral, apertural and ventral views. (7) Suture line at H = 16.6 mm, × 2. PIMUZ30319.
          Paratype.
          Loc. Nam72, base Ceratite Marls, Nammal Nala, *Ambites radiatus* Regional Zone; middle Dienerian.
4–11:   (8–10) Lateral, apertural and ventral views. (11) Suture line at H = 14.3 mm, × 2. PIMUZ30320.
          Paratype.
          Loc. Nam72, base Ceratite Marls, Nammal Nala, *Ambites radiatus* Regional Zone; middle Dienerian.

12–19: *Ambites subradiatus* n. sp.
12–14:   Lateral, apertural and ventral views. PIMUZ30321. Holotype.
          Loc. Amb54, base Ceratite Marls, Amb, *Ambites discus* Regional Zone; middle Dienerian.
15, 16:   Lateral and ventral views. PIMUZ30322. Paratype.
          Loc. Nam50, base Ceratite Marls, Nammal Nala, *Ambites discus* Regional Zone; middle Dienerian.
17:       Suture line at H = 11.1 mm, × 2. PIMUZ30323. Paratype.
          Loc. Amb54, base Ceratite Marls, Amb, *Ambites discus* Regional Zone; middle Dienerian.
18, 19:   Lateral and apertural views. PIMUZ30324. Paratype.
          Loc. Nam52, base Ceratite Marls, Nammal Nala, *Ambites discus* Regional Zone; middle Dienerian.

20–23: *Ambites?* sp. indet.
20–23:   (20–22) Lateral, apertural and ventral views. (23) Suture line at H = 13.4 mm, × 2. PIMUZ30325.
          Loc. Nam382, base Ceratite Marls, Nammal Nala, *Ambites discus* Regional Zone; middle Dienerian.
24, 25:   Lateral and ventral views. PIMUZ30326.
          Loc. Nam380, base Ceratite Marls, Nammal Nala, *Ambites discus* Regional Zone; middle Dienerian.

26–33: *Ambites superior* (Waagen, 1895)
26–28:   Lateral, apertural and ventral views. PIMUZ30330.
          Loc. Nam53, base Ceratite Marls, Nammal Nala, *Ambites superior* Regional Zone; middle Dienerian.
29–31:   Lateral, apertural and ventral views. PIMUZ30331.
          Loc. Nam371, base Ceratite Marls, Nammal Nala, *Ambites superior* Regional Zone; middle Dienerian.
32, 33:   Polished cross section and cross section. PIMUZ30332.
          Loc. Nam302, base Ceratite Marls, Nammal Nala, *Ambites superior* Regional Zone; middle Dienerian.

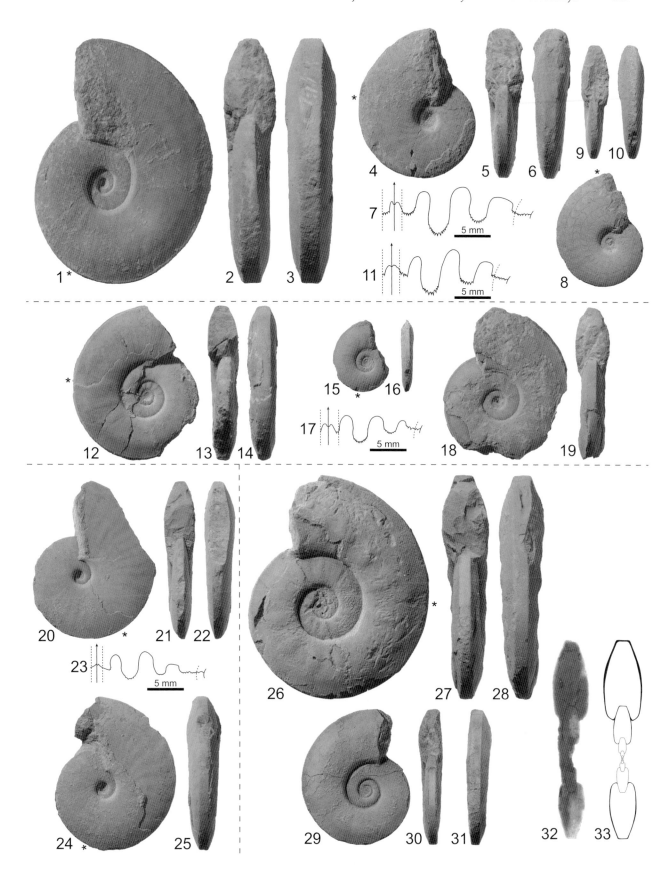

Plate 11 All figures natural size unless otherwise indicated; asterisks indicate the position of the last septum.

1–17:   ***Ambites superior* (Waagen, 1895)**
1–3:   Lateral, apertural and ventral views. PIMUZ30333.
        Loc. Nam302, base Ceratite Marls, Nammal Nala, *Ambites superior* Regional Zone; middle Dienerian.
4–6:   Lateral, apertural and ventral views. PIMUZ30334.
        Loc. Nam53, base Ceratite Marls, Nammal Nala, *Ambites superior* Regional Zone; middle Dienerian.
7, 8:   Lateral and ventral views. PIMUZ30335.
        Loc. Nam53, base Ceratite Marls, Nammal Nala, *Ambites superior* Regional Zone; middle Dienerian.
9:      Suture line at H = 20.7 mm, × 1.5 (mirrored image). PIMUZ30336.
        Loc. Nam53, base Ceratite Marls, Nammal Nala, *Ambites superior* Regional Zone; middle Dienerian.
10:     Suture line at H = 18.5 mm, × 1.5. PIMUZ30337.
        Loc. Nam53, base Ceratite Marls, Nammal Nala, *Ambites superior* Regional Zone; middle Dienerian.
11:     Suture line at H = 18.3 mm, × 1.5 (mirrored image). PIMUZ30338.
        Loc. Nam53, base Ceratite Marls, Nammal Nala, *Ambites superior* Regional Zone; middle Dienerian.
12:     Suture line at H = 13.9 mm, × 1.5 (mirrored image). PIMUZ30339.
        Loc. Nam53, base Ceratite Marls, Nammal Nala, *Ambites superior* Regional Zone; middle Dienerian.
13, 14: Lateral and ventral views. PIMUZ30340.
        Loc. Nam53, base Ceratite Marls, Nammal Nala, *Ambites superior* regional Zone; middle Dienerian.
15–17: Lateral, apertural and ventral views. PIMUZ30341.
        Loc. Nam371, base Ceratite Marls, Nammal Nala, *Ambites superior* Regional Zone; middle Dienerian.

Plate 12   All figures natural size unless otherwise indicated; asterisks indicate the position of the last septum)

1–3:   *Ambites superior* (Waagen, 1895)
1–3:   Lateral, apertural and ventral views. PIMUZ30342.
         Loc. Nam308, base Ceratite Marls, Nammal Nala, *Ambites superior* Regional Zone; middle Dienerian.

4–19:   *Ambites lilangensis* (von Krafft, 1909)
4–7:   (4–6) Lateral, apertural and ventral views. (7) Suture line at H = 19.7 mm, × 1.5.
         PIMUZ30343.
         Loc. Nam100, base Ceratite Marls, Nammal Nala, *Ambites lilangensis* Regional Zone; middle Dienerian.
8–11:   (8–10) Lateral, apertural and ventral views. (11) Suture line at H = 13.5 mm, × 1.5. PIMUZ30344.
         Loc. Nam344, base Ceratite Marls, Nammal Nala, *Ambites lilangensis* Regional Zone; middle Dienerian.
12–15:   (12–14) Lateral, apertural and ventral views. (15) Suture line at H = 22.4 mm, × 1.5. PIMUZ30345.
         Loc. Nam100, base Ceratite Marls, Nammal Nala, *Ambites lilangensis* Regional Zone; middle Dienerian.
16–19:   (16–18) Lateral, apertural and ventral views. (19) Suture line at H = 17.4 mm, × 1.5. PIMUZ30346.
         Loc. Nam344, base Ceratite Marls, Nammal Nala, *Ambites lilangensis* Regional Zone; middle Dienerian.

Plate 13   All figures natural size unless otherwise indicated; asterisks indicate the position of the last septum.

1–6:   ***Ambites lilangensis* (von Krafft, 1909)**
1–3:   Lateral, apertural and ventral views. PIMUZ30347.
       Loc. Nam100, base Ceratite Marls, Nammal Nala, *Ambites lilangensis* Regional Zone ; middle Dienerian.
4–6:   Lateral, apertural and ventral views. PIMUZ30348.
       Loc. Nam100, base Ceratite Marls, Nammal Nala, *Ambites lilangensis* Regional Zone ; middle Dienerian.

7–9:   ***Ambites* cf. *A. impressus* (Waagen, 1895)**
       Lateral, apertural and ventral views. PIMUZ30349.
       Loc. Nam344, base Ceratite Marls, Nammal Nala, *Ambites lilangensis* Regional Zone ; middle Dienerian.

10–26: *Ambites bjerageri* n. sp.
10, 11: Lateral and ventral views. PIMUZ30350. Paratype
       Loc. Nam503, base Ceratite Marls, Nammal Nala, *Ambites lilangensis* Regional Zone; middle Dienerian.
12–15: (12–14) Lateral, apertural and ventral views. (15) Suture line at H = 7.7 mm, × 3. PIMUZ30351. Holotype.
       Loc. Nam100, base Ceratite Marls, Nammal Nala, *Ambites lilangensis* Regional Zone; middle Dienerian.
16–18: Lateral, apertural and ventral views. PIMUZ30352. Paratype.
       Loc. Nam92, base Ceratite Marls, Nammal Nala, *Ambites lilangensis* Regional Zone; middle Dienerian.
19–21: Lateral, apertural and ventral views. PIMUZ30353. Paratype.
       Loc. Nam501, base Ceratite Marls, Nammal Nala, *Ambites lilangensis* Regional Zone; middle Dienerian.
22–24: Lateral, apertural and ventral views. PIMUZ30354. Paratype.
       Loc. Nam501, base Ceratite Marls, Nammal Nala, *Ambites lilangensis* Regional Zone; middle Dienerian.
25, 26: Lateral and ventral views. PIMUZ30355. Paratype.
       Loc. Nam344, base Ceratite Marls, Nammal Nala, *Ambites lilangensis* Regional Zone; middle Dienerian.

27–35: *Vavilovites* cf. *V. sverdrupi* (Tozer, 1963)
27–29: (27, 28) Lateral and ventral views. (29) Suture line at H = 23.2 mm. PIMUZ30356.
       Loc. Nam396, base Ceratite Marls, Nammal Nala, *Vavilovites* cf. *V. sverdrupi* Regional Zone; late Dienerian.
30, 31: Lateral and ventral views. PIMUZ30357.
       Loc. Nam396, base Ceratite Marls, Nammal Nala, *Vavilovites* cf. *V. sverdrupi* Regional Zone; late Dienerian.
32, 33: Lateral and ventral views. PIMUZ30358.
       Loc. Nam316, base Ceratite Marls, Nammal Nala, *Vavilovites* cf. *V. sverdrupi* Regional Zone; late Dienerian.
34, 35: Lateral and ventral views. PIMUZ30359.
       Loc. Nam318, base Ceratite Marls, Nammal Nala, *Vavilovites* cf. *V. sverdrupi* Regional Zone; late Dienerian.

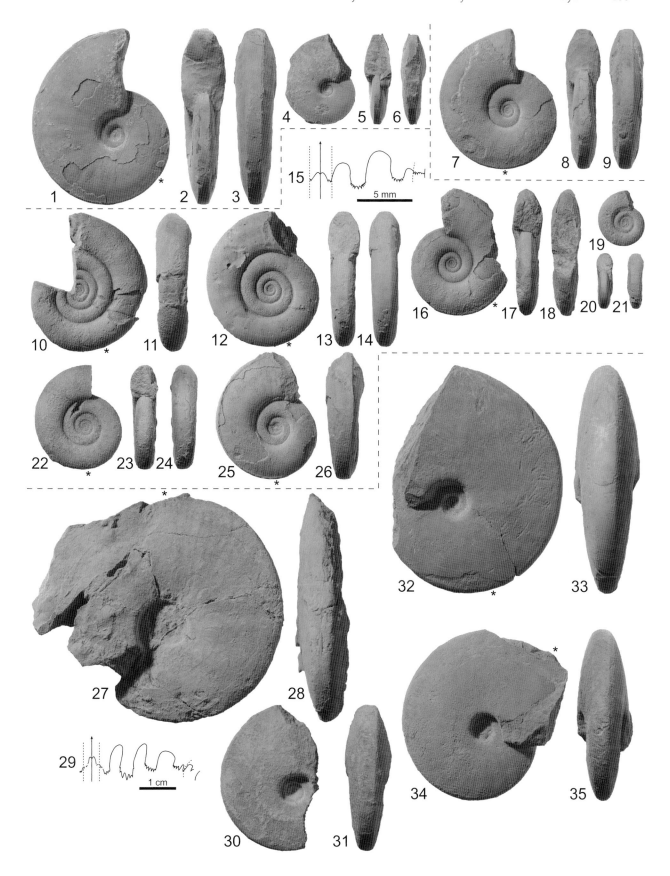

Plate 14   All figures natural size unless otherwise indicated; asterisks indicate the position of the last septum.

1–36:   ***Koninckites vetustus* Waagen, 1895**

1, 2:   Lateral views. PIMUZ30360.

3, 4:   Apertural and ventral views. Specimen with partially preserved body chamber, but position of the last septum unknown.
Loc. Nam71, base Ceratite Marls, Nammal Nala, *Koninckites vetustus* Regional Zone; late Dienerian.

5–7:   Lateral, apertural and ventral views. PIMUZ30361.
Loc. Nam347, base Ceratite Marls, Nammal Nala, *Koninckites vetustus* Regional Zone; late Dienerian.

8, 9:   Lateral and ventral views. PIMUZ30362. × 2.
Loc. War104, base Ceratite Marls, Wargal, *Koninckites vetustus* Regional Zone; late Dienerian.

10, 11: Lateral and ventral views. PIMUZ30363. × 2.
Loc. Nam305, base Ceratite Marls, Nammal Nala, *Koninckites vetustus* Regional Zone; late Dienerian.

12–14: Lateral, apertural and ventral views. PIMUZ30364.
Loc. Nam305, base Ceratite Marls, Nammal Nala, *Koninckites vetustus* Regional Zone; late Dienerian.

15–17: Lateral, apertural and ventral views. PIMUZ30365.
Loc. Nam349, base Ceratite Marls, Nammal Nala, *Koninckites vetustus* Regional Zone; late Dienerian.

18, 19: (18, 19) Lateral, (20) apertural and ventral (21) views. PIMUZ30366.
Loc. Nam312, base Ceratite Marls, Nammal Nala, *Koninckites vetustus* Regional Zone; late Dienerian.

22–24: Lateral, apertural and ventral views. PIMUZ30367.
Loc. War104, base Ceratite Marls, Wargal, *Koninckites vetustus* Regional Zone; late Dienerian.

25–27: Lateral, apertural and ventral views. PIMUZ30368.
Loc. War104, base Ceratite Marls, Wargal, *Koninckites vetustus* Regional Zone; late Dienerian.

28–30: Lateral, apertural and ventral views. PIMUZ30369.
Loc. War104, base Ceratite Marls, Wargal, *Koninckites vetustus* Regional Zone; late Dienerian.

31, 32: Polished cross section and cross section. PIMUZ30370.
Loc. Nam305, base Ceratite Marls, Nammal Nala, *Koninckites vetustus* Regional Zone; late Dienerian.

33, 34: Polished cross section and cross section. PIMUZ30371.
Loc. Nam305, base Ceratite Marls, Nammal Nala, *Koninckites vetustus* Regional Zone; late Dienerian.

35, 36: Polished cross section and cross section. PIMUZ30372.
Loc. Nam305, base Ceratite Marls, Nammal Nala, *Koninckites vetustus* Regional Zone; late Dienerian.

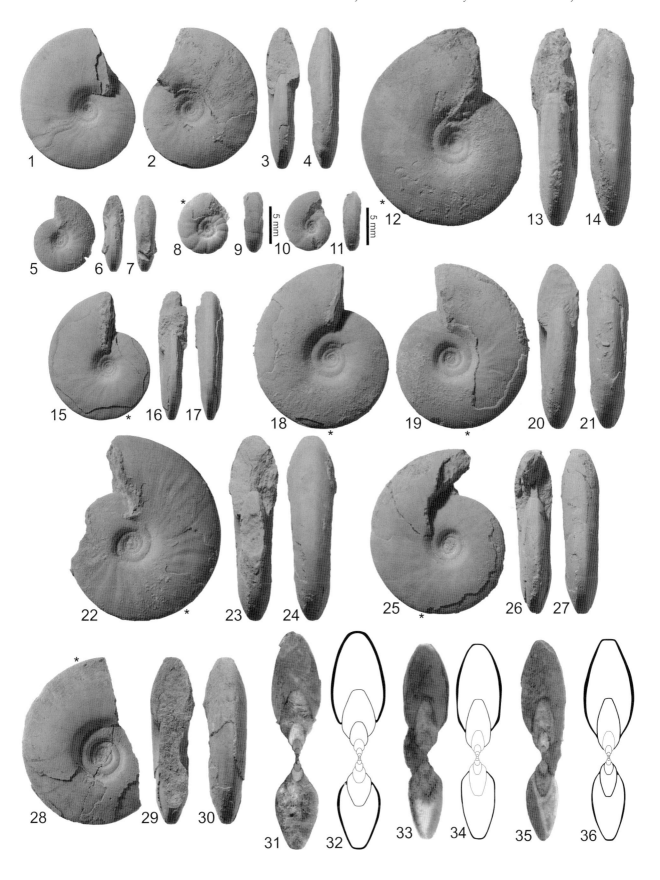

Plate 15   All figures natural size unless otherwise indicated; asterisks indicate the position of the last septum.

1–20:   *Koninckites vetustus* Waagen, 1895.
1, 2:   Lateral and ventral views. GSI7161. Lectotype. (Photo by B. Kummel).
        Lower Ceratite Limestone, Chiddru, precise bed and locality unknown.
3–6:   (3–5) Lateral, apertural and ventral views. (6) Suture line at H = 18.5 mm, × 1.5 (mirrored image).
        PIMUZ30373.
        Loc. Nam63, base Ceratite Marls, Nammal Nala, *Koninckites vetustus* Regional Zone; late Dienerian.
7–10:   (7–9) Lateral, apertural and ventral views. (10) Suture line at H = 16.6 mm, × 1.5. PIMUZ30374.
        Loc. Nam305, base Ceratite Marls, Nammal Nala, *Koninckites vetustus* Regional Zone; late Dienerian.
11–14: (11–13) Lateral, apertural and ventral views. (14) Suture line at H = 17 mm, × 1.5 (mirrored image).
        PIMUZ30375.
        Loc. Nam83, base Ceratite Marls, Nammal Nala, *Koninckites vetustus* Regional Zone; late Dienerian.
15:     Suture line at H = 18.7 mm. × 1.5 (mirrored image). PIMUZ30376.
        Loc. Nam305, base Ceratite Marls, Nammal Nala, *Koninckites vetustus* Regional Zone; late Dienerian.
16–20: (16–19) Lateral, apertural and ventral views, × 0.7. (20) Suture line at H = 30.3 mm. PIMUZ30377.
        Loc. Nam305, base Ceratite Marls, Nammal Nala, *Koninckites vetustus* Regional Zone; late Dienerian.

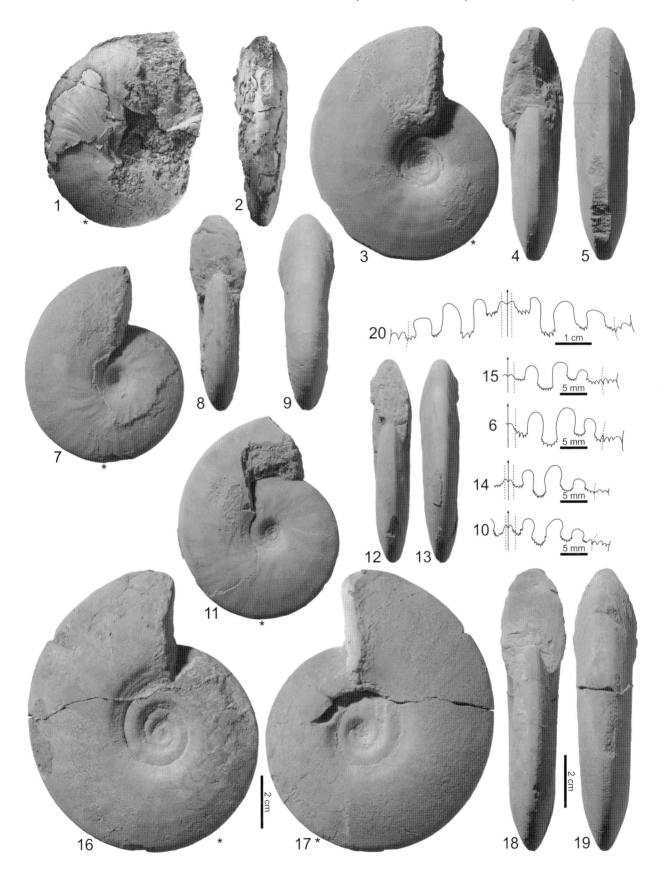

Plate 16   All figures natural size; asterisks indicate the position of the last septum.

1–15:   *Koninckites vetustus* Waagen, 1895

1–3:   Lateral, apertural and ventral views. PIMUZ30378.
        Loc. War104, base Ceratite Marls, Wargal, *Koninckites vetustus* Regional Zone; late Dienerian.

4–6:   Lateral, apertural and ventral views. PIMUZ30379.
        Loc. Nam312, base Ceratite Marls, Nammal Nala, *Koninckites vetustus* Regional Zone; late Dienerian.

7–9:   Lateral, apertural and ventral views. PIMUZ30380.
        Loc. Chi51, top Lower Ceratite Limestone, Chiddru, *Koninckites vetustus* Regional Zone; late Dienerian.

10–12: Lateral, apertural and ventral views. PIMUZ30381.
        Loc. Nam71, base Ceratite Marls, Nammal Nala, *Koninckites vetustus* Regional Zone; late Dienerian.

13–15: Lateral, apertural and ventral views. PIMUZ30382.
        Loc. Nam312, base Ceratite Marls, Nammal Nala, *Koninckites vetustus* Regional Zone; late Dienerian.

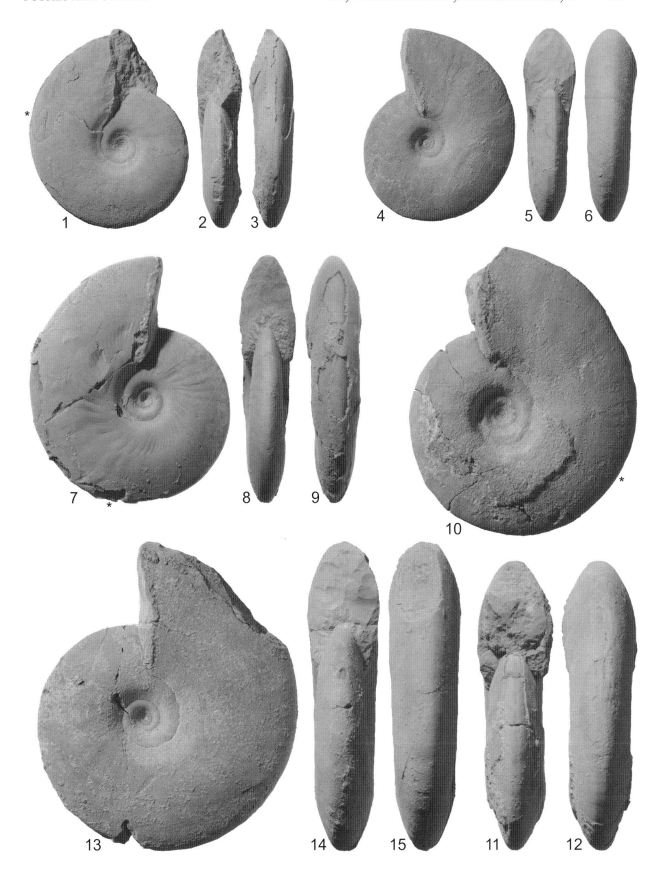

Plate 17 All figures natural size unless otherwise indicated; asterisks indicate the position of the last septum.

1–4: *Koninckites vetustus* **Waagen, 1895**

  (1, 2) Lateral, (3) apertural and (4) ventral views. PIMUZ30383.
  Loc. Nam83, base Ceratite Marls, Nammal Nala, *Kingites davidsonianus* Regional Zone; late Dienerian.

5– 37: *Koninckites khoorensis* **(Waagen, 1895)**

5, 6: Lateral and ventral views. PIMUZ30384. × 1.5.
  Loc. Nam101d, base Ceratite Marls, Nammal Nala, *Kingites davidsonianus* Regional Zone; late Dienerian.

7–9: Lateral, apertural and ventral views. PIMUZ30385. (See also Pl. 18, fig. 9). × 1.5.
  Loc. Nam61, base Ceratite Marls, Nammal Nala, *Kingites davidsonianus* regional Zone; late Dienerian.

10–12: Lateral, apertural and ventral views. PIMUZ30386. (See also Pl. 18, fig. 7).
  Loc. Nam101d, base Ceratite Marls, Nammal Nala, *Kingites davidsonianus* Regional Zone; late Dienerian.

13, 14: Lateral and ventral views. PIMUZ30387. × 1.5.
  Loc. Nam101d, base Ceratite Marls, Nammal Nala, *Kingites davidsonianus* Regional Zone; late Dienerian.

15–17: Lateral, apertural and ventral views. PIMUZ30388. × 1.5.
  Loc. Nam67, base Ceratite Marls, Nammal Nala, *Kingites davidsonianus* Regional Zone; late Dienerian.

18–20: Lateral, apertural and ventral views. PIMUZ30389.
  Loc. Nam67, base Ceratite Marls, Nammal Nala, *Kingites davidsonianus* Regional Zone; late Dienerian.

21, 22: Lateral and ventral views. PIMUZ30390.
  Loc. Nam59, base Ceratite Marls, Nammal Nala, *Kingites davidsonianus* Regional Zone; late Dienerian.

23–25: Lateral, apertural and ventral views. PIMUZ30391. (See also Pl. 18, fig. 8).
  Loc. Nam61, base Ceratite Marls, Nammal Nala, *Kingites davidsonianus* Regional Zone; late Dienerian.

26–28: Lateral, apertural and ventral views. PIMUZ30392.
  Loc. Nam61, base Ceratite Marls, Nammal Nala, *Kingites davidsonianus* Regional Zone; late Dienerian.

29–31: Lateral, apertural and ventral views. PIMUZ30393. (See also Pl. 18, fig. 6).
  Loc. Nam101d, base Ceratite Marls, Nammal Nala, *Kingites davidsonianus* Regional Zone; late Dienerian,

32–34: Lateral, apertural and ventral views. PIMUZ30394.
  Loc. Nam101d, base Ceratite Marls, Nammal Nala, *Kingites davidsonianus* Regional Zone; late Dienerian.

35–37: Lateral, apertural and ventral views. PIMUZ30395.
  Loc. Nam61, base Ceratite Marls, Nammal Nala, *Kingites davidsonianus* Regional Zone; late Dienerian.

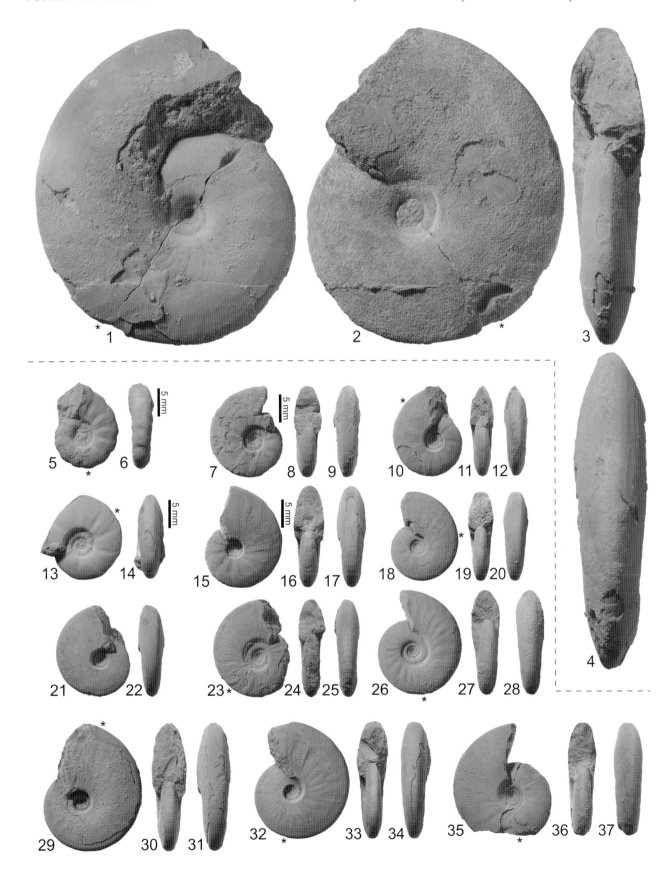

Plate 18   All figures natural size unless otherwise indicated; asterisks indicate the position of the last septum.

1– 31: *Koninckites khoorensis* (Waagen, 1895)
1:      Suture line at H = 28.7 mm. PIMUZ30396.
        Loc. Nam61, base Ceratite Marls, Nammal Nala, *Kingites davidsonianus* Regional Zone; late Dienerian.
2:      Suture line at H = 30.3 mm (mirrored image. PIMUZ30397. (See also Pl. 20, figs 4–6).
        Loc. Nam59, base Ceratite Marls, Nammal Nala, *Kingites davidsonianus* Regional Zone; late Dienerian.
3:      Suture line at H = 22.4 mm, × 1.2. (mirrored image). PIMUZ30398. (See also Pl. 20, figs 1–3).
        Loc. Nam724, base Ceratite Marls, Nammal Nala, *Kingites davidsonianus* Regional Zone; late
        Dienerian.
4:      Suture line at H = 18.9 mm, × 1.5. (mirrored image). PIMUZ30399. (See also Pl. 20, figs 7, 8).
        Loc. Nam59, base Ceratite Marls, Nammal Nala, *Vavilovites* cf. *V. sverdrupi* Regional Zone; late
        Dienerian.
5:      Suture line at H = 17.8 mm, × 1.5. PIMUZ30400. (See also Pl. 19, figs 8, 9).
        Loc. Nam59, base Ceratite Marls, Nammal Nala, *Kingites davidsonianus* Regional Zone; late Dienerian.
6:      Suture line at H = 17.7 mm, × 1.5. PIMUZ30393. (See also Pl. 17, figs 29–31).
        Loc. Nam101d, base Ceratite Marls, Nammal Nala, *Kingites davidsonianus* Regional Zone; late
        Dienerian.
7:      Suture line at H = 9.9 mm, × 2 (mirrored image). PIMUZ30386. (See also Pl. 17, figs 10–12).
        Loc. Nam101d, base Ceratite Marls, Nammal Nala, *Kingites davidsonianus* Regional Zone; late
        Dienerian.
8:      Suture line at H = 7.5 mm, × 2 (mirrored image). PIMUZ30391. (See also Pl. 17, figs 23–25).
        Loc. Nam61, base Ceratite Marls, Nammal Nala, *Kingites davidsonianus* Regional Zone; late Dienerian).
9:      Suture line at H = 7.3 mm, × 2. PIMUZ30385. (See also Pl. 17, figs 7–9).
        Loc. Nam61, base Ceratite Marls, Nammal Nala, *Kingites davidsonianus* Regional Zone; late Dienerian.
10, 11: Polished cross section and cross section. PIMUZ30401.
        Loc. Nam61, base Ceratite Marls, Nammal Nala, *Kingites davidsonianus* Regional Zone; late Dienerian.
12, 13: Polished cross section and cross section. PIMUZ30402.
        Loc. Nam61, base Ceratite Marls, Nammal Nala, *Kingites davidsonianus* Regional Zone; late Dienerian.
14, 15: Polished cross section and cross section. PIMUZ30403.
        Loc. Nam336, base Ceratite Marls, Nammal Nala, *Kingites davidsonianus* Regional Zone; late
        Dienerian.
16–18: Asymmetric specimen with a strong ornamentation on the left side. Lateral and ventral views.
        PIMUZ30404.
        Loc. Nam502, base Ceratite Marls, Nammal Nala, *Kingites davidsonianus* Regional Zone; late
        Dienerian.
19–21: Lateral, apertural and ventral views. PIMUZ30405.
        Loc. Nam101d, base Ceratite Marls, Nammal Nala, *Kingites davidsonianus* Regional Zone; late
        Dienerian.
22, 23: Lateral and ventral views. PIMUZ30406.
        Loc. Nam304, base Ceratite Marls, Nammal Nala, *Kingites davidsonianus* Regional Zone; late
        Dienerian.
24–26: Lateral, apertural and ventral views. PIMUZ30407.
        Loc. Nam61, base Ceratite Marls, Nammal Nala, *Kingites davidsonianus* Regional Zone; late Dienerian.
27, 28: Lateral and ventral views. PIMUZ30408.
        Loc. Nam101d, base Ceratite Marls, Nammal Nala, *Kingites davidsonianus* Regional Zone; late
        Dienerian.
29–31: (29, 30) Lateral and ventral views. (31) Suture line at H = 13.3 mm, × 2.
        PIMUZ30409.
        Loc. Nam59, base Ceratite Marls, Nammal Nala, *Kingites davidsonianus* Regional Zone; late Dienerian.

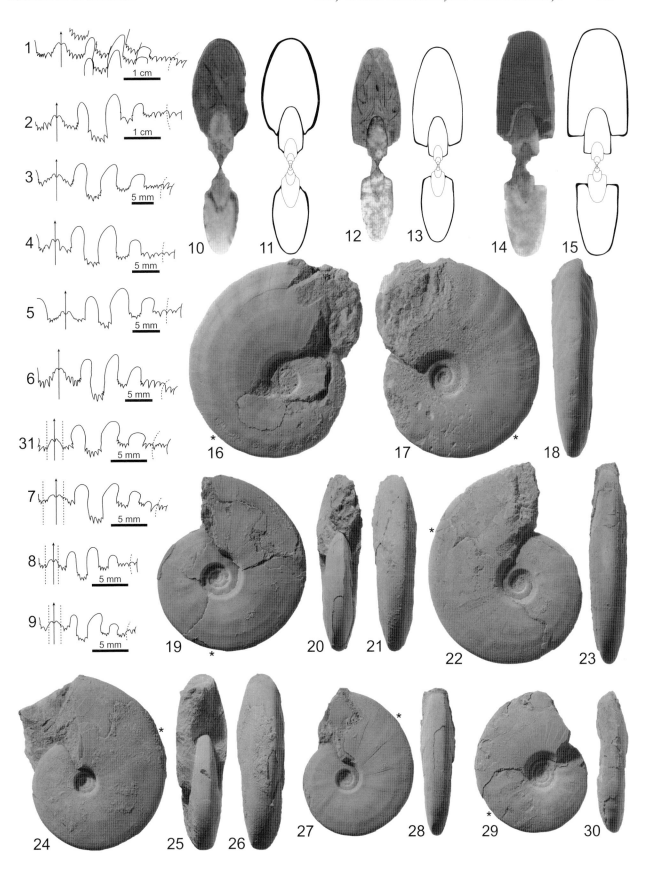

Plate 19 All figures natural size; asterisks indicate the position of the last septum.

1–12: ***Koninckites khoorensis* (Waagen, 1895)**
1, 2: Lateral and ventral views. PIMUZ30410.
Loc. Nam130, base Ceratite Marls, Nammal Nala, *Kingites davidsonianus* Regional Zone; late Dienerian.
3, 4: Ventral and lateral views. PIMUZ30411.
Loc. Nam337, base Ceratite Marls, Nammal Nala, *Kingites davidsonianus* Regional Zone; late Dienerian.
5–7: Lateral and ventral views. PIMUZ30412.
Loc. Nam59, base Ceratite Marls, Nammal Nala, *Kingites davidsonianus* Regional Zone; late Dienerian.
8, 9: Lateral and ventral views. PIMUZ30400. (See also Pl. 18, fig. 5).
Loc. Nam59, base Ceratite Marls, Nammal Nala, *Kingites davidsonianus* Regional Zone; late Dienerian.
10–12: Lateral, apertural and ventral views. PIMUZ30413.
Loc. Nam61, base Ceratite Marls, Nammal Nala, *Kingites davidsonianus* Regional Zone; late Dienerian.

Plate 20   All figures natural size; asterisks indicate the position of the last septum.

1–13:   *Koninckites khoorensis* (Waagen, 1895)
1–3:   Lateral, apertural and ventral views. PIMUZ30398. (See also Pl. 18, fig. 3).
         Loc. Nam724, base Ceratite Marls, Nammal Nala, *Vavilovites* cf. *V. sverdrupi* Regional Zone; late
         Dienerian.
4–6:   Lateral, apertural and ventral views. PIMUZ30397. (See also Pl. 18, fig. 2).
         Loc. Nam59, base Ceratite Marls, Nammal Nala, *Kingites davidsonianus* Regional Zone; late Dienerian.
7, 8:   Lateral and ventral views. PIMUZ30399. (See also Pl. 18, fig. 4).
         Loc. Nam59, base Ceratite Marls, Nammal Nala, *Kingites davidsonianus* Regional Zone; late Dienerian.
9, 10:   Lateral and ventral views. PIMUZ30414.
         Loc. Nam346, base Ceratite Marls, Nammal Nala, *Kingites davidsonianus* Regional Zone; late
         Dienerian.
11–13:   Lateral, apertural and ventral views. PIMUZ30415.
         Loc. Nam502, base Ceratite Marls, Nammal Nala, *Kingites davidsonianus* Regional Zone; late
         Dienerian.

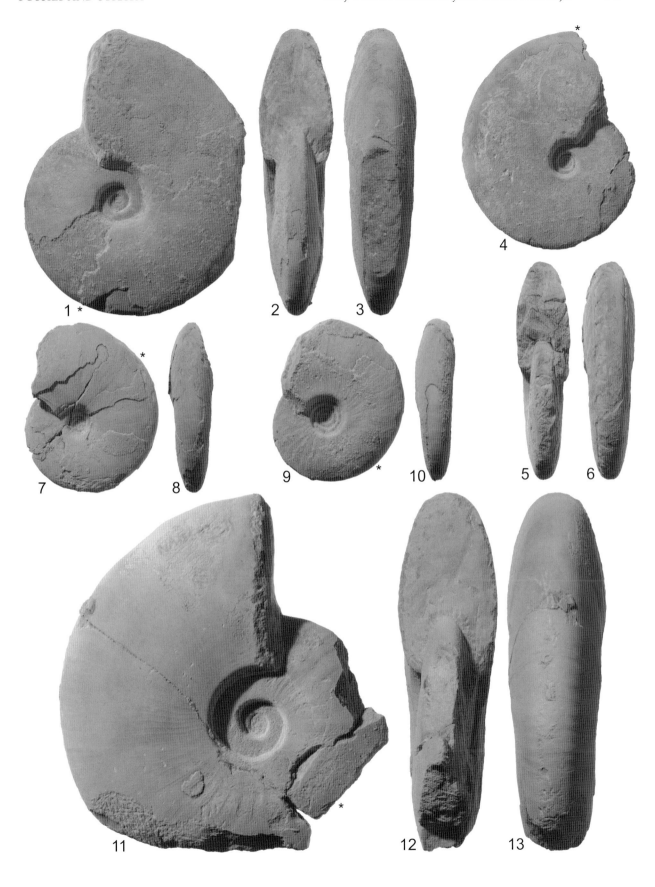

Plate 21   All figures natural size unless otherwise indicated; asterisks indicate the position of the last septum.

1–6: *Radioceras truncatum* **(Spath, 1934)**
1–4:   (1–3) Lateral, apertural and ventral views. (4) Suture line at H = 15.5 mm, × 1.5 (mirrored image). PIMUZ30416.
        Loc. Nam350, base Ceratite Marls, Nammal Nala, *Awanites awani* Regional Zone; late Dienerian.
5, 6:   Lateral and ventral views. PIMUZ30417.
        Loc. War104, base Ceratite Marls, Wargal, *Koninckites vetustus* Regional Zone; late Dienerian.

7–13: *Pashtunites kraffti* **(Spath, 1934) n. gen.**
7–9:   Lateral, apertural and ventral views. PIMUZ30418.
        Loc. Nam83, base Ceratite Marls, Nammal Nala, *Koninckites vetustus* Regional Zone; late Dienerian.
10–13:   (10–12) Lateral, apertural and ventral views. (13) Suture line at H = 21.3 mm. PIMUZ30419.
        Loc. Nam83, base Ceratite Marls, Nammal Nala, *Koninckites vetustus* Regional Zone; late Dienerian.

14–23: *Awanites awani* **n. gen. et n. sp.**
14–16:   Lateral, apertural and ventral views. PIMUZ30420. Paratype.
        Loc. Nam350, base Ceratite Marls, Nammal Nala, *Awanites awani* Regional Zone; late Dienerian.
17–19:   Lateral, apertural and ventral views. PIMUZ30421. Paratype.
        Loc. Nam350, base Ceratite Marls, Nammal Nala, *Awanites awani* Regional Zone; late Dienerian.
20–23:   (20–22) Lateral, apertural and ventral views. (23) Suture line at H = 15.8 mm, × 1.5 (mirrored image). PIMUZ30422. Holotype.
        Loc. Nam350, base Ceratite Marls, Nammal Nala, *Awanites awani* Regional Zone; late Dienerian.

24–29: *Koiloceras sahibi* **n. sp.**
24–26:   Lateral, apertural and ventral views. PIMUZ30423. Holotype.
        Loc. Chi51, top Lower Ceratite Limestone, Chiddru, *Koninckites vetustus* Regional Zone; late Dienerian.
27–29:   Lateral, apertural and ventral views. PIMUZ30424. Paratype.
        Loc. War104, base Ceratite Marls, Wargal, *Koninckites vetustus* Regional Zone; late Dienerian.

30–32: *Xenodiscoides?* **sp. indet.**
        Lateral, apertural and ventral views. PIMUZ30425.
        Loc. Chi51, top Lower Ceratite Limestone, Chiddru, *Koninckites vetustus* Regional Zone; late Dienerian.

33–35: *Shamaraites?* **sp. indet.**
        Lateral, apertural and ventral views. × 2. PIMUZ30426.
        Loc. War104, base Ceratite Marls, Wargal, *Koninckites vetustus* Regional Zone; late Dienerian.

36–39: *Bukkenites sakesarensis* **n. sp.**
36, 37:   Lateral and apertural views. PIMUZ30427. Paratype.
        Loc. Amb104, base Lower Ceratite Limestone, Amb, *Gyronites dubius* Regional Zone; early Dienerian.
38, 39:   Lateral views. PIMUZ30428. Paratype.
        Loc. Amb104, base Lower Ceratite Limestone, Amb, *Gyronites dubius* Regional Zone; early Dienerian.

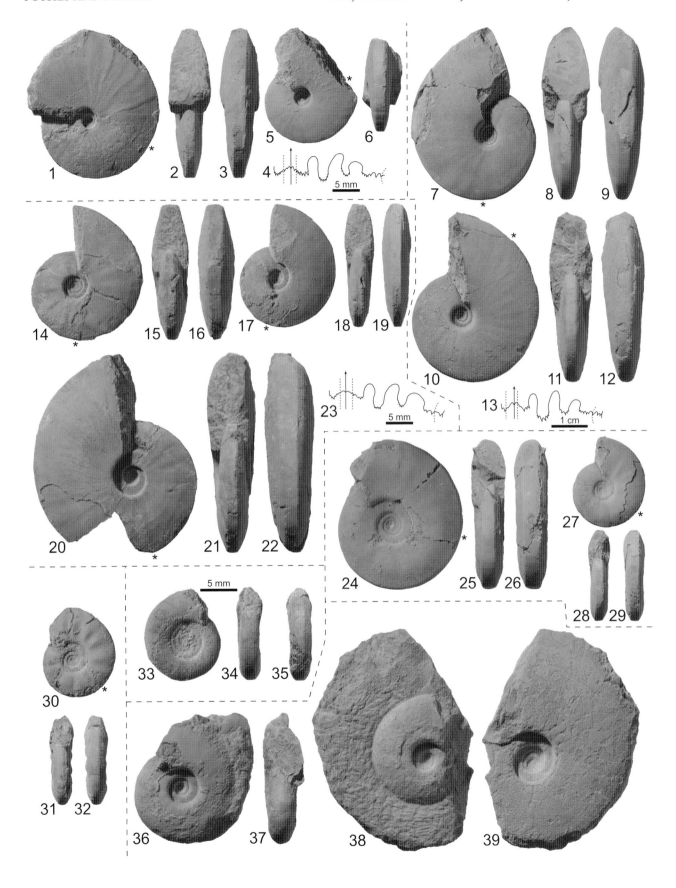

Plate 22   All figures natural size; asterisks indicate the position of the last septum.

1–13:   *Bukkenites sakesarensis* **n. sp.**
1–3:   Lateral, apertural and ventral views. PIMUZ30429. Holotype.
         Loc. Amb104, base Lower Ceratite Limestone, Amb, *Gyronites dubius* Regional Zone; early Dienerian.
4–6:   Lateral, apertural and ventral views. PIMUZ30430. Paratype.
         Loc. Amb104, base Lower Ceratite Limestone, Amb, *Gyronites dubius* Regional Zone; early Dienerian.
7–9:   Lateral, apertural and ventral views. Robust variant. PIMUZ30431. Paratype.
         Loc. Amb104, base Lower Ceratite Limestone, Amb, *Gyronites dubius* Regional Zone; early Dienerian.
10–13:   (10, 11) Lateral and apertural views. (12) Suture line at H = 21.8 mm.
         (13) Dorsal part of the external suture line 120° further than the one drawn in 4c (H not measurable).
         PIMUZ30432. Paratype.
         Loc. Amb104, base Lower Ceratite Limestone, Amb, *Gyronites dubius* Regional Zone; early Dienerian.

14, 15:   *Proptychites lawrencianus* **(de Konninck, 1863)** *sensu* **Waagen, 1895.**
         Polished cross section and cross section. PIMUZ30433.
         Loc. Nam301, base Ceratite Marls, Nammal Nala, *Ambites lilangensis* Regional Zone; middle Dienerian.

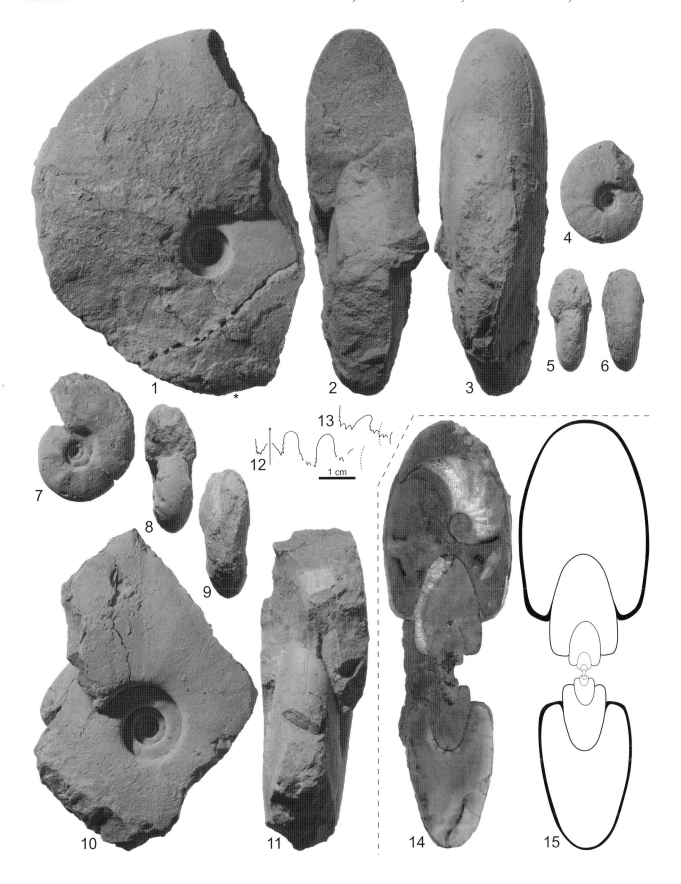

Plate 23   All figures natural size unless otherwise indicated; asterisks indicate the position of the last septum.

1–11:   ***Proptychites lawrencianus* (de Konninck, 1863) *sensu* Waagen, 1895**

1–4:   (1–3) Lateral, apertural and ventral views, × 0.5. (4) Suture line at H = 56 mm. PIMUZ30434.
       Loc. Nam300, floated block from base Ceratite Marls, Nammal Nala, *Ambites lilangensis* Regional
       Zone; middle Dienerian.

5–7:   Lateral, apertural and ventral views. Inner whorls of a larger broken specimen. PIMUZ30435.
       Loc. Nam100, base Ceratite Marls, Nammal Nala, *Ambites lilangensis* Regional Zone; middle Dienerian.

8:     Suture line at H = 16.1 mm, × 1.5. From the inner whorls of a large broken specimen. PIMUZ30436.
       Loc. Nam503, base Ceratite Marls, Nammal Nala, *Ambites lilangensis* Regional Zone; middle Dienerian.

9–11: Lateral, apertural and ventral views, × 0.7. PIMUZ30437.
       Loc. Nam100, base Ceratite Marls, Nammal Nala, *Ambites lilangensis* Regional Zone; middle Dienerian.

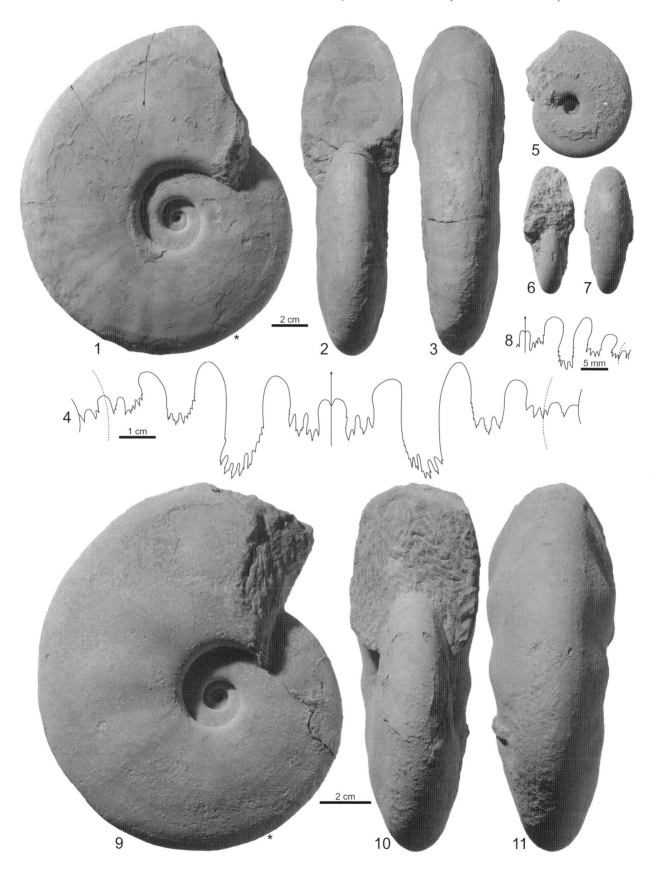

Plate 24 All figures natural size unless otherwise indicated; all specimens are incomplete phragmocones.

1–4: ***Proptychites wargalensis* n. sp.**
(1–3) Lateral, apertural and ventral views. (4) Suture line at H = 25.6 mm (mirrored image). PIMUZ30438. Holotype.
Loc. War5, middle Lower Ceratite Limestone, Wargal, *Gyronites plicosus* Regional Zone; early Dienerian.

5–14: ***Proptychites oldhamianus* Waagen, 1895**
5–7:   Lateral, apertural and ventral views. PIMUZ30439.
Loc. Nam378, top Lower Ceratite Limestone, Nammal Nala, *Gyronites frequens* Regional Zone; early Dienerian.
8–10:   Lateral, apertural and ventral views. PIMUZ30440.
Loc. Nam335, middle Lower Ceratite Limestone, Nammal Nala, *Gyronites plicosus* Regional Zone; early Dienerian.
11–14: (11–13) Lateral, apertural and ventral views. (14) Suture line at H = 10.8 mm, × 2. PIMUZ30441.
Loc. Nam335, middle Lower Ceratite Limestone, Nammal Nala, *Gyronites plicosus* Regional Zone; early Dienerian.

15–17: ***Proptychites ammonoides* Waagen, 1895**
Lateral, apertural and ventral views. PIMUZ30442.
Loc. Nam384, base Ceratite Marls, Nammal Nala, *Ambites radiatus* Regional Zone; middle Dienerian.

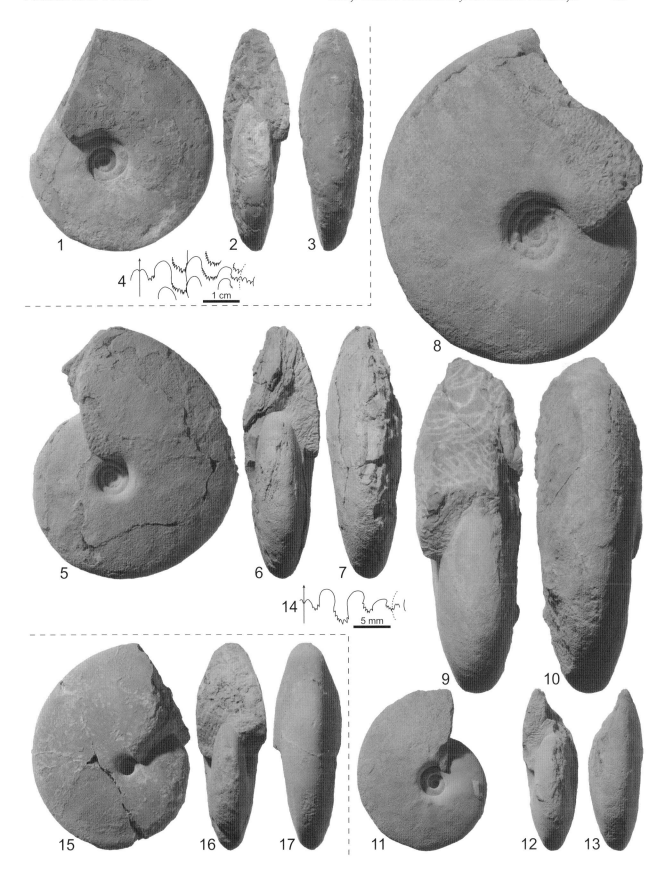

Plate 25   All figures natural size unless otherwise indicated; asterisks indicate the position of the last septum.

1–15:   *Proptychites ammonoides* Waagen, 1895.

1–5:   (1–4) Lateral, apertural and ventral views, × 0.6. (5) Suture line at H = 62.7 mm. (mirrored image). PIMUZ30443.

Loc. Nam380, base Ceratite Marls, Nammal Nala, *Ambites discus* Regional Zone; middle Dienerian.

6:    Suture line at H = 42.7 mm. PIMUZ30444.

Loc. Nam53, base Ceratite Marls, Nammal Nala, *Ambites superior* Regional Zone; middle Dienerian.

7–9:   (7, 8) Lateral and ventral views. (9) Suture line at H = 15.1 mm, × 1.5. PIMUZ30445.

Loc. Nam50, base Ceratite Marls, Nammal Nala, *Ambites discus* Regional Zone; middle Dienerian.

10–14:   (10–13) Lateral, apertural and ventral views, × 0.5. (14) Suture line at H = 85 mm. Specimen with two encrusting bivalves (b) with both valves preserved. PIMUZ30446.

Loc. Nam400, floated block from base Ceratite Marls, Nammal Nala, *Ambites discus* Regional Zone; middle Dienerian.

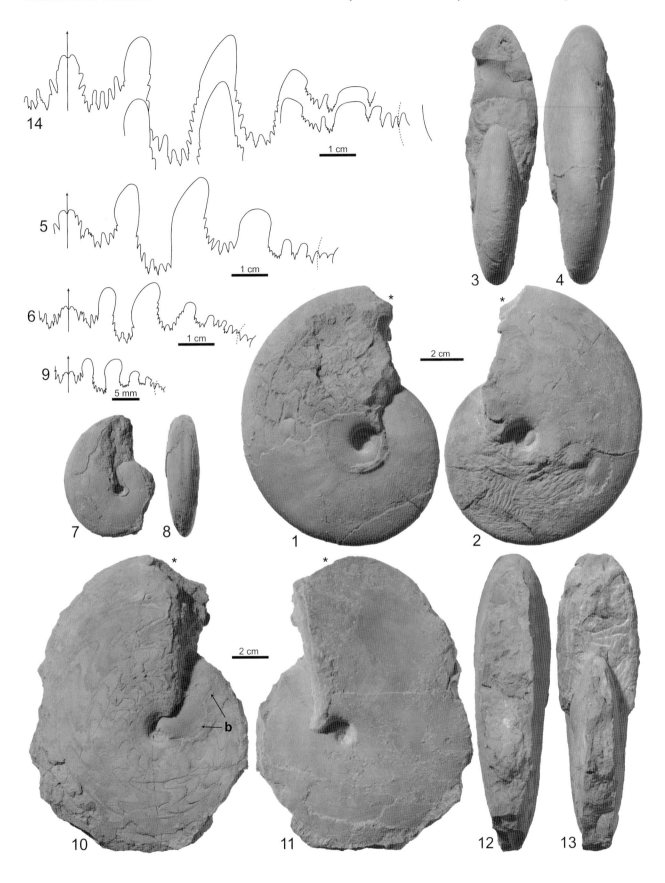

Plate 26   All figures natural size unless otherwise indicated.

1, 2:   *Proptychites ammonoides* **Waagen, 1895.**
Lateral and ventral views, × 0.6. PIMUZ30447.
Loc. Nam53, base Ceratite Marls, Nammal Nala, *Ambites superior* Regional Zone; middle Dienerian.

3–8:   *Proptychites* **cf. *P. pagei* Ware *et al.*, 2011.**
3–7:   (3–6) Lateral, apertural and ventral views. (7) Suture line at H = 53 mm. Left side (3) without shell, showing the umbilicus. Right side (4) with shell, umbilicus occluded. PIMUZ30448.
Loc. Nam501, base Ceratite Marls, Nammal Nala, *Ambites lilangensis* Regional Zone; middle Dienerian.
8:   Slightly weathered suture line at H = 48.1 mm. PIMUZ30451.
Loc. Nam345, base Ceratite Marls, Nammal Nala, *Ambites lilangensis* Regional Zone; middle Dienerian.

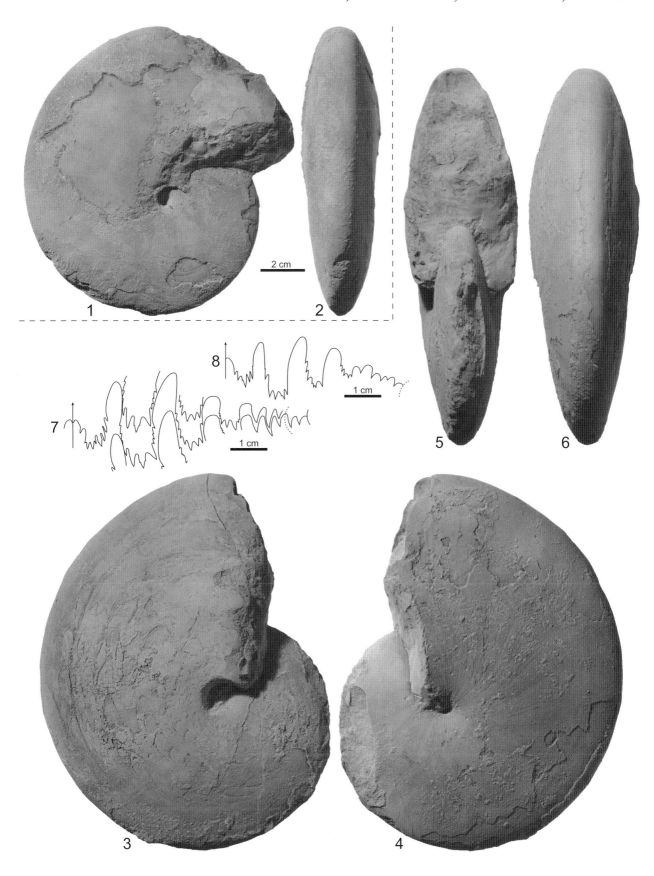

Plate 27   All figures natural size unless otherwise indicated; asterisks indicate the position of the last septum.

1–9:   *Mullericeras spitiense* **(von Krafft, 1909).**
1–3:   Lateral, apertural and ventral views. PIMUZ30500.
        Loc. Nam396, base Ceratite Marls, Nammal Nala, *Vavilovites* cf. *V. sverdrupi* Regional Zone; late Dienerian.
4–7:   (4–6) Lateral, apertural and ventral views. (7) Suture line at H = 21 mm, × 1.5. PIMUZ30452.
        Loc. Nam344, base Ceratite Marls, Nammal Nala, *Ambites lilangensis* Regional Zone; middle Dienerian.
8, 9:   Lateral and ventral views. PIMUZ30496.
        Loc. Nam396, base Ceratite Marls, Nammal Nala, *Vavilovites* cf. *V. sverdrupi* Regional Zone; late Dienerian.

10–28: *Mullericeras shigetai* **n. sp.**
10–13: (10–12) Lateral, apertural and ventral views. (13) Suture line at H = 29 mm, × 1.5 (mirrored image). PIMUZ30453. Holotype.
        Loc. Nam381, base Ceratite Marls, Nammal Nala, *Ambites radiatus* Regional Zone; middle Dienerian.
14:     Suture line at H = 20.6 mm, × 1.5. PIMUZ30454. Paratype.
        Loc. Nam382, base Ceratite Marls, Nammal Nala, *Ambites discus* Regional Zone; middle Dienerian.
15–17: (15–16) Lateral and ventral views. (17) Suture line at H = 15.3 mm, × 3 (mirrored image). PIMUZ30457. Paratype.
        Loc. Nam364, base Ceratite Marls, Nammal Nala, *Ambites discus* Regional Zone; middle Dienerian.
18–20: Lateral, apertural and ventral views. PIMUZ30455. Paratype.
        Loc. Nam381, base Ceratite Marls, Nammal Nala, *Ambites radiatus* Regional Zone; middle Dienerian.
21–24: (21–23) Lateral, apertural and ventral views. (24) Suture line at H = 22.1 mm, × 1.5 (mirrored image). PIMUZ30458. Paratype.
        Loc. Amb53, top Lower Ceratite Limestone, Amb, *Ambites atavus* Regional Zone; middle Dienerian.
25–28: (25–27) Lateral, apertural and ventral views. (28) Suture line at H = 20.2 mm, × 1.5. PIMUZ30456. Paratype.
        Loc. Nam527, base Ceratite Marls, Nammal Nala, *Ambites discus* Regional Zone; middle Dienerian.

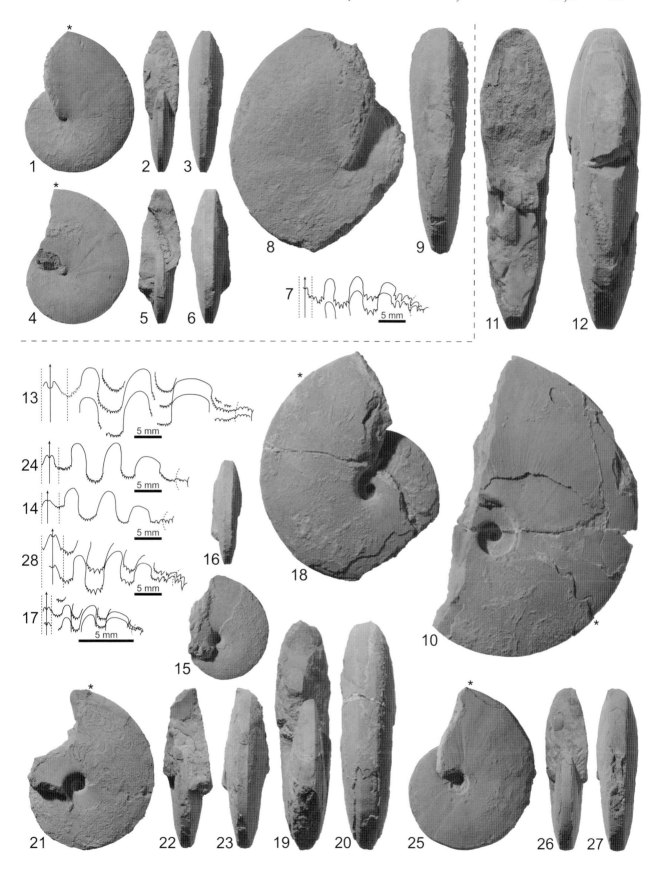

Plate 28 All figures natural size unless otherwise indicated; asterisks indicate the position of the last septum.

1–9: *Mullericeras indusense* **n. sp.**
1:      Suture line at H = 17.3 mm, × 1.5. PIMUZ30459. Paratype.
        Loc. Nam53, base Ceratite Marls, Nammal Nala, *Ambites superior* Regional Zone; middle Dienerian.
2–4:   (2, 3) Lateral and ventral views. (4) Suture line at H = 20.6 mm, × 1.5. PIMUZ30460. Holotype.
        Loc. Nam302, base Ceratite Marls, Nammal Nala, *Ambites superior* Regional Zone; middle Dienerian.
5–7:   Lateral, apertural and ventral views. PIMUZ30461. Paratype.
        Loc. Nam302, base Ceratite Marls, Nammal Nala, *Ambites superior* Regional Zone; middle Dienerian.
8, 9:   (8) Lateral view. (9) Suture line at H = 22.6 mm (mirrored image), × 1.5. PIMUZ30462. Paratype.
        Loc. Nam53, base Ceratite Marls, Nammal Nala, *Ambites superior* Regional Zone; middle Dienerian.

10–18: *Mullericeras niazii* **n. sp.**
10–13: (10–12) Lateral, apertural and ventral views. (13) Suture line at H = 22.4 mm, × 1.5 (mirrored
        image). PIMUZ30464. Holotype.
        Loc. Nam100, base Ceratite Marls, Nammal Nala, *Ambites lilangensis* Regional Zone; middle Dienerian.
14–16: Lateral, apertural and ventral views. PIMUZ30465. Paratype.
        Loc. Nam100, base Ceratite Marls, Nammal Nala, *Ambites lilangensis* Regional Zone; middle Dienerian.
17, 18: Lateral and ventral views. PIMUZ30466. Paratype.
        Loc. Nam100, base Ceratite Marls, Nammal Nala, *Ambites lilangensis* Regional Zone; middle Dienerian.

19–31: *Ussuridiscus varaha* **(Diener, 1895)**
19–21: Lateral, apertural and ventral views. PIMUZ30470.
        Loc. Amb104, base Lower Ceratite Limestone, Amb, *Gyronites dubius* Regional Zone; early Dienerian.
22–24: Lateral, apertural and ventral views. PIMUZ30468.
        Loc. Amb104, base Lower Ceratite Limestone, Amb, *Gyronites dubius* Regional Zone; early Dienerian.
25–27: Lateral, apertural and ventral views. PIMUZ30467.
        Loc. Amb104, base Lower Ceratite Limestone, Amb, *Gyronites dubius* Regional Zone; early Dienerian.
28–30: Lateral, apertural and ventral views. PIMUZ30469.
31:     Suture line at H = 19.8 mm, × 1.5 (mirrored image).
        Loc. Amb104, base Lower Ceratite Limestone, Amb, *Gyronites dubius* Regional Zone; early Dienerian.

32–41: *Ussuridiscus ensanus* **(von Krafft, 1909)**
32–35: (32, 33) Lateral, (34) apertural and (35) ventral views. PIMUZ30474.
        Loc. War100, from dipslope near top Lower Ceratite Limestone, Wargal, precise age unknown
        (*Gyronites plicosus* or *Gyronites frequens* Regional Zones; early Dienerian.
36–38: Lateral, apertural and ventral views. PIMUZ30472.
        Loc. Nam377, middle Lower Ceratite Limestone, Nammal Nala, *Gyronites plicosus* Regional Zone; early
        Dienerian.
39–41: Lateral, apertural and ventral views. PIMUZ30471.
        Loc. Nam393, middle Lower Ceratite Limestone, Nammal Nala, *Gyronites plicosus* Regional Zone; early
        Dienerian.

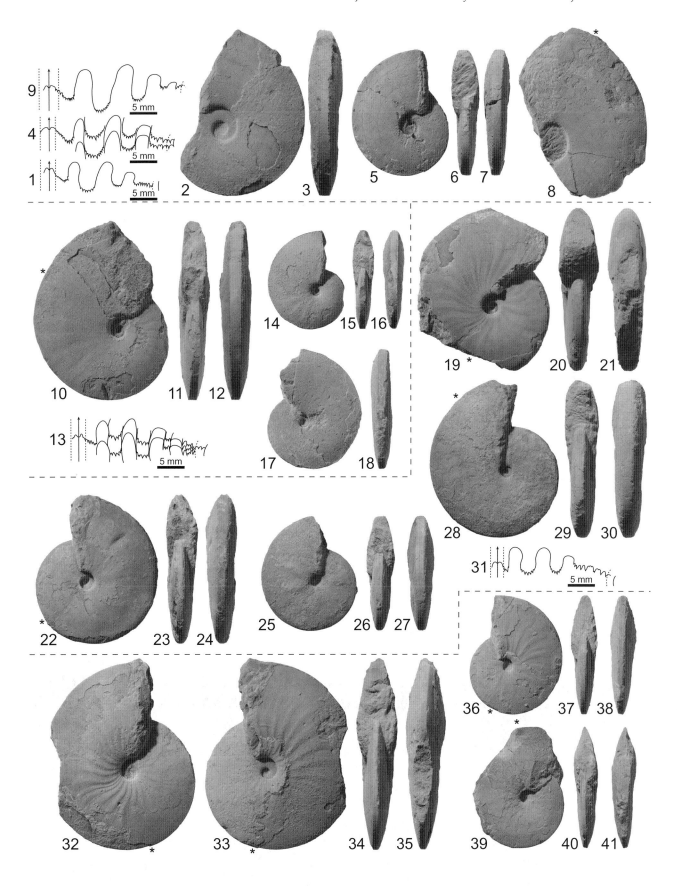

Plate 29   All figures natural size unless otherwise indicated; asterisks indicate the position of the last septum.

1–15:   *Ussuridiscus ensanus* **(von Krafft, 1909)**
1–3:   Lateral (1, 2) and ventral (3) views. PIMUZ30473.
          Loc. Nam378, top Lower Ceratite Limestone, Nammal Nala, *Gyronites frequens* Regional Zone; early
          Dienerian.
4–7:   (4–6) Lateral, apertural and ventral views. (7) Suture line at H = 20.8 mm, × 1.5. PIMUZ30475.
          Loc. Nam377, middle Lower Ceratite Limestone, Nammal Nala, *Gyronites plicosus* Regional Zone; early
          Dienerian.
8:       Suture line at H = 17 mm, × 1.5 (mirrored image). PIMUZ30476.
          Loc. War100, from dipslope near top Lower Ceratite Limestone, Wargal, precise age unknown
          (*Gyronites plicosus* or *Gyronites frequens* Regional Zones; early Dienerian.
9:       Suture line at H = 16.3 mm, × 1.5 (mirrored image). PIMUZ30477.
          Loc. War100, from dipslope near top Lower Ceratite Limestone, Wargal, precise age unknown
          (*Gyronites plicosus* or *Gyronites frequens* Regional Zones; early Dienerian.
10–14: (10, 11) Lateral, (12) apertural and (13) ventral views. (14) Suture line at H = 16 mm, × 1.5.
          PIMUZ30478.
          Loc. War100, from dipslope near top Lower Ceratite Limestone, Wargal, precise age unknown
          (*Gyronites plicosus* or *Gyronites frequens* Regional Zones; early Dienerian.

15–21:   *Ussuridiscus ornatus* **n. sp.**
15–17: Lateral, apertural and ventral views. PIMUZ30479. Paratype.
          Loc. War6, middle Lower Ceratite Limestone, Wargal, *Gyronites plicosus* Regional Zone; early
          Dienerian.
18–21: (18–20) Lateral, apertural and ventral views. (21) Suture line at H = 14.1 mm, × 2 (mirrored image).
          PIMUZ30481. Holotype.
          Loc. Nam377, middle Lower Ceratite Limestone, Nammal Nala, *Gyronites plicosus* Regional Zone; early
          Dienerian.

22–27:   *Ussuridiscus?* **sp. indet.**
22–24: (22, 23) Lateral and ventral views. (24) Suture line at H = 16.9 mm, × 1.5 (mirrored image).
          PIMUZ30483.
          Loc. Nam391, base Lower Ceratite Limestone, Nammal Nala, ?latest Griesbachian – earliest Dienerian.
25–27: (25, 26) Lateral and apertural views. (27) Suture line at H = 11.5 mm, × 2 (mirrored image).
          PIMUZ30484.
          Loc. Nam391, base Lower Ceratite Limestone, Nammal Nala, ?latest Griesbachian – earliest Dienerian.

28–31:   *Ussuridiscus ventriosus* **n. sp.**
          (28–30) Lateral, apertural and ventral views. (31) Suture line at H = 14.9 mm, × 1.5. PIMUZ30482.
          Holotype.
          Loc. Nam377, middle Lower Ceratite Limestone, Nammal Nala, *Gyronites plicosus* Regional Zone; early
          Dienerian.

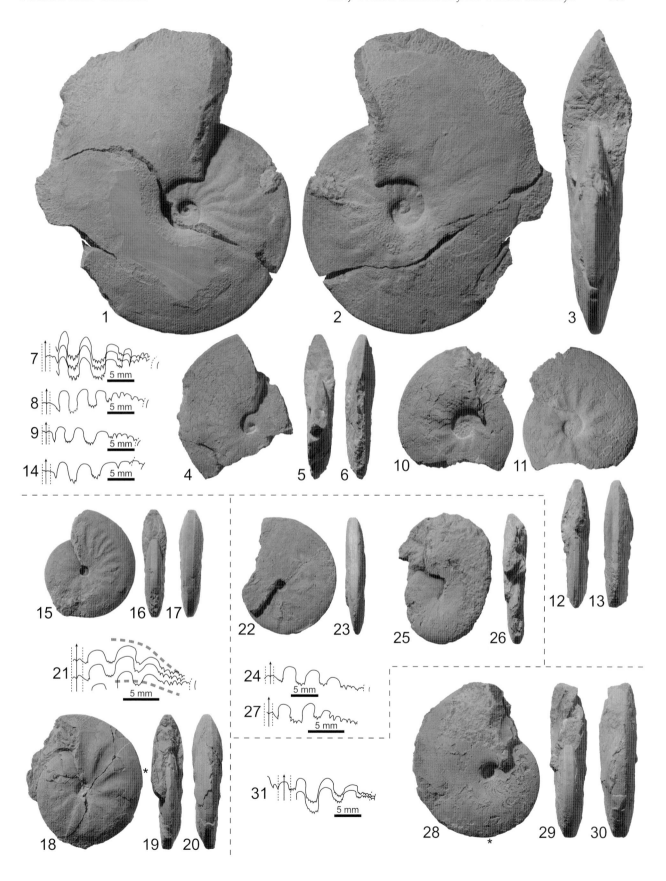

Plate 30   All figures natural size unless otherwise indicated; asterisks indicate the position of the last septum.

1–3:   ***Kingites davidsonianus* (de Koninck, 1863)**
1–3:   Lateral, apertural and ventral views. PIMUZ30486.
   Loc. Nam346, base Ceratite Marls, Nammal Nala, *Kingites davidsonianus* Regional Zone; late Dienerian.
4, 5:   Lateral and apertural views. GSI7155. Holotype of *Kingites lens* Waagen, 1895. (Photo by B. Kummel). Ceratite Marls, Wargal, bed and locality unknown.
6–9:   (6–8) Lateral, apertural and ventral views. (9) Suture line at H = 25 mm (mirrored image). PIMUZ30485.
   Loc. Nam61, base Ceratite Marls, Nammal Nala, *Kingites davidsonianus* Regional Zone; late Dienerian.
10–12:   (10, 11) Lateral and ventral views. (12) Suture line at H = 21.6 mm (mirrored image). Specimen with two juvenile specimens of *Kon. khoorensis* (k) in its body chamber. PIMUZ30492.
   Loc. Nam61, base Ceratite Marls, Nammal Nala, *Kingites davidsonianus* Regional Zone; late Dienerian.
13:   Suture line at H = 27.2 mm (mirrored image). PIMUZ30487.
   Loc. Nam61, base Ceratite Marls, Nammal Nala, *Kingites davidsonianus* Regional Zone; late Dienerian.
14:   Suture line at H = 21.2 mm (mirrored image). PIMUZ30488.
   Loc. Nam315, base Ceratite Marls, Nammal Nala, *Kingites davidsonianus* Regional Zone; late Dienerian.
15–18:   (15–17) Lateral, apertural and ventral views. (18) Suture line at H = 14.3 mm, × 1.5 (mirrored image). PIMUZ30489.
   Loc. Nam61, base Ceratite Marls, Nammal Nala, *Kingites davidsonianus* Regional Zone; late Dienerian.

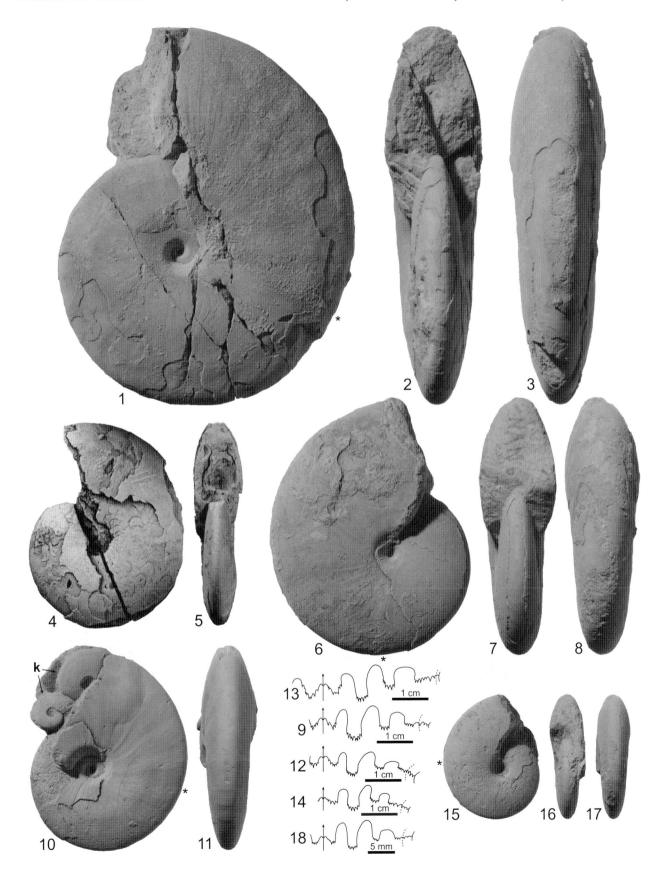

Plate 31   All figures natural size unless otherwise indicated; asterisks indicate the position of the last septum.

1–4:   ***Kingites davidsonianus* (de Koninck, 1863)**

1, 2:   Lateral and ventral views. With its lower jaw (j) preserved at the beginning of its body chamber. PIMUZ30490.
Loc. Nam101d, base Ceratite Marls, Nammal Nala, *Kingites davidsonianus* Regional Zone; late Dienerian.

3, 4:   Polished cross section (3) and cross section (4). PIMUZ30491.
Loc. Nam313, base Ceratite Marls, Nammal Nala, *Kingites davidsonianus* Regional Zone; late Dienerian.

5–9:   ***Kingites korni* Brühwiler *et al.*, 2010**

(5–8) Lateral, apertural and ventral views. (9) Suture line at H = 25 mm. Inner mould of a complete specimen. PIMUZ30493.
Loc. Nam70, base Ceratite Marls, Nammal Nala, *Koninckites vetustus* regional Zone; late Dienerian.

10–19: *Clypites typicus* Waagen, 1895

10–12:   Lateral, apertural and ventral views. PIMUZ30494.

13:   Suture line at H = 38 mm, × 1 (mirrored image).
Loc. Nam130, base Ceratite Marls, Nammal Nala, *Kingites davidsonianus* Regional Zone; late Dienerian.

14–17:   (14–16) Lateral, apertural and ventral views. (17) Suture line at H = 21.2 mm, × 1.5 (mirrored image). PIMUZ30495.
Loc. Nam61, base Ceratite Marls, Nammal Nala, *Kingites davidsonianus* Regional Zone; late Dienerian.

18:   Suture line at H = 26.3 mm, × 1.5 (mirrored image). PIMUZ30498.
Loc. Nam319, base Ceratite Marls, Nammal Nala, *Kingites davidsonianus* Regional Zone; late Dienerian.

19:   Suture line at H = 26.4 mm, × 1.5 PIMUZ30499.
Loc. Nam61, base Ceratite Marls, Nammal Nala, *Kingites davidsonianus* Regional Zone; late Dienerian.

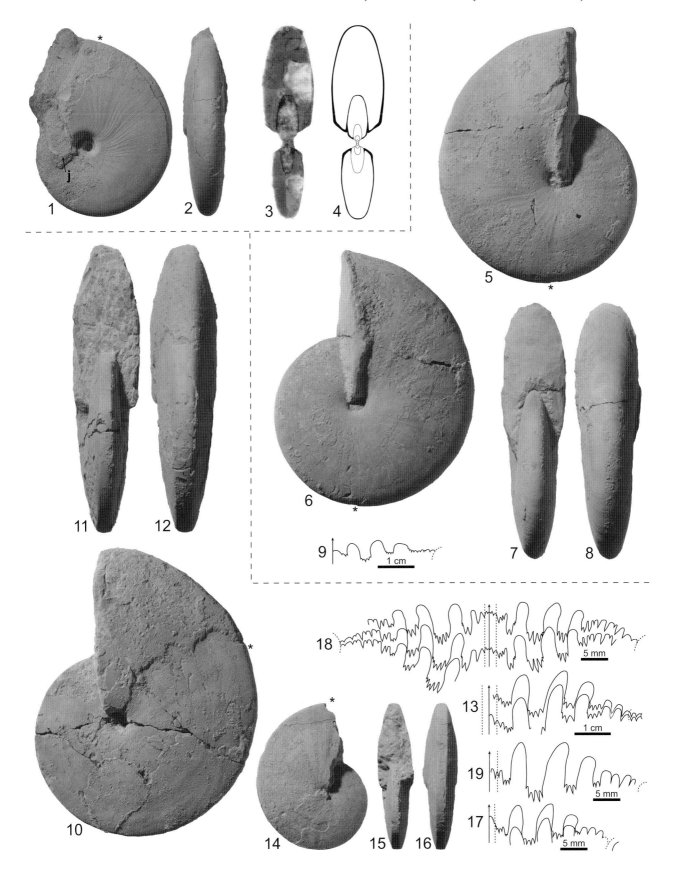

Plate 32 All figures natural size unless otherwise indicated; asterisks indicate the position of the last septum.

1–4: ***Clypites typicus* Waagen, 1895.**
1, 2:   Lateral and ventral views. PIMUZ30497.
         Loc. Nam130, base Ceratite Marls, Nammal Nala, *Kingites davidsonianus* Regional Zone; late
         Dienerian.
3, 4:   Lateral and apertural views. PIMUZ30501.
         Loc. Nam83, base Ceratite Marls, Nammal Nala, *Koninckites vetustus* Regional Zone; late Dienerian.

5–16: ***Pseudosageceras simplelobatum* n. sp.**
         (5–7) Lateral, apertural and ventral views. (8) Suture line at H = 27.4 mm, × 1.5 (mirrored image).
         PIMUZ30502. Paratype.
         Loc. Nam63, base Ceratite Marls, Nammal Nala, *Koninckites vetustus* Regional Zone; late Dienerian.
9:       Suture line at H = 40.8 mm (mirrored image). PIMUZ30503. Paratype.
         Loc. Nam61, base Ceratite Marls, Nammal Nala, *Kingites davidsonianus* Regional Zone; late Dienerian.
10–13:   (10–12) Lateral, apertural and ventral views. × 2. (13) Suture line at H = 9 mm, × 3 (mirrored
         image). PIMUZ30504. Paratype.
         Loc. Nam67, base Ceratite Marls, Nammal Nala, *Kingites davidsonianus* regional Zone; late Dienerian.
14–16:   Lateral, apertural and ventral views. PIMUZ30505. Paratype.
         Loc. Chi51, top Lower Ceratite Limestone, Chiddru, *Koninckites vetustus* regional Zone; late Dienerian.

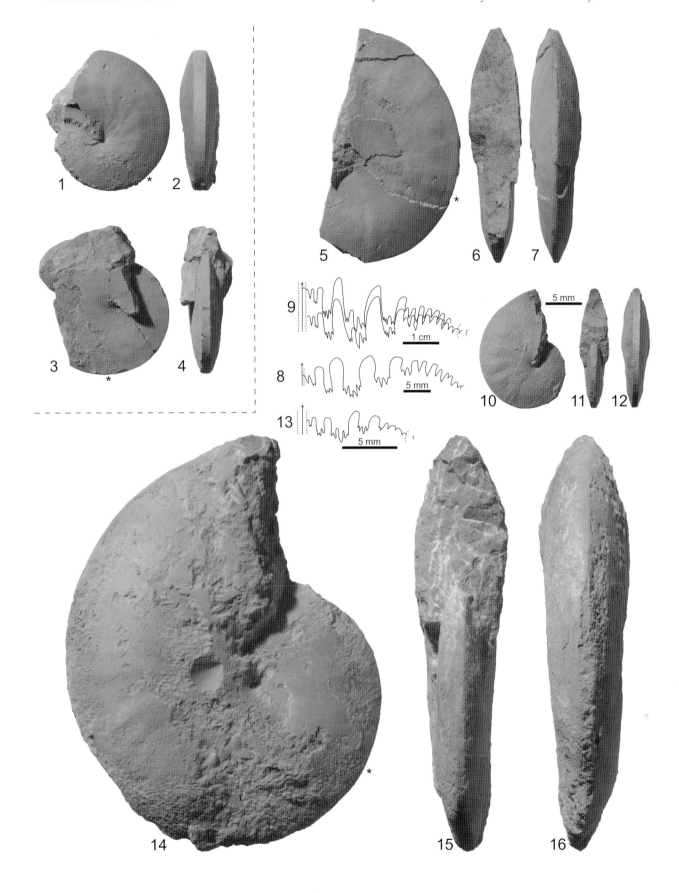

Plate 33   All figures natural size unless otherwise indicated asterisks indicate the position of the last septum.

1–10:   ***Pseudosageceras simplelobatum* n. sp.**

1–3:   Lateral, apertural and ventral views. PIMUZ30506. Holotype.
       Loc. Nam346, base Ceratite Marls, Nammal Nala, *Kingites davidsonianus* Regional Zone; late
       Dienerian.

4, 5:   Polished cross section (4) and cross section (5). PIMUZ30507. Paratype.
        Loc. Nam336, base Ceratite Marls, Nammal Nala, *Kingites davidsonianus* Regional Zone; late
        Dienerian.

6, 7:   Lateral and ventral views. PIMUZ30508. Paratype.
        Loc. Nam346, base Ceratite Marls, Nammal Nala, *Kingites davidsonianus* Regional Zone; late
        Dienerian.

8–10:   Lateral, apertural and ventral views. PIMUZ30509. Paratype.
        Loc. Nam350, base Ceratite Marls, Nammal Nala, *Awanites awani* Regional Zone; late Dienerian.

11–15: ***Subacerites friski* n. gen. et n. sp.**
       Lateral (11, 12), apertural (13) and ventral (14) views. (15) Suture line at H = 19.1 mm, × 1.5
       (mirrored image). PIMUZ30510. Holotype.
       Loc. Amb3, base Ceratite Marls, Amb, *Kingites davidsonianus* Regional Zone; late Dienerian.

16–19: Gen. et sp. indet.
       (16–18) Lateral, apertural and ventral views, × 2. (19) Suture line at H = 8.5 mm, × 3 PIMUZ30511.
       Loc. Amb104, base Lower Ceratite Limestone, Amb, *Gyronites dubius* Regional Zone; early Dienerian.

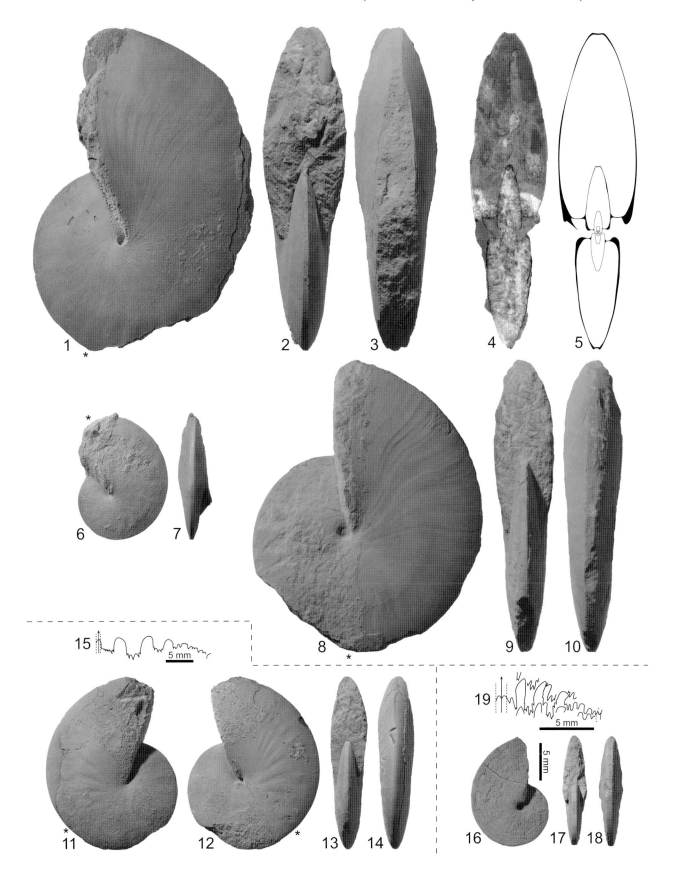

Dienerian (Early Triassic) ammonoids from Spiti, Himachal Pradesh, India

by

David Ware, Hugo Bucher, Thomas Brühwiler
and Leopold Krystyn

Acknowledgements
Financial support for the publication of this issue of
Fossils and Strata was provided by the Lethaia Foundation

Contents

Dienerian (Early Triassic) ammonoids from Spiti, Himachal Pradesh, India

DAVID WARE, HUGO BUCHER, THOMAS BRÜHWILER AND LEOPOLD KRYSTYN

FOSSILS AND STRATA

THE LETHAIA FOUNDATION

Ware, D., Bucher, H., Brühwiler, T. & Krystyn L. 2018: Dienerian (Early Triassic) ammonoids from Spiti, Himachal Pradesh, India. Fossils and Strata, No 63, pp. 179–241. doi: 10.1111/let.12274

The results of a high resolution bedrock controlled sampling of Mud, Guling and Lalung in the Spiti District, India are presented. These areas yielded abundant and rather well preserved Dienerian ammonoids, which compare well with the revised ammonoid successions from the Salt Range (Pakistan). The Dienerian ammonoid faunas from both regions are remarkably similar and the new threefold subdivision of the Dienerian (early, middle and late) proposed in the Salt Range also applies to Spiti. Moreover, 10 out of the 12 Dienerian regional zones defined in the Salt Range are recognized in Spiti, with the same associations of characteristic species. Thus the initial biostratigraphical scheme established in the Salt Range can be reproduced laterally and is valid throughout most of the Northern Indian Margin. The four new species *Gyronites levilatus*, *Gyronites bullatus*, *Ambites nyingmai* and *Vavilovites meridialis* are introduced. □ *Ammonoidea, biostratigraphy, Dienerian, Early Triassic, India, Spiti.*

David Ware ✉ [david.ware@mfn-berlin.de], Museum für Naturkunde, Leibniz Institute for Evolution and Biodiversity Science, Invalidenstrasse 43 10115 Berlin, Germany; Hugo Bucher [hugo.fr.bucher@pim.uzh.ch], Thomas Brühwiler [bruehwiler@pim.uzh.ch], Paläontologisches Institut und Museum der Universität Zürich, Karl Schmid-Strasse 4 CH-8006 Zürich, Switzerland; Leopold Krystyn [leopold.krystyn@univie.ac.at], Institut für Paläontologie, Althanstraße 14 1090 Wien, Austria

Introduction

The Northern Indian Margin is known since the end of the 19th century as a classical region for the study of Early Triassic ammonoid faunas. In a companion work (Ware *et al.* 2018), Dienerian ammonoids of the Salt Range (Pakistan) were thoroughly revised. A new high resolution biostratigraphical scheme was proposed for the Salt Range and correlation with other recently established biozonations (Tozer 1994 for Arctic Canada; Dagys & Ermakova 1996 for northern Siberia; Shigeta & Zakharov 2009 for Primorye, Russia) was discussed. The Spiti region (Himachal Pradesh, India) is a classical area for studying Early Triassic ammonoids of the Northern Indian Margin. Ammonites were described in two monographs first by Diener (1897) followed by von Krafft & Diener (1909). These two pioneering works unfortunately lack a detailed stratigraphic context, and many new species are only indicated as coming from the 'lower division of the lower Trias', a very vague indication, which does not provide the basis for an accurate and precise biozonation. Until very recently, no other papers concerning Early Triassic ammonoids from Spiti have been published. In the mid-1990s, Leopold Krystyn started the investigation of various sections in Spiti, which led him to propose a section near Mud as a Global Boundary Stratotype Section and Point (GSSP) candidate for the Induan–Olenekian boundary (equivalent to the Dienerian–Smithian boundary; Krystyn *et al.* 2007a, b). Only a few ammonoids were illustrated and described in these studies, as the focus was on the boundary interval. After the work in India by Krystyn *et al.* (2007a,b) the research group at the University of Zürich conducted two intensive seasons of fieldwork in 2008 and 2009. This lead to the discovery of typical Smithian ammonoids below the boundary (Brühwiler *et al.* 2010) proposed by Krystyn *et al.* (2007a,b), thus questioning the proposed definition of the Induan–Olenekian boundary (see the foreword, Ware & Bucher 2018a for details). Recently, Brühwiler *et al.* (2012) revised and described the middle and upper Smithian ammonoids from Spiti district.

The present work focuses on the abundant and fairly well preserved ammonoid faunas from a detailed bedrock controlled sampling in the Mud, Guling and Lalung areas of the Spiti District. The great similarity with the Salt Range, Pakistan, allows the authors to establish a detailed and laterally reproducible but informal biozonation for the Dienerian of the Northern Indian Margin. Based on these results, a formal biozonation for the Northern Indian Margin was constructed by Ware *et al.* (2015).

DOI 10.1111/let.12274 © 2018 Lethaia Foundation. Published by John Wiley & Sons Ltd

Geological framework

The Spiti District is situated in the central Himalayas in Himachal Pradesh, northern India, about 400 km North of New Delhi (Fig. 1). It includes the valley formed by the Spiti River and its tributaries. During the Early Triassic, it was situated in the southern Tethys on the northern Gondwana margin, at a palaeolatitude of ca. 30°S (Fig. 1A).

Stratigraphy

In the Spiti District, the Lower Triassic and Anisian sedimentary rocks are referred to the Mikin Formation (Bhargava *et al.* 2004). It disconformably overlies the Wuchiapingian to Changhsingian (Upper Permian) Kuling Shales and is overlain by the Upper Ladinian (Middle Triassic) Kaza Formation. The Mikin Formation is divided into the Lower Limestone Member, the Limestone and Shale Member, the Niti Limestone and the Himalayan Muschelkalk. The Dienerian is represented by the upper part of the Lower Limestone Member and the shale interval at the base of the Limestone and Shale Member.

Based on their ammonoid content, the Lower Limestone Member and Limestone and Shale Member were subdivided into a certain number of lithological units named 'beds', the names and number of which vary depending on the author (such as '*Otoceras* beds', '*Meekoceras* beds', etc.). These names have also changed depending on the stratigraphic resolution and taxonomy used by different authors. For example, the Dienerian shale interval at the base of the Limestone and Shale Member was included in the '*Otoceras* beds' by Diener (1897), then separated from the '*Otoceras* beds' and referred to the '*Meekoceras* beds' by von Krafft & Diener (1909), renamed '*Gyronites* beds' by Bhargava *et al.* (2004), and finally renamed '*Ambites* beds' by Brühwiler *et al.* (2010). To avoid any confusion with previous works, and considering that lithological units should be named independently from their faunal content to avoid problems linked with possible diachronism of the lithological units, these names are not used here (except in reference to previous studies).

Lithology and ammonoid preservation

Dienerian ammonoids of Spiti are found in three different types of facies, one being typical of the Lower Limestone Member while the other two are from the shale interval at the base of the Limestone

and Shale Member. The lithological description of the Mikin Formation was published in Bhargava *et al.* (2004).

Lower Limestone Member

The Lower Limestone Member consists of a thin (0.4–1.5 m thick) series of massive shelly calcarenite beds. It is often designated in the literature as the '*Otoceras* beds' (e.g. von Krafft & Diener 1909; Bhargava *et al.* 2004) and is Griesbachian to middle Dienerian. The Dienerian strata in the upper part of the Lower Limestone Member consist of thin, hard coarse-grained coquinoid limestone beds very rich in ammonoids, bivalves and unidentifiable shell fragments. The Dienerian part of this unit can generally easily be differentiated in the field from its Griesbachian part by a minor facies change, the Griesbachian part consisting of massive and coarser grained limestone beds without intercalated shale in contrast to the Dienerian succession, which is composed of limestone and shale. This difference however could not be observed in the Mud Top Section and in the Guling River Section. The ammonoids are often poorly preserved and broken, with only a preserved portion of the body chamber, and are frequently distorted. They are occasionally phosphatized, with concentration of iron oxides where the shell material has been dissolved. In the Lower Limestone Member, all specimens are usually small sized, inclusive of nuclei of larger individuals, thus suggesting some mechanical sorting. These layers were deposited in a shallow basinal environment with low sedimentation rates, so the stratigraphic record is occasionally affected by condensation. In two sections, Lalung ridge 1 (LMH-3, fig. 11) and Guling Village (LMH-3, fig. 5), the upper bed of the Lower Limestone Member contains both the genus *Gyronites*, with species typical of the *Gyronites frequens* Regional Zone (uppermost lower Dienerian), and the genus *Ambites*, with species typical of the *Ambites atavus* Regional Zone (lowermost earliest middle Dienerian). Such condensed occurrences are excluded from the construction of the biochronological scheme. In the Mud Bottom section, the uppermost bed of the Lower Limestone Member is different. It is lenticular, and consists of fine grained limestone with well preserved and undistorted ammonoids.

Limestone and Shale Member

The Limestone and Shale Member is a ca. 15 m thick interval composed of alternating limestone and shale beds. It is classically divided into three

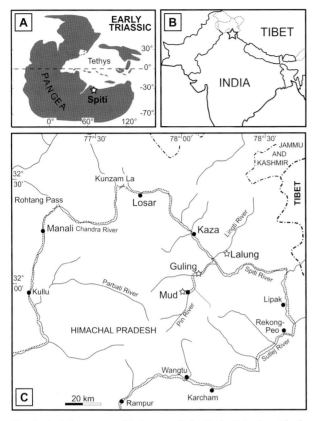

Fig. 1. A, Palaeogeographical map of the Early Triassic with the palaeoposition of Spiti (modified after Brayard *et al.* 2006). B, Map of India showing the location of Spiti. C, Location map of the three studied areas (open stars) in Spiti District (modified after Brühwiler *et al.* 2012).

often flattened, an indication of high sedimentation rates. Body chambers show geopetal structures, their upper half being filled with sparitic calcite and bitumen. These early diagenetic concretions may also yield complete, still articulated fish skeletons (Romano *et al.* 2016). Such complete fishes, associated with the abundance of pyrite and organic matter, indicates that these layers were deposited under anoxic conditions at the sediment–seawater interface. Throughout this interval, ammonoids are very frequently encrusted by bivalves similar to the ones previously described in Ware *et al.* (2011, 2018), especially the involute forms, independently of their taxonomic affinities. These bivalves are generally encrusting the umbilicus on both sides of the shell (see for example Pl. 5, fig. 1; Pl. 6, fig. 6; Pl. 7, figs 3, 6, 7; Pl. 8, figs 1, 4, 6, 7; Pl. 10, fig. 8), sometimes inducing a departure from normal umbilical coiling. Bivalves encrusting the whorl flanks of ammonoids are not unusual (see for example Pl. 6, fig. 1). Some large ammonoid specimens are completely encrusted by hundreds of these bivalves (see for example Pl. 7, fig. 1; Pl. 11, fig. 1), except on the last portion of the body chamber, an additional indication that these bivalves encrusted the shell *in vivo*. The bivalves encrusting the specimen figured on Plate 7, figure 1, were described in detail by Hautmann *et al.* (2017), and identified as Ostreidae. Although generally too poorly preserved to be identified, it can reasonably be assumed that these encrusting bivalves all belong to the same species.

Present work

Most of the specimens presented herein were collected during two fieldwork seasons in 2008 and 2009. To complete this study, a few specimens collected previously by Leopold Krystyn in Lalung and Guling were also included. Finally, following the discovery of complete fishes in 2009, further field investigations were carried out in 2010 by Winand Brinkmann and associates from the Paläontologisches Institut und Museum, University of Zürich, to find additional fishes in the Dienerian shale interval, and additional ammonoids were collected in order to adequately date the fish material.

Study area

Three areas were investigated, from South to North: Mud, Guling and Lalung (Fig. 1C). Mud and Guling are situated along the Pin River, a southern tributary of the Spiti River, while Lalung is situated along the Lingti River, a northern

parts (e.g. Bhargava *et al.* 2004): the 'Gyronites beds', the '*Flemingites* beds' and the '*Parahedenstroemia* beds'. Its age ranges from the middle Dienerian to the end of the Smithian. Its lower (Dienerian) part consists of a black shale interval rich in organic matter and pyrite (corresponding to the '*Gyronites* beds' of Bhargava *et al.* 2004). It contains only a few limestone beds in Mud and Guling, whereas in Lalung limestone beds are much more abundant. Ammonoids in the shales are crushed, heavily distorted and their shell is partially dissolved, usually too poorly preserved to be identified even at the family level. The limestone beds display two different types of facies. A first facies mostly consists of tempestites in which the fossils are very unevenly distributed. These beds are often discontinuous and show some grain size sorting. Ammonoids generally have their body chamber partially broken and are often imbricated and distorted. A second facies yielding ammonoids consists of large (up to ca. 50 cm in diameter) early diagenetic limestone concretions. In this case, the specimens are generally complete, undistorted, and with preserved peristome. The upper side of the phragmocone is tributary of the Spiti River. The Geological Survey of India was the first to explore these three classic localities in the late 19th century under the direction of Karl Griesbach. Several sections were sampled in detail in each of these localities. Although strata of Early Triassic age are extensively exposed throughout the Spiti region, they are very often difficult to reach, thus limiting the number of sections studied here.

Mud area

Lower Triassic strata can easily be reached on the northern flank of a wide side valley NW of Mud village. This area is of primary importance for Early Triassic stratigraphy because Krystyn et al. (2007a,b) proposed the succession as a candidate GSSP for the Dienerian–Smithian boundary. Two sections have been sampled (Fig. 2A), the first one corresponding to the section M05 of Krystyn et al. (2007a; Fig. 2C).

Mud Bottom section
The Mud Bottom section is situated slightly above the candidate GSSP (section M04), at an altitude of ca. 4130 m (coordinates: N31°57′58.4″; E78°01′21.4″). This section is much better exposed than the GSSP candidate. The stratigraphic log and ammonoid distribution are given in Figure 3.

Mud Top section
This composite section consists of three partly overlapping and closely spaced (within 100 m) sections. These sections can easily be correlated on the basis of lithology, thus allowing the construction of a single composite section as presented here (Fig. 4). These partial sections are at ca. 4750 m elevation, around the following coordinates: N31°58′36.2″; E78°00′23.3″. These sections are much thinner than the Mud Bottom section, especially in the case of the Lower Limestone Member (Fig. 2B), where the Griesbachian is represented by a ca. 15 cm thick bed, whereas it corresponds to a 1 m thick interval, which can be subdivided into 4 beds in the Mud Bottom section. The stratigraphic log and ammonoid distribution are given in Figure 4.

Guling area

Around Guling, the Lower Triassic strata are folded, forming a syncline with easily accessible sections. Three sections were investigated and sampled. The first two sections are very similar in thickness. However, the comparison of their respective bases is obscured by small scale faulting parallel to bedding in the river section. The upper parts of the sections are otherwise nearly identical. The Tilling section is much thinner than the two other sections of this area, but the succession of beds of the interval which has been sampled is identical.

Guling Village section
This section is located about 100 m north of Guling, at ca. 3650 m elevation (coordinates: N32°02′44.3″; E78°05′26.2″). The stratigraphic log and ammonoid distribution are given in Figure 5.

Here, the Dienerian shale interval was covered by scree, so a trench was dug to sample it. Consequently, the lateral continuity of the layers could not be assessed. The upper part of the Lower Limestone Member contains more shales than in other sections, so the boundary between the Lower Limestone Member and Limestone and Shale Member is not well defined and not shown precisely on the logs. It is however marked by a change in facies, and the topmost bed of the Lower Limestone Member corresponds to a condensed horizon containing the *Gyronites frequens* and the *Ambites atavus* regional zones.

Guling River section
The second easily accessible section (Fig. 6) was sampled on the other side of the river, south of Guling at ca. 3740 m elevation (coordinates: N32°02′7.1″; E78°05′10.9″). It is the only section where the two parts of the Lower Limestone Member (Griesbachian and early Dienerian) do not show any change in facies. The limestone beds at the base of the Limestone and Shale Member (Fig. 6) are not laterally continuous, and have obviously been interrupted by small scale low angle faults. The stratigraphic log and ammonoid distribution are shown in Figure 7.

Tilling section
The Tilling section is situated further south in the continuity of the Guling River section. It is situated north of the village of Tilling at ca. 4030 m elevation (coordinates: N32°00′46.2″; E78°05′04.5″) and has been briefly sampled during a field work session conducted in 2010 with the aim of collecting fish fossils (published in Romano et al. 2016). Only a few ammonoids were collected in order to date the shale interval with fish fossils. The base of this section has not been sampled. The stratigraphic log and ammonoid distribution are presented in Figure 8.

Lalung area

Extensive exposures of Lower Triassic strata can be seen near the Village of Lalung, but most of these are

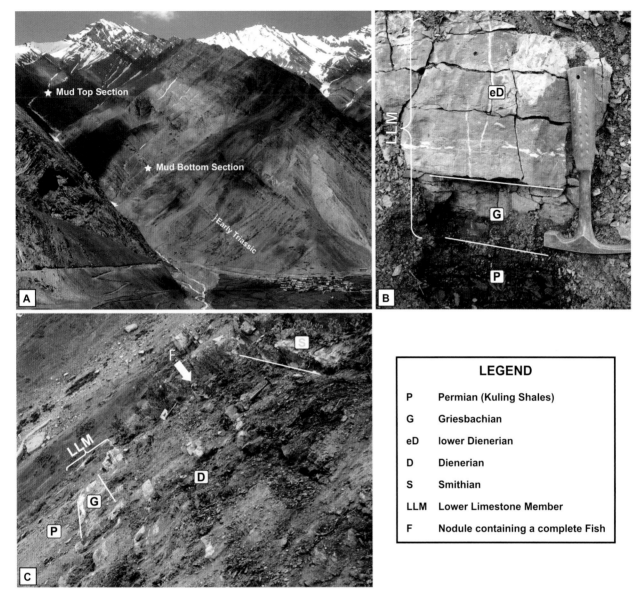

Fig. 2. **A,** General view of the exposures near Mud, with location of the two studied sections. **B,** Close-up view of the Permian-Triassic boundary and of the Lower Limestone Member of Mud Top Section. **C,** Photo of Mud Bottom Section (section M05 of Krystyn *et al.* 2007a). Note the thickness difference of the Lower Limestone Member compared with the Top Section.

inaccessible. As a consequence, only three sections, all situated in a small gulley north of Lalung, could be investigated. In this locality, the Lower Limestone Member is much thicker than at Mud and Guling, while the basal shale interval of the Limestone and Shale Member is much thinner and with more abundant limestone beds than in the two other localities.

Lalung Cliff sections
The Lalung Cliff section (Fig. 9) is situated on the southern side of this gulley, below a small cliff formed by the Niti Limestone. It is the most complete section and is situated at ca. 3860 m elevation

(coordinates: N32°08′57.8″; E78°14′36.5″). The stratigraphic log and ammonoid distribution are given in Figure 10.

Lalung Ridge 1+2 sections
Two additional and small sections situated only ca. 20 m apart on a ridge on the northern side of the same gulley were also investigated. Here, only the Lower Limestone Member and a few beds at the very base of the Limestone and Shale Member are exposed, the rest of the section being covered by scree. These two sections are intensively weathered, and even the bedding plane cannot be seen easily. As a consequence,

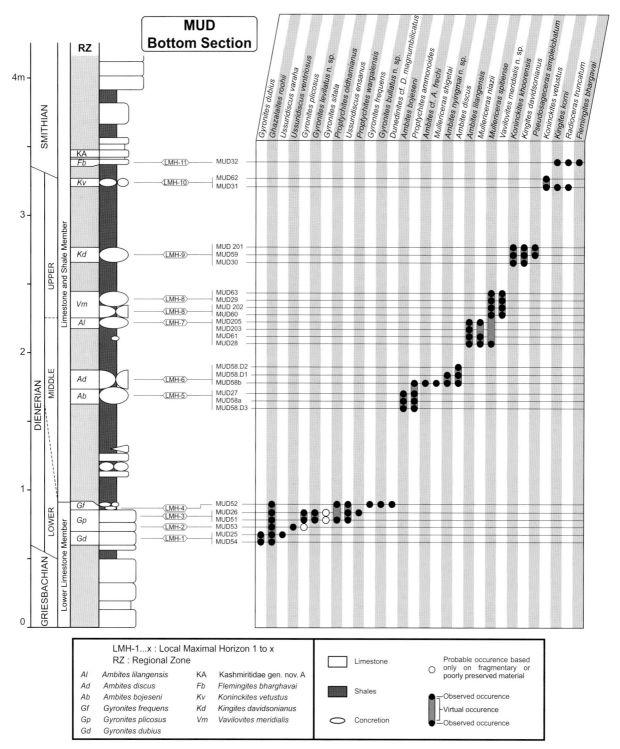

Fig. 3. Section Mud Bottom: lithostratigraphy, ammonoid occurrences and biostratigraphy.

these two sections could not be directly correlated on the basis of lithology, thus preventing
the construction of a composite section. These
two sections are situated at ca. 3840 m elevation
(coordinates: N32°09′05.1″; E78°14′37.4″ and
N32°09′04.8″; E78°14′38.1″). The stratigraphic

logs and ammonoid distribution are given in Figures 11, 12. In the Ridge 1 section, the topmost
bed of the Lower Limestone Member is a condensed horizon, containing a mixed assemblage
of the *Gyronites frequens* and *A. atavus* associations, like in the Guling Village section.

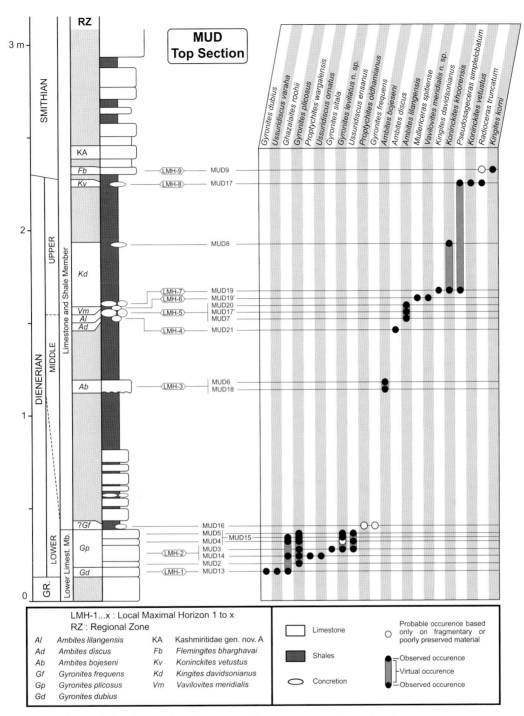

Fig. 4. Composite section Mud Top: lithostratigraphy, ammonoid occurrences and biostratigraphy.

Biostratigraphy

The very detailed bed-rock-controlled sampling on which this study is based allowed us establishing a succession of 10 different regional zones for the Spiti District (Figs 13, 14). These regional zones are in very good agreement with those described in the Salt Range (Ware *et al.* 2018).

Only two regional zones first identified in the Salt Range have not been subsequently recognized in Spiti. The correlation of the regional zones of the Salt Range with those of the Spiti district is straightforward. This new biostratigraphical scheme (Fig. 13) considerably improves the biostratigraphic resolution of the Dienerian part of the succession in Spiti, which was so far

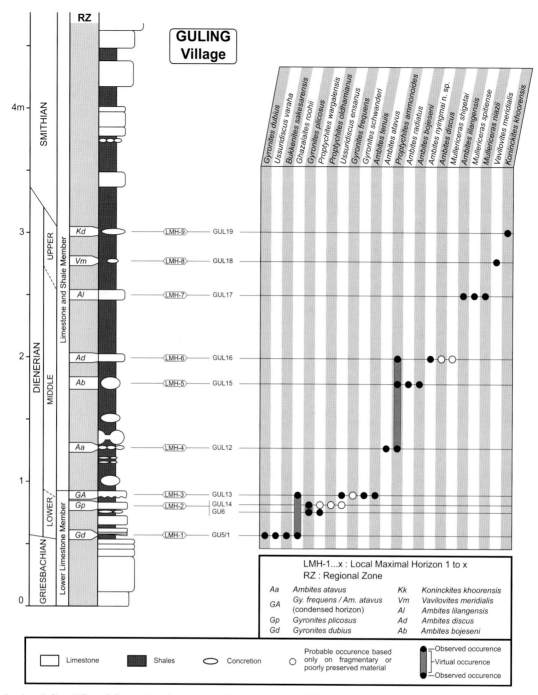

Fig. 5. Section Guling Village: lithostratigraphy, ammonoid occurrences and biostratigraphy.

divided into five zones (Krystyn *et al.* 2004, 2007a; Brühwiler *et al.* 2010). Most of the regional zones described here directly correspond to local maximal horizons, with only two exceptions: the *Gyronites plicosus* Regional Zone and the *Kingites davidsonianus* Regional Zone (see below).

The comparison of this biostratigraphical scheme with those of other regions has been given in the revision of the Dienerian of Salt Range (Ware *et al.* 2018). Figures 15, 16 show the ranges of Dienerian ammonoid species and genera from the Spiti district. For each regional zone, the list of co-occurring species is given with the list of characteristic species and pairs of species in braces. Additionally, the number of specimens of each species in each zone is given in brackets.

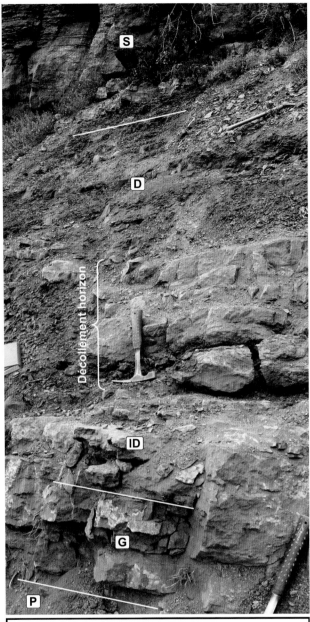

Fig. 6. Photo of the section Guling River. Note the décollement horizon, with a thick layer of irregular limestone beds replaced laterally by shales.

Early Dienerian ammonoid faunas

Early Dienerian faunas are restricted to the upper part of the Lower Limestone Member. Each regional zone is characterized and largely dominated by a different species of *Gyronites*. In this interval, the sedimentation rate is very low, so the stratigraphic record is occasionally affected by hiatuses and condensation. Indeed, the *Gyronites dubius* Regional Zone is missing in the section Guling River, where beds belonging to the *Gyronites plicosus* Regional Zone rest directly onto beds of Griesbachian age. Also, in two sections (Guling River and Lalung Ridge 1), the topmost bed of the Lower Limestone Member is condensed, including a mixture of species characteristic of the uppermost lower Dienerian *Gyronites frequens* Regional Zone and of the lowermost middle Dienerian *Ambites atavus* Regional Zone.

Gyronites dubius Regional Zone

Co-occurring species. – {*Bukkenites sakesarensis* (*n* = 4)}, *Ghazalaites roohii* (*n* = 7), {*Gyronites dubius* (*n* = 18)}, {*Ussuridiscus varaha* (*n* = 3)}.

Occurrence in the investigated sections

This regional zone has been recognized in most of the investigated sections. However, it is missing in Guling River. In Lalung Cliff and Lalung Ridge 2, the beds that might correspond stratigraphically to the zone contained only poorly preserved ammonoids that could not be identified - even at the family level.

Distribution

The very same faunal association has been found in Amb in the Salt Range (Ware *et al.* 2018). It corresponds to UA-zone DI-1 (Ware *et al.* 2015).

Gyronites plicosus Regional Zone

Co-occuring species

Ghazalaites roohii (*n* = 25), {*Gyronites levilatus* (*n* = 21)}, {*Gyronites plicosus* (*n* = 147)}, {*Gyronites sitala* (*n* = 13)}, *Proptychites oldhamianus* (*n* = 5), {*Proptychites wargalensis* (*n* = 4)}, *Ussuridiscus ensanus* (*n* = 13), {*Ussuridiscus ornatus* (*n* = 4)}, {*Ussuridiscus ventriosus* (*n* = 1)}.

Occurrence in the investigated sections

This regional zone has been recognized in every section investigated here.

Distribution

The same regional zone has been recognized in the Salt Range, except for *Gyronites levilatus*, which has so far only been found in Spiti. It corresponds to UA-zone DI-2 (Ware *et al.* 2015).

Remarks

This association corresponds to two 'Local Maximal Horizons' in the Mud Top (LMH-2 and LMH-3, fig. 4) and Mud Bottom (LMH-2 and LMH-3, fig. 3) sections. However, these 'Local Maximal Horizons'

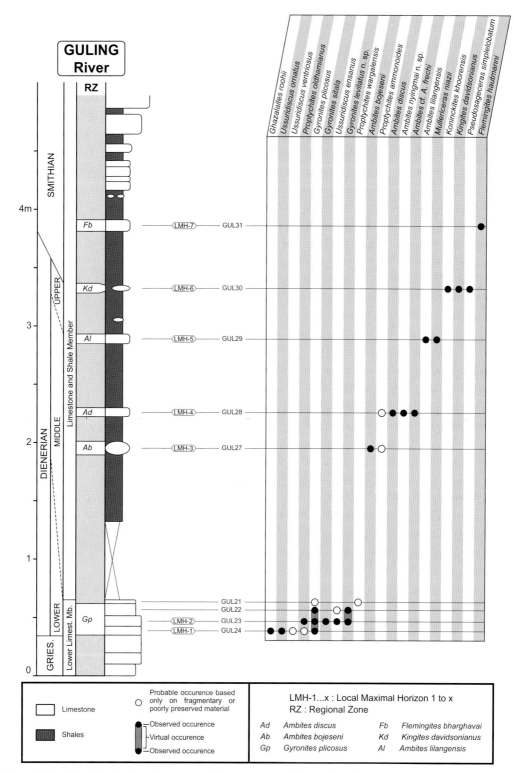

Fig. 7. Section Guling River: lithostratigraphy, ammonoid occurrences and biostratigraphy.

are only differentiated by rare species represented in the collections only by a few specimens. Hence, they are presently lumped into one regional zone. With nine species, it is the most diverse Dienerian fauna in the Spiti District.

Gyronites frequens Regional Zone

Co-occurring species
{*Dunedinites* cf. *D. magnumbilicatus* (*n* = 1)}, *Ghazalaites roohii* (*n* = 4), {*Gyronites frequens*

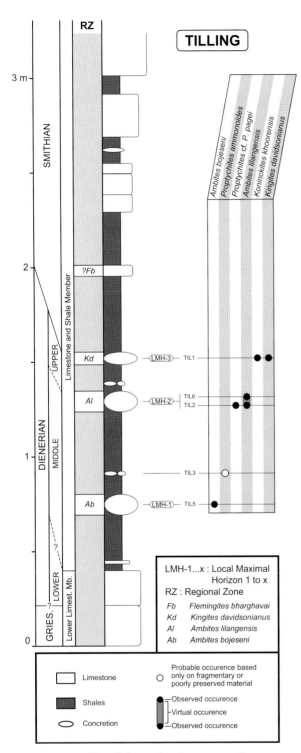

Fig. 8. Section Tilling: lithostratigraphy, ammonoid occurrences and biostratigraphy.

(*n* = 140)}, {*Gyronites schwanderi* (*n* = 1)}, {*Gyronites bullatus* (*n* = 2)}, *Proptychites oldhamianus* (*n* = 6) *Ussuridiscus ensanus* (*n* = 21).

Occurrence in the investigated sections

This regional zone corresponds in every section to the topmost bed(s) of the Lower Limestone Member, except in Guling River, where the section is interrupted by a fault breccia. In Guling Village (LMH-3, fig. 5) and Lalung Ridge 1 (LMH-3, fig. 11), it is condensed and mixed with the earliest middle Dienerian fauna. The rare *Gyronites bullatus* and *Dunedinites* cf. *D. magnumbilicatus* have been recorded from a lens with very well preserved ammonoids at the top of the Lower Limestone Member in the Mud Bottom section.

Distribution

This regional zone is an obvious correlative of the *Gyronites frequens* Regional Zone of the Salt Range, Pakistan irrespective of the two rare species mentioned above. It corresponds to UA-zone DI-3 (Ware *et al.* 2015).

Middle Dienerian ammonoid faunas

Middle Dienerian ammonoid faunas occur in the lower half of the shale interval at the base of the Limestone and Shale Member. Each zone is exclusively characterized by a different species of *Ambites*, each of these being very abundant. Many very well preserved and complete specimens were found in the early diagenetic limestone concretions of this interval.

Specimens derived from tempestitic layers are not as well preserved as they are often broken and distorted. As indicated above, the base of this interval is occasionally condensed and may locally include the uppermost part of the lower Dienerian.

Ambites atavus Regional Zone

Co-occurring species

{*Ambites atavus* (*n* = 36)}, {*Ambites tenuis* (*n* = 1)}, *Mullericeras shigetai* (*n* = 8), *Proptychites ammonoides* (*n* = 6).

Occurrence in the investigated sections

This regional zone has only been recognized in Lalung and in Guling Village sections. The corresponding interval in Mud is represented by shale and limestone beds in which only unidentifiable ammonoids were found. In Guling Village and Lalung Ridge 1, it is condensed with the *Gyronites frequens* Regional Zone.

Fig. 9. Photo of the section Lalung Cliff. Note the abundance of limestone beds compared with the other sections. The Griesbachian–Dienerian Boundary is here not well defined as no Griesbachian or earliest Dienerian faunas were found. Hence, it is based on the lithological change only.

Distribution
This regional zone is exactly the same as that recognized in the Salt Range, Pakistan. It corresponds to UA-zone DI-4 (Ware *et al.* 2015).

Ambites bojeseni Regional Zone

Co-occuring species
{*Ambites bojeseni* (*n* = 83)}, {*Ambites radiatus* (*n* = 2)}, *Mullericeras shigetai* (*n* = 0, virtual occurrence only), *Proptychites ammonoides* (*n* = 18).

Occurrence in the investigated sections
This regional zone has been recognized in all sections except Lalung Cliff.

Correlation
Sharing exactly the same association of species, it obviously correlates with the *Ambites radiatus* regional Zone in the Salt Range, Pakistan. The only difference is that *Ambites radiatus* is the most abundant species in the Salt Range. It corresponds to UA-zone DI-5 (Ware *et al.* 2015).

Remarks
No specimen of *Mullericeras shigetai* could be recovered from these beds, so this species is only virtually present (*i.e.* present in the underlying and overlying beds) in this regional zone.

Ambites discus Regional Zone

Co-occuring species
{*Ambites discus* (*n* = 72)}, {*Ambites* cf. *A. frechi* (*n* = 2)}, {*Ambites nyingmai* (*n* = 54)}, {*Ambites subradiatus* (*n* = 1)}, *Mullericeras shigetai* (*n* = 4), *Proptychites ammonoides* (*n* = 10).

Occurrence in the investigated sections
This regional zone has been recognized in all sections except Tilling.

Correlation
It corresponds exactly to the *Ambites discus* Regional Zone in the Salt Range, Pakistan, with the same association of characteristic species, with one additional new species (*Ambites nyingmai*). It corresponds to UA-zone DI-6 (Ware *et al.* 2015).

Remarks
Ambites cf. *A. frechi* is represented by two immature specimens which identification remains uncertain. Thus, a direct correlation with the fauna from British Columbia where this species was originally described remains doubtful.

Ambites lilangensis Regional Zone

Co-occuring species
{*Ambites lilangensis* (*n* = 137)}, {*Mullericeras niazii* (*n* = 14)}, *Mullericeras spitiense* (*n* = 3), {*Proptychites* cf. *P. pagei* (*n* = 1)}.

Occurrence in the investigated sections
This regional zone has been recognized in all investigated sections.

Distribution
This is the lateral correlative to the *Ambites lilangensis* Regional Zone in Salt Range. It corresponds to UA-zone DI-8 (Ware *et al.* 2015).

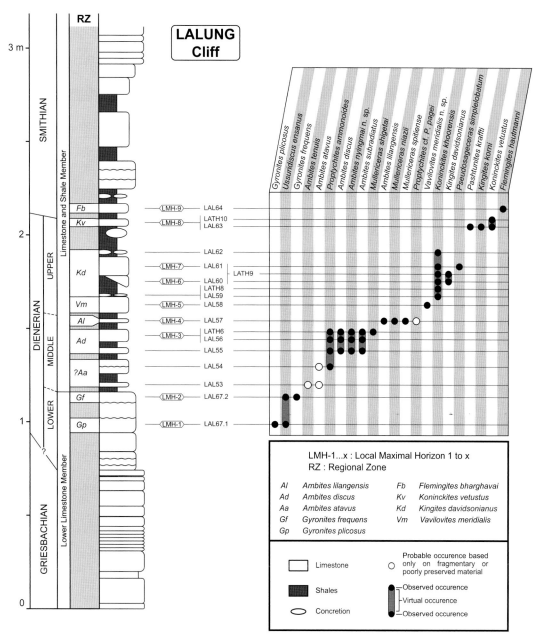

Fig. 10. Section Lalung Cliff: lithostratigraphy, ammonoid occurrences and biostratigraphy.

Remarks

Proptychites lawrencianus is quite abundant in the Salt Range, but is here represented only by one specimen from a float block in Lalung. It probably derived from these beds as this species is restricted to this zone in the Salt Range.

Late Dienerian ammonoid faunas

The late Dienerian faunas are distributed throughout the upper half of the shale interval at the base of the Limestone and Shale Member. Each regional zone is characterized by a different paranoritid. Large numbers of very well preserved and complete specimens were found in the concretions of this interval. Ammonoids from tempestitic layers are often fragmentary and distorted. The youngest fauna of the Dienerian has been partly described by Brühwiler *et al.* (2010).

Vavilovites meridialis Regional Zone

Co-occuring species
Mullericeras spitiense (*n* = 9), {*Vavilovites meridialis* (*n* = 47)}.

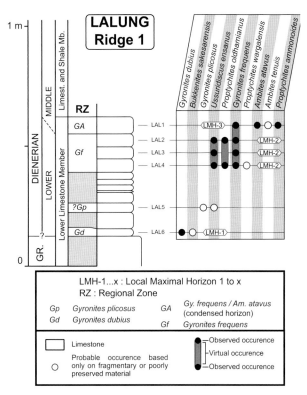

Fig. 11. Section Lalung Ridge 1: lithostratigraphy, ammonoid occurrences and biostratigraphy.

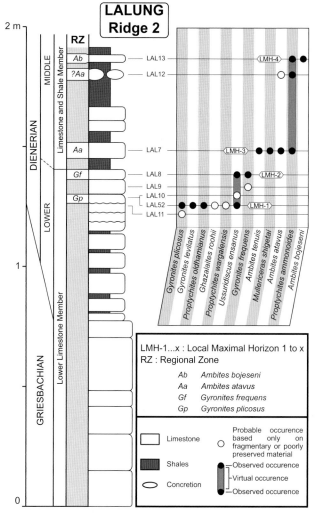

Fig. 12. Section Lalung Ridge 2: lithostratigraphy, ammonoid occurrences and biostratigraphy.

Occurrence in the investigated sections
Except for Tilling and Guling River sections this regional zone has been recognized in all sections.

Distribution
This regional zone corresponds to the same association as that of the *Vavilovites* cf. *V. sverdrupi* Regional Zone of Ware *et al.* (2018) in the Salt Range, which is here renamed *Vavilovites meridialis* regional zone. In the Salt Range, the species *Koninckites khoorensis* and *Pseudosageceras simplelobatum* co-occur with the two mentioned species, but have not yet been documented from this regional zone in Spiti. It corresponds to UA-zone DI-9 (Ware *et al.* 2015).

Kingites davidsonianus Regional Zone

Co-occurring species
{*Kingites davidsonianus* (n = 64)}, {*Koninckites khoorensis* (n = 219)}, *Pseudosageceras simplelobatum* (n = 8).

Occurrence in the investigated sections
This regional zone has been recognized in all investigated section.

Correlation
The same regional zone has been found in the Salt Range, where it also contains *Clypites typicus* and

Subacerites friski. It corresponds to UA-zone DI-10 (Ware *et al.* 2015).

Remarks
This zone corresponds to two 'Local Maximal Horizons' in Lalung Cliff (LMH-6 and LMH-7, fig. 10), where *Kingites davidsonianus* and *Pseudosageceras simplelobatum* occur in two successive beds. These 'Local Maximal Horizons' are here lumped together as these two species are known to occur together in other sections. In Spiti, *Koninckites khoorensis* occurs exclusively in this zone. However, it ranges through two zones in the Salt Range (see Ware *et al.* 2018). Hence *Kingites davidsonianus* provides the best index name for this zone.

Koninckites vetustus Regional Zone

Co-occurring species
{*Kingites korni* (n = 3)}, {*Koninckites vetustus* (n = 26)}, {*Pashtunites kraffti* (n = 6)},

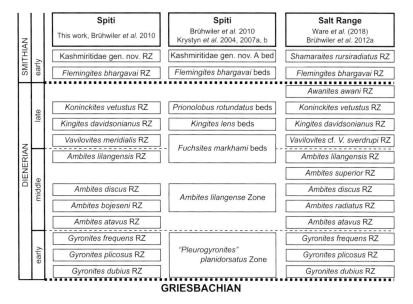

		Spiti This work, Brühwiler *et al.* 2010	Spiti Brühwiler *et al.* 2010 Krystyn *et al.* 2004, 2007a, b	Salt Range Ware *et al.* (2018) Brühwiler *et al.* 2012a
SMITHIAN	early	Kashmiritidae gen. nov. RZ	Kashmiritidae gen. nov. A bed	*Shamaraites rursiradiatus* RZ
		Flemingites bhargavai RZ	*Flemingites bhargavai* beds	*Flemingites bhargavai* RZ
DIENERIAN	late			*Awanites awani* RZ
		Koninckites vetustus RZ	*Prionolobus rotundatus* beds	*Koninckites vetustus* RZ
		Kingites davidsonianus RZ	*Kingites lens* beds	*Kingites davidsonianus* RZ
		Vavilovites meridialis RZ	*Fuchsites markhami* beds	*Vavilovites* cf. *V. sverdrupi* RZ
		Ambites lilangensis RZ		*Ambites lilangensis* RZ
	middle			*Ambites superior* RZ
		Ambites discus RZ	*Ambites lilangense* Zone	*Ambites discus* RZ
		Ambites bojeseni RZ		*Ambites radiatus* RZ
		Ambites atavus RZ		*Ambites atavus* RZ
	early	*Gyronites frequens* RZ	"*Pleurogyronites*" *planidorsatus* Zone	*Gyronites frequens* RZ
		Gyronites plicosus RZ		*Gyronites plicosus* RZ
		Gyronites dubius RZ		*Gyronites dubius* RZ

GRIESBACHIAN

Fig. 13. Biostratigraphic subdivisions of the Dienerian and earliest Smithian of Spiti and correlation with the previously established zonation for this area and the zonation established for the Salt Range (RZ: regional zone).

Pseudosageceras simplelobatum ($n = 2$), *Radioceras truncatum* ($n = 6$).

Occurrence in the investigated sections
This regional zone has been recognized in the sections near Mud and Lalung, but has not been recovered from Guling. It corresponds to UA-zone DI-11 (Ware *et al.* 2015).

Distribution
The same faunal association has been found in the Salt Range, with the two additional species *Clypites typicus* and *Koiloceras sahibi*.

Conclusion

The high-resolution bed-rock controlled sampling of different sections near Mud, Guling and Lalung yielded abundant and well-preserved Dienerian ammonoids. Their comparison with the newly established taxonomy and biostratigraphic scheme of the Salt Range (Ware *et al.* 2018) enabled us to recognize 10 out of the 12 different ammonoid regional zones defined in the Salt Range, each with almost identical species contents. The only differences concern a few rare taxa and relative abundance of the different species. The similarity of faunas of the Salt Range and Spiti is very remarkable, and shows that the succession of faunal associations established in the Salt Range is laterally reproducible. The correlation of this new biozonation with those of other regions is discussed in detail in Ware *et al.* (2015, 2018).

Systematic palaeontology

By Ware, D. *and* Bucher, H.

Systematic descriptions follow the principles and classification established for Dienerian ammonoids of the Salt Range by Ware & Bucher (2018b). Many species were described in detail in this work and their description is not repeated here. Only some remarks are made when the material from Spiti allows us to give some more details concerning the morphology. Intraspecific variability of the shell proportions is quantified by the four classic geometrical parameters of the ammonoid shell: diameter (D), whorl height (H), whorl width (W) and umbilical diameter (U). For new species, provided that at least 4 specimens were measurable, the three last parameters (H, W and U) are plotted against the diameter in absolute values as well as relative to the diameter (H/D, W/D and U/D). For all the species already described by Ware & Bucher (2018b), only the relative values are plotted and compared with the populations from the Salt Range. Measurements of species for which <4 measurable specimens are available are given in Table 1. Sample numbers are reported on the stratigraphic sections (Figs 3–5, 7–8, 10–12). Synonymy lists and taxa in open nomenclature are annotated following the recommendations of Matthews (1973) and Bengtson (1988).

Repository. – All figured specimens are curated in the collections of the Paläontologisches Institut und Museum der Universität Zürich (abbreviated PIMUZ).

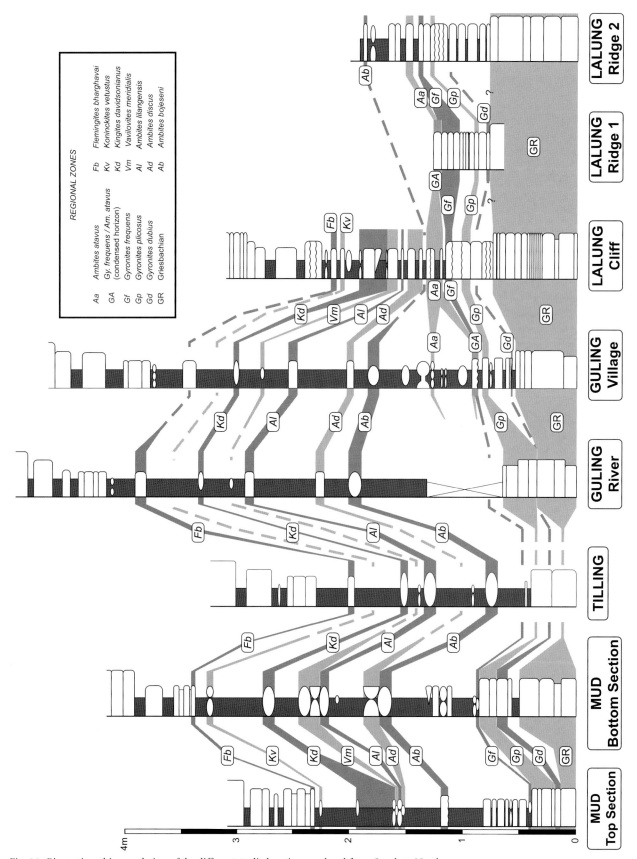

Fig. 14. Biostratigraphic correlation of the different studied sections, ordered from South to North.

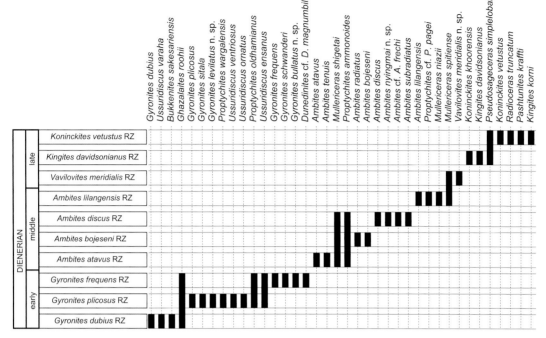

Fig. 15. Range chart showing the biostratigraphical distribution of Dienerian ammonoid species in Spiti (RZ: regional zone).

Class Cephalopoda Cuvier, 1797

Subclass Ammonoidea Agassiz, 1847

Order Ceratitida Hyatt, 1884

Superfamily Meekocerataceae Waagen, 1895

Family Ophiceratidae Arthaber, 1911

Genus *Ghazalaites* Ware & Bucher, 2018b

Type species. – Ghazalaites roohii Ware & Bucher, 2018b.

Ghazalaites roohii Ware & Bucher, 2018b

Plate 1, figures 1–7; Figure 17

v 2018b *Ghazalaites roohii* n. gen., n. sp. Ware & Bucher, pp. 38–39, fig. 15, pl. 2, figs 12–14, 15–17 (holotype), 18–24 (*cum syn.*).

Material. – 36 specimens.

Measurements. – See Figure 17.

Occurrence. – Guling River, sample Gul24 ($n = 1$); Guling Village, samples GU5/1 ($n = 1$) and Gul13 ($n = 1$); Mud Bottom Section, samples Mud25 ($n = 2$), Mud26 ($n = 4$), Mud51 ($n = 1$), Mud52 ($n = 4$), Mud53 ($n = 3$) and Mud54 ($n = 1$); Mud Top Section, samples Mud3 ($n = 1$), Mud4 ($n = 3$), Mud5 ($n = 1$), Mud13 ($n = 3$), Mud14 ($n = 9$) and Mud15 ($n = 1$). Some poorly preserved specimens from sample Lal52 from Lalung Ridge 2 are tentatively assigned to this species.

Family Gyronitidae Waagen, 1895

Genus *Gyronites* Waagen, 1895

Type species. – Gyronites frequens Waagen, 1895.

Gyronites frequens Waagen, 1895

Plate 1, figures 8–15; Figure 18

1895 *Gyronites frequens* n. gen., n. sp. Waagen, pp. 292–294, pl. 38, figs 1, 2 (lectotype), 3, 4, pl. 40, fig. 4.

v 2018b *Gyronites frequens* Waagen, 1895; Ware & Bucher, pp. 41–44, figs 16–18, pl. 3, figs 1–5, pl. 4, figs 1–17 (*cum syn.*).

Material. – 53 specimens.

Measurements. – See Figure 18.

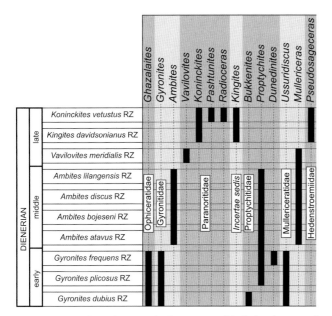

Fig. 16. Range chart showing the biostratigraphical distribution of Dienerian ammonoid genera (grouped by families) in Spiti (RZ: regional zone).

Occurrence. – Lalung Cliff, sample Lal67.2 ($n = 5$); Lalung Ridge 1, samples Lal1 ($n = 3$), Lal2 ($n = 13$), Lal3 ($n = 6$) and Lal4 ($n = 6$); Lalung Ridge 2, sample Lal8 ($n = 8$); Mud Bottom Section, sample Mud52 ($n = 12$). Some poorly preserved specimens from samples Mud16 from Mud Top Section, Gul13 from Guling Village and Lal9 from Lalung Ridge 2 are tentatively assigned to this species.

Gyronites dubius (von Krafft, 1909)

Plate 1, figures 16–18

1909   *Meekoceras dubium* n. sp. von Krafft , pp. 50, 51, pl. 24, figs 6–10, 11 (lectotype), 12–14.

v 2018b   *Gyronites dubius* von Krafft, 1909; Ware & Bucher, pp. 44–45, figs 16, 19, pl. 4, figs 18–35 (*cum syn.*).

Material. – 18 specimens.

Measurements. – See Table 1.

Occurrence. – Guling Village, sample GU5/1 ($n = 1$); Lalung Ridge 1, sample Lal6 ($n = 1$); Mud Bottom Section, samples Mud25 ($n = 7$) and Mud54 ($n = 2$); Mud Top Section, sample Mud13 ($n = 7$).

Gyronites plicosus Waagen, 1895;

Plate 2, figures 1–9

1895   *Gyronites plicosus* n. gen., n. sp. Waagen, pp. 298–300, pl. 38, fig. 11 (holotype).

v 2018b   *Gyronites plicosus* Waagen, 1895; Ware & Bucher, pp. 46–48, figs 16, 20, 21, pl. 5, figs 4–28 (*cum syn.*).

Material. – 147 specimens.

Measurements. – See Table 1.

Occurrence. – Guling River, samples Gul22 ($n = 7$) and Gul23 ($n = 17$) Gul24 ($n = 9$); Guling Village, samples GU6 ($n = 1$) and Gul14 ($n = 1$); Lalung Cliff, sample Lal52 ($n = 10$); Lalung Ridge2, sample Lal67.1 ($n = 5$); Mud Bottom Section, samples Mud26 ($n = 26$), Mud51 ($n = 8$); Mud Top Section, samples Mud2 ($n = 2$), Mud3 ($n = 29$), Mud4 ($n = 9$), Mud5 ($n = 3$), Mud14 ($n = 6$), Mud15 ($n = 14$). Poorly preserved specimens from sample Mud53 from Mud Bottom Section, Gul21 from Guling River, Lal5 from Lalung Ridge 1 and Lal11 from Lalung Ridge 2 are tentatively assigned to this species.

Gyronites sitala (Diener, 1897)

Plate 2, figures 10–12

1897   *Danubites sitala* n. sp. Diener, pp. 49, 50, pl. 15, figs 12 (lectotype), 13.

v 2018b   *Gyronites sitala* (Diener, 1897); Ware & Bucher, p. 48, fig. 16, pl. 5, figs 29–33.

Material. – 12 specimens.

Measurements. – No measurable specimen available.

Occurrence. – Guling River, sample Gul23 ($n = 7$); Mud Top section, sample Mud3 ($n = 5$). Some poorly preserved specimens from sample Mud26 and Mud51 from Mud Bottom section are tentatively assigned to this species.

Gyronites levilatus n. sp.

Plate 2, figures 13–21

Table 1. Measurements of Dienerian ammonoids from Spiti for which <4 specimens were measurable.

Genus	Species	Specimen number	Section	Age	D	H	W	U
Gyronites	*dubius*	PIMUZ30835	Mud Bottom	Gd	26.5	8.9	6.4	11.5
Gyronites	*plicosus*	PIMUZ30836	Mud Bottom	Gp	28.5	9.2	6.2	12.4
Gyronites	*plicosus*	PIMUZ30837	Mud Bottom	Gp	26	7.7	5.4	12.8
Gyronites	*plicosus*	PIMUZ30838	Mud Bottom	Gp	31.1	9.9	7	13.4
Gyronites	*levilatus*	PIMUZ30842	Mud Top	Gp	28.8	12.6	7.2	8.4
Gyronites	*levilatus*	PIMUZ30845	Mud Bottom	Gp	41.1	16.7	7.6	12.2
Gyronites	*bullatus*	PIMUZ30847	Mud Bottom	Gf	10.6	3.7	2.8	4.4
Ambites	*radiatus*	PIMUZ30865	Guling Village	Ab	27.4	11.9	6.7	7.3
Ambites	cf.*frechi*	PIMUZ30872	Mud Bottom	Ad	25.6	10.8	6	7.8
Proptychites	*oldhamianus*	PIMUZ30904	Mud Bottom	Gf	32.8	15.1	12.9	7.2
Proptychites	*oldhamianus*	PIMUZ30903	Mud Bottom	Gf	56.8	28.9	18.9	8.9
Dunedinites	cf. *magnumbilicatus*	PIMUZ30913	Mud Bottom	Gf	17.3	7.3	11.1	5.8
Mullericeras	*spitiense*	PIMUZ30917	Mud Bottom	Al	33.9	20.3	8.6	0
Mullericeras	*spitiense*	PIMUZ30916	Mud Bottom	Vm	56.8	33.9	14.3	0
Mullericeras	*spitiense*	PIMUZ30918	Mud Bottom	Vm	53.8	32.9	15	0
Mullericeras	*shigetai*	PIMUZ30919	Mud Bottom	Ad	25.9	13.8	6.8	3.8
Mullericeras	*shigetai*	PIMUZ30920	Mud Bottom	Ad	31.1	17	7.3	3.3
Kingites	*korni*	PIMUZ27863	Mud Bottom	Kv	58.3	35.1	15.3	0
Kingites	*korni*	PIMUZ27864	Mud Bottom	Kv	32.9	20.6	7.7	0

Abbreviations: D, Diameter; H, whorl height; W, whorl width; U, umbilical diameter.
For the age abbreviations see Figure 14.

Derivation of name. – From the Latin *levis*, meaning smooth, and *latus*, meaning flanks.

Holotype. – Specimen PIMUZ30844 (Pl. 2, figs 19–21).

Type locality. – Mud Top Section, Himachal Pradesh, India.

Type horizon. – Sample Mud3, ca. 20 cm above base of the Lower Limestone Member, *Gyronites plicosus* Regional Zone, lower Dienerian.

Diagnosis. – Compressed *Gyronites* with very shallow umbilicus and smooth flanks.

Material. – 21 specimens.

Description. – Moderately evolute (U/D ≈ 29%) and compressed (W/D ≈ 22%, W/H ≈ 51%) platyconic shell with a broad tabulate venter. Flanks slightly convex, with maximum width at mid-flanks. The flanks bend suddenly just before the umbilical seam, joining the previous whorl with an obtuse angle, and without forming any differentiated umbilical wall. The flanks are perfectly smooth. Suture line typical of *Gyronites*, with a narrow ventral lobe, broad lateral lobes and saddles, the lobes having few small indentations at their base and a very short auxiliary series.

Measurements. – See Table 1.

Discussion. – This species is very similar to *Gyronites frequens*, which was originally diagnosed as nonornamented. However, *Gyronites frequens* has been documented by Ware & Bucher (2018b) to be ribbed, sometimes even rather strongly. The original material of *Gyronites frequens* described by Waagen (1895) appears to be smooth because of strong weathering. *Gyronites levilatus* is also more involute than *Gyronites frequens*, and its indistinct umbilical wall is a unique trait among all representative of this genus. Some species of *Ambites* look superficially close to the species here described, but they differ clearly by their bottleneck shaped venter, the presence of ornamentation on the flanks, and their suture line which has a broader ventral lobe and a longer auxiliary series.

Occurrence. – Guling River, samples Gul22 (*n* = 1) and Gul23 (*n* = 2); Lalung Ridge 2, sample Lal52 (*n* = 2); Mud Bottom Section, samples Mud26 (*n* = 6) and Mud51 (*n* = 6); Mud Top Section, samples Mud3 (*n* = 1), Mud5 (*n* = 1) and Mud15 (*n* = 2). Some poorly preserved specimens from sample Mud4 of Mud Top Section are tentatively assigned to this species.

Gyronites bullatus n. sp.

Plate 2, figures 22–28

Derivation of name. – Refers to the umbilical bullae on inner whorls.

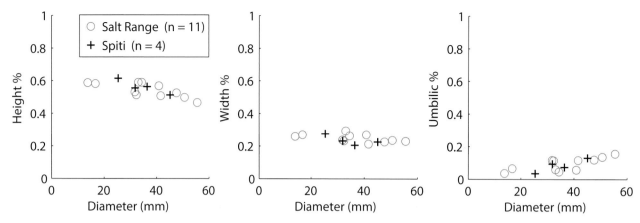

Fig. 17. Ghazalaites roohii Ware & Bucher 2018b. Scatter diagrams of H/D, W/D, and U/D (D = conch diameter, H = whorl height, W = whorl width, U = umbilical diameter).

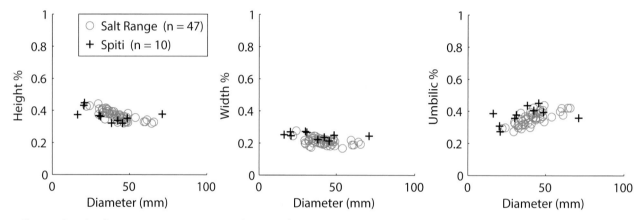

Fig. 18. Gyronites frequens Waagen, 1895. Scatter diagrams of H/D, W/D, and U/D (abbreviations as in Fig. 17).

Holotype. – Specimen PIMUZ30846 (Pl. 2, figs 22–25).

Type locality. – Mud Bottom Section (section M05 of Krystyn *et al.* 2007a), Himachal Pradesh, India.

Type horizon. – Sample Mud 52, top of the Lower Limestone Member, *Gyronites frequens* regional Zone, lower Dienerian.

Diagnosis. – *Gyronites* with very evolute inner whorls bearing distant umbilical bullae which are replaced by weak sub-radial ribs on the more compressed adult body chamber.

Material. – Two specimens.

Description. – Evolute sub-platyconic shell characterized by a pronounced change of morphology at maturity. Inner whorls very evolute (U/D = 42%) with a slightly compressed whorl section (W/H = 76%). Flanks flat and slightly convergent towards the tabulate venter, imparting the whorl

section a sub-trapezoidal shape. Umbilical wall rather high and vertical, grading into a narrowly rounded umbilical shoulder. Flanks with distant umbilical bullae. On the body chamber, the whorl section becomes slightly more compressed (W/H = 70%) with slightly convex sub-parallel flanks and a slightly oblique umbilical wall. The umbilical bullae are replaced by more closely spaced, slightly sigmoid, weak, blunt ribs. Due to the incompleteness of the holotype, the transition between the juvenile and adult stage could not be observed. The suture line could only be observed on the juvenile part of the shell. It is typical of *Gyronites* with a second lateral saddle larger than the others, lobes with few small indentations at their base and no auxiliary series.

Measurements. – See Table 1.

Discussion. – This species is characterized by the presence of umbilical bullae in its inner whorls. The absence of auxiliary series is due to the small size at which the suture line could be observed.

At larger size, it is probable that a short auxiliary series with a few small indentations should appear. The inner whorls resemble robust variants of *Gyronites dubius*, but the latter differs by its more trapezoidal whorl section with a higher, better individualized umbilical wall. The body chamber resembles those of *Gyronites frequens* and of smooth variants of *Gyronites plicosus*, but the inner whorls clearly differ from those of these two species.

Occurrence. – Mud Bottom section, sample Mud52 ($n = 2$).

Gyronites schwanderi Ware & Bucher, 2018b

Plate 2, figures 29, 30

v 2018b *Gyronites schwanderi* n. sp. Ware & Bucher, pp. 48–49, fig. 16, pl. 5, figs 34, 35, 36–38 (holotype).

Material. – Four specimens.

Measurements. – No measurable specimen available.

Occurrence. – Guling Village, sample Gul13 ($n = 4$).

Genus *Ambites* Waagen, 1895

Type species. – *Ambites discus* Waagen, 1895.

Discussion. – This genus has been revised in detail by Ware & Bucher (2018b), but the very fine preservation of the shell on some specimens from Spiti allows the addition of the two following observations: (1) very thin sinuous lirae following the trajectory of the growth lines occur on the flanks; (2) the protruding ventro-lateral shoulders are underlined by a very fine strigation. Additionally, when some bivalves encrust the flanks of individuals belonging to this genus, they cast a fine strigation, which is generally not preserved on the shell (for example, see Pl. 4, fig. 3 and Pl. 6, fig. 2). This feature has not been observed in other genera, even when the specimens have encrusting bivalves on their flank.

Ambites discus Waagen, 1895

Plate 3, figures 1–9; Figure 19

1895 *Ambites discus* n. sp. Waagen, pp. 152–154, pl. 21, figs 4, 5 (lectotype).

v 2018b *Ambites discus* Waagen, 1895; Ware & Bucher, pp. 51–53, figs. 22–24, pl. 6, figs 1–31, pl. 7, figs 1–17 (*cum syn.*).

Material. – 72 specimens.

Measurements. – See Figure 19.

Occurrence. – Guling River, sample Gul28 ($n = 5$); Lalung Cliff, samples Lal55 ($n = 23$), Lal56 ($n = 6$) and Lath6 ($n = 15$); Mud Bottom Section, samples Mud58b ($n = 14$), Mud58.D1 ($n = 3$) and Mud58.D2 ($n = 1$); Mud Top Section, sample Mud21 ($n = 5$). Some additional poorly preserved specimens from sample Gul16 from Guling Village are tentatively assigned to this species.

Ambites atavus (Waagen, 1895)

Plate 3, figures 10–15

1895 *Prionolobus atavus* n. sp. Waagen, pp. 309, 310, pl. 34, fig. 4 (lectotype), pl. 35, fig. 4.

v 2018b *Ambites atavus* (Waagen, 1895); Ware & Bucher, pp. 53–55, figs 22, 24, 25, pl. 8, figs 1–26, pl. 9, figs 1, 2 (*cum syn.*).

Material. – 39 specimens.

Measurements. – No measurable specimen available.

Occurrence. – Guling Village, sample Gul12 ($n = 1$); Lalung Ridge 1, sample Lal1 ($n = 3$); Lalung Ridge2, sample Lal7 ($n = 35$). Some poorly preserved specimens from sample Lal53 and Lal54 from Lalung Cliff are tentatively assigned to this species.

Ambites tenuis Ware & Bucher, 2018b

Plate 2, figures 31–34

v 2018b *Ambites tenuis* n. sp. Ware & Bucher, pp. 55–56, fig. 22, pl. 9, figs 3–7 (holotype), 8–10.

Material. – Two specimens.

Measurements. – No measurable specimen available.

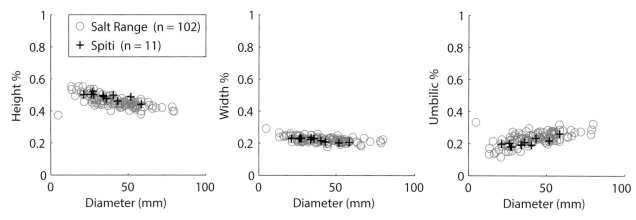

Fig. 19. *Ambites discus* Waagen, 1895. Scatter diagrams of H/D, W/D, and U/D (abbreviations as in Fig. 17).

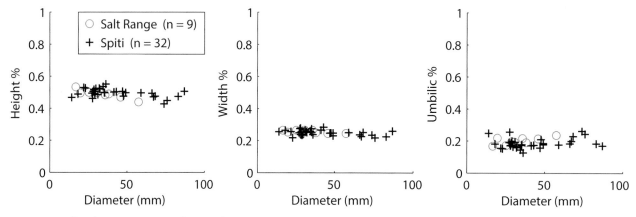

Fig. 20. *Ambites bojeseni* Ware & Bucher, 2018b. Scatter diagrams of H/D, W/D, and U/D (abbreviations as in Fig. 17).

Occurrence. – Guling Village, sample Gul13 (*n* = 1) and Lalung Ridge2, sample Lal7 (*n* = 1). Some poorly preserved specimens from samples Lal53 from Lalung Cliff and Lal1 from Lalung Ridge 1 are tentatively assigned to this species.

Ambites bojeseni Ware & Bucher, 2018b

Plate 3, figures 16–18; Plate 4, figures 1–15; Plate 5, figures 1–4; Figure 20

v 2018b *Ambites bojeseni* n. sp. Ware & Bucher, pp. 57–58, figs 22, 29, pl. 9, figs 29–31, pl. 10, figs 1–3 (holotype), 4–11.

Material. – 83 specimens.

Measurements. – See Figure 20.

Discussion. – Compared to the Salt Range material, more abundant and better preserved specimens were found in Spiti, which allows the addition of the two following points to the description provided by Ware & Bucher (2018b). First, the slight allometry detected for the specimens from the Salt Range was in fact a sample bias. For specimens from Spiti, the relative umbilical size varies between 12% and 26% for diameters above 25 mm (Fig. 20), without any clear trend towards greater evolution with size, except for large adult specimens with umbilical egression of the last half whorl. Second, at large size, the suture line tends to have more numerous indentations at the base of the lobes, and evolute variants have a suture line with considerably shallower lobes and shorter saddles.

This more abundant material also allows us to demonstrate that *Ambites bojeseni* is generally more involute and compressed than *Ambites lilangensis*. At a diameter of more than 25 mm, the means of the different parameters are as follow: for *Ambites bojeseni*, x̄(U/D) = 18.6%, x̄(W/D) = 24.9%, and x̄(W/H) = 49.7%, whereas for *Ambites lilangensis*, x̄(U/D) = 20.1%, x̄(W/D) = 26.2% and x̄(W/H) = 54.4%.

Occurrence. – Guling River, sample Gul27 (*n* = 6); Guling Village, sample Gul15 (*n* = 25); Lalung Ridge2, sample Lal13 (*n* = 2); Mud Bottom Section, samples Mud27 (*n* = 29), Mud58a (*n* = 9) and Mud58.D3 (*n* = 1); Mud Top Section, samples Mud6 (*n* = 3) and Mud18 (*n* = 7); Tilling, sample Til5 (*n* = 1).

Ambites radiatus (Brühwiler, Brayard, Bucher & Guodun, 2008)

Plate 5, figures 5–9

v 2008   *Pleurambites radiatus* n. sp. Brühwiler, Brayard, Bucher & Guodun, p. 1168, pl. 5, figs 1 (holotype), 2, 3.

v 2018b *Ambites radiatus* (Brühwiler, Brayard, Bucher & Guodun, 2008); Ware & Bucher, pp. 56–57, figs 22, 27, 28, pl. 9, figs 11–28.

Material. – Two specimens.

Measurements. – See Table 1.

Discussion. – This species is very rare in Spiti, only two incomplete small specimens were found in Guling.

Occurrence. – Guling Village, sample Gul15 (*n* = 2).

Ambites nyingmai n. sp.

Plate 5, figures 10–23; Figure 21

Derivation of name. – Named after the Nyingma school of Buddhism, whose last few Buchen Lamas still live in the Pin Valley.

Holotype. – Specimen PIMUZ30867 (Pl. 5, figs 10–13).

Type locality. – Mud Bottom section (section M05 of Krystyn *et al.* 2007b), Himachal Pradesh, India.

Type horizon. – Sample Mud 58b, middle part of the shale interval at the base of the Limestone and Shale Member, upper part of bed 8 of Krystyn *et al.* (2007a,b), *Ambites discus* Regional Zone, middle Dienerian.

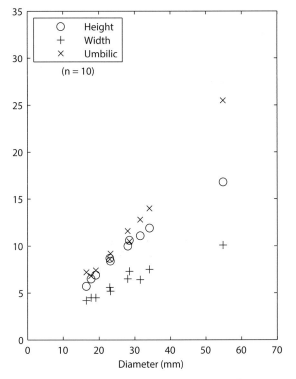

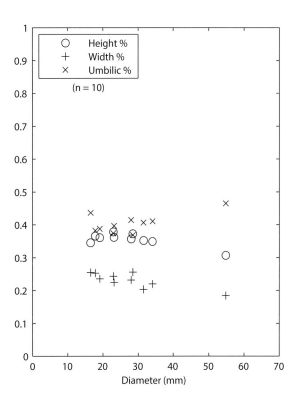

Fig. 21. Ambites nyingmai n. sp. Scatter diagrams of H/D, W/D, and U/D (abbreviations as in Fig. 17).

Diagnosis. – Very evolute and compressed *Ambites* with either no or only faint ornamentation.

Material. – 54 specimens.

Description. – This species is characterized by its very evolute (for D > 25 mm, U/D ≈ 42%), compressed (for D > 25 mm, W/D ≈ 22% and W/H ≈ 63%) and sub-platyconic shape, its flanks being slightly convex with maximal width at inner third. According to the graphs (Fig. 21), the shell tends to become more evolute and compressed with growth, but more measurable specimens would be necessary to test this possible allometry. The bottleneck shaped venter is present but subdued, visible only on the outer shell of well-preserved specimens and absent on the internal mould. The umbilical wall is slightly oblique, poorly individualized and grading into a broadly rounded umbilical shoulder. Some specimens have low and broad radial folds on the flanks while others are nearly smooth, with weak folds parallel to growth lines. The suture line features very deep lobes and elongated saddles, the second lateral saddle being very broad whereas the third one is very small. The auxiliary series is very short with a few small indentations and no differentiated auxiliary lobe.

Measurements. – See Figure 21.

Discussion. – Adult specimens of this species closely resemble evolute variants of *Ambites superior* (Waagen, 1895), from which they differ only by their much more evolute inner whorls. *Ambites nyingmai* is otherwise very distinct from any other species of *Ambites*. It can also be mistaken with smooth evolute variants of *Gyronites frequens*, but its bottleneck shaped venter, although poorly marked, and its more evolute inner whorls clearly differentiate it from this species. The Smithian flemingitid *Rohillites* Waterhouse, 1996 has a very close shell geometry but differs by its typically rursiradiate ribbing.

Occurrence. – Guling River, sample Gul28 (*n* = 2); Guling Village, sample Gul16 (*n* = 6); Lalung Cliff, samples Lal55 (*n* = 9), Lal56 (*n* = 8) and Lath6 (*n* = 10); Mud Bottom section, samples Mud58b (*n* = 16) and Mud58.D1 (*n* = 3).

Ambites subradiatus Ware & Bucher, 2018b

Plate 5, figures 30, 31

v 2018b *Ambites subradiatus* n. sp. Ware & Bucher, pp. 58–59, figs 22, 30, pl. 10, figs 12–14 (holotype), 15–19 (*cum syn.*).

Material. – Three specimens.

Measurements. – No measurable specimen available.

Occurrence. – Lalung Cliff, samples Lal55 (*n* = 1), Lal56 (*n* = 1) and Lath6 (*n* = 1).

Ambites cf. *A. frechi* (Tozer, 1994)

Plate 5, figures 26–29

? 1994 *Pleurambites frechi* n. gen., n. sp. Tozer, p. 68, fig. 14, pl. 13, figs 1, 2, 3 (holotype), 6.

Material. – Two specimens.

Description. – The small complete phragmocone is a moderately evolute (U/D = 30.5%), thick (W/D = 23.4%, W/H = 55.6%), and platyconic shell. Its flanks are very slightly convex with maximal width at inner third of the whorl section. The venter is tabulate with strongly prominent ventro-lateral shoulders. The umbilical wall is vertical and rather high, differentiated by a broadly rounded umbilical shoulder. It is characterized by strong, slightly convex blunt ribs, which fade towards the venter. The last preserved half whorl bears 11 ribs. The suture line has relatively deep lobes and elongated saddles, the second lateral saddle being the largest, slightly bent towards the umbilicus, whereas the third saddle is the smallest, with a rounded tip. The auxiliary series is rather short, with irregularly spaced denticules.

Measurements. – See Table 1.

Discussion. – Tozer (1994) created the genus *Pleurambites* for this ribbed species, but as already discussed by Ware & Bucher (2018b), the presence of ribs in a species is not sufficient to justify its separation from smooth forms by means of another genus. Therefore, *Pleurambites* is here synonymized with *Ambites*. Our specimens are

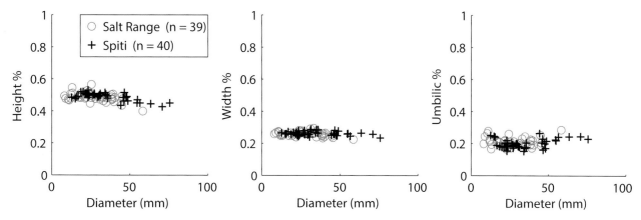

Fig. 22. *Ambites lilangensis* (von Krafft 1909). Scatter diagrams of H/D, W/D, and U/D (abbreviations as in Fig. 17).

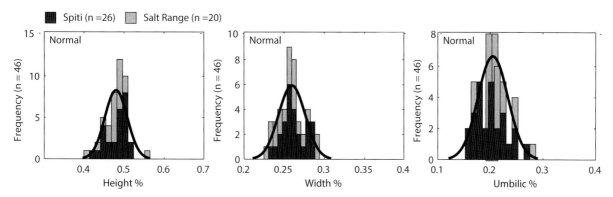

Fig. 23. *Ambites lilangensis* (von Krafft, 1909). Cumulative histograms of H/D, W/D, and U/D (abbreviations as in Fig. 17). Because of growth allometries, specimens of <25 mm in diameters have been excluded from this analysis. The label 'normal' on each histogram indicates that a Lilliefors test performed on the total population (Spiti plus Salt Range) does not allow rejecting the null hypothesis of normality at a confidence interval of 95%.

slightly more involute and compressed than those of Tozer (1994). However, our description is here based on two small phragmocones only, one of them being strongly distorted and poorly preserved. We therefore cannot test whether this difference is due to the size difference and/or intraspecific variability, and thus leave them in open nomenclature.

Occurrence. – Guling River, sample Gul28 (*n* = 1) and Mud Bottom section, sample Mud58b (*n* = 1).

Ambites lilangensis (von Krafft, 1909)

Plate 5, figures 24, 25; Plate 6, figures 1–24;
Figures 22, 23

p 1909   *Meekoceras lilangense* n. sp. von Krafft,
          pp. 23–25, pl. 1, figs 1, 2 (lectotype), 3,
          5–7.

v 2018b   *Ambites lilangensis* von Krafft, 1909; Ware
          & Bucher, pp. 62–63, figs 22, 33, pl. 12, figs
          4–19, pl. 13, figs 1–6 (*cum syn.*).

Emended diagnosis. – Rather involute (U/D ≈ 20%) and thick (W/H ≈ 54%, W/D ≈ 26%) *Ambites* with a very slightly convex venter. Suture line with lobes generally having only a few deep indentations and a poorly differentiated auxiliary lobe.

Material. – 137 specimens.

Description. – A rather complete description of this species has already been provided by Ware & Bucher (2018b) and is not repeated here. However, it was based only on incomplete specimens, and the better preserved specimens found in Spiti allow the addition of the following points. Its growth is only slightly allometric. Most sub-adult specimens (between 25 and 50 mm in diameter) have a relative umbilical diameter of <20%, whereas all specimens of more than 50 mm in diameter have a relative umbilical diameter of more than 20%, showing a slight umbilical egression on the adult body chamber. Instead of being perfectly flat like in other species of *Ambites*, the venter is often very slightly arched, some specimens having even a low blunt median ridge on the venter, imparting a very low roof shape. A few specimens do not have the bottlenecked venter typical of the genus. Most specimens have a more complex suture line than other *Ambites*, with less numerous but deeper indentations and a frequent auxiliary lobe. However, one specimen (Pl. 6, fig. 7) has a simpler suture line, with more numerous and smaller indentations on the lobes, and a narrow ventral lobe with pointed branches. This illustrates the strong variability of the suture line.

Measurements. – See Figures 22, 23.

Discussion. – Von Krafft (1909) indicated that this species was present in five different layers within the 'Meekoceras beds' of Lalung. However, as already discussed by Ware & Bucher (2018b), two of the specimens he illustrated were misidentified, and actually correspond to *Ambites discus*, thus questioning the alleged long stratigraphic distribution of this species. Here, *Ambites lilangensis* only occurs in one layer at each locality, inclusive of Lalung. As mentioned previously, *Ambites lilangensis* and *Ambites bojeseni* are very close to each other, but the former tends to be more evolute and depressed than the latter. This especially applies to complete adult specimens, the relative umbilical diameter of which is always above 20% for *Ambites lilangensis*, whereas it is often (but not always) under 20% for *Ambites bojeseni*. Large and evolute variants of *Ambites bojeseni* are very similar to adult specimens of *Ambites lilangensis*, as the former differs only by its slightly thinner whorl section and venter. This underlines the importance of analysing large enough samples for a clear and consistent taxonomy for these smooth forms.

Occurrence. – Guling River, sample Gul29 ($n = 11$); Guling Village, sample Gul17 ($n = 17$); Lalung Cliff, sample Lal57 ($n = 17$); Mud Bottom Section, samples Mud28 ($n = 25$), Mud61 ($n = 15$), Mud203 ($n = 3$) and Mud205 ($n = 4$); Mud Top Section, samples Mud7 ($n = 11$), Mud17′ ($n = 6$) and Mud20 ($n = 6$); Tilling, samples Til2 ($n = 13$) and Til6 ($n = 9$).

Family Paranoritidae Spath, 1930

Genus *Vavilovites* Tozer, 1971

Type species. – *Paranorites sverdrupi* Tozer, 1963.

Vavilovites meridialis n. sp.

Plate 7, figures 1–24; Plate 8, figures 1–5; Figure 24

   non 1897 *Meekoceras markhami* n. sp. Diener, pp. 75–77, pl. 6, figs 4 (lectotype), 6.

       1901 *Meekoceras noetlingi* n. sp. von Krafft in Griesbach; p. 30.

   v 1901 *Meekoceras noetlingi* von Krafft, 1901; Noetling, p. 466.

   v 1904 *Meekoceras noetlingi* von Krafft, 1901; Noetling, p. 546.

       1909 *Meekoceras markhami* Diener, 1897; von Krafft & Diener, pp. 20–23, pl. 11, figs 1–5, pl. 12, figs 1–3, pl. 13, figs 1–5, pl. 15, figs 4–5.

   v 2011 *Vavilovites* sp. indet. Ware, Jenks, Hautmann & Bucher., pp. 176, 177, fig. 19.

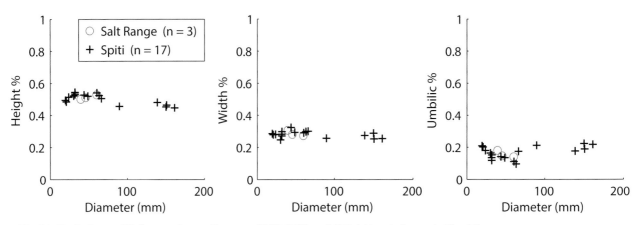

Fig. 24. Vavilovites meridialis n. sp. Scatter diagrams of H/D, W/D, and U/D (abbreviations as in Fig. 17).

v 2012 *Paranorites* sp. indet. Brühwiler,
        Bucher & Krystyn, pp. 137, 139, fig.
        14AB–AE.

v 2018b *Vavilovites* cf. *V. sverdrupi* Tozer,
         1963; Ware & Bucher, pp. 66–67, fig.
         35, pl. 13, figs 27–35.

Derivation of name. – From the Latin 'meridialis', from the south.

Holotype. – Specimen PIMUZ30883 (Pl. 7, fig. 1).

Type locality. – Mud Bottom section (section M05 of Krystyn *et al.* 2007a), Himachal Pradesh, India.

Type horizon. – Sample Mud 63, upper part of the shale interval at the base of the Limestone and Shale Member, bed 9 of Krystyn *et al.* (2007a,b), *Vavilovites meridialis* Regional Zone, late Dienerian.

Diagnosis. – *Vavilovites* with sharply tabulate venter on innermost whorls (at D < 3 cm), becoming progressively arched to well rounded on outer whorls when reaching a diameter larger than ca. 70 mm.

Material. – 47 specimens.

Description. – Their large (mature specimens exceed 15 cm in diameter), moderately involute (adults with U/D ≈ 20%) discoidal shell shows a substantial change in whorl section during ontogeny. The inner whorls (up to a diameter of ca. 30 mm) are moderately involute (U/D ≈ 20%) and sub-platyconic, with a maximal width at inner third of the whorl section and a tabulate venter sometimes showing a slightly bottleneck shaped venter close to that of inner whorls of *Ambites lilangensis*. The flanks carry some weak broad folds following the trajectory of growth lines. The next whorls (up to 10 cm) become more involute (U/D ≈ 14%). Their maximal width shifts progressively towards the umbilicus, giving the whorl section a more trapezoidal shape. The venter becomes broadly arched, and the ventral shoulders fade simultaneously. The ornamentation disappears completely. Adult specimens (D > 10 cm) become more evolute (U/D ≈ 20%) and their maximal width is at the inner fourth to inner fifth of the whorl height. Their flanks converge strongly towards the venter and sometimes show a slight concavity before reaching the perfectly rounded venter, giving the whorl section a sub-lanceolate shape. During the entire growth, the umbilical wall is vertical to slightly overhanging, with a gently rounded shoulder. The suture line is typical of Paranoritidae, with a vaguely differentiated auxiliary lobe and a broad flattened third lateral saddle.

Measurements. – See Figure 24.

Discussion. – This species displays a large intraspecific variability, from thin involute variants to thick and more evolute variants, but this is difficult to evaluate because of ubiquitous *in vivo* encrusting bivalves. Every specimen of more than 2 cm in diameter has bivalves encrusting both sides of the umbilicus, often inducing a deviation of the coiling and thus making measurements unsuitable for statistical analysis (on the graph, Fig. 24, only specimens without any obvious growth disturbance were included). The few large specimens (D > 10 cm) were all found to be crowded with hundreds of such epibionts on the whole surface of the shell, with exception of the last quarter of whorl of the body chamber.

This species clearly differs from *Vavilovites sverdrupi* by its more involute coiling and its rounded venter on adult whorls. It also resembles *Vavilovites subtriangularis* Vavilov (1976) which also has a rounded venter on the outer whorls, but has a thinner whorl section and less triangular whorl section with more convex flanks and a broader venter. The inner whorls of thin variants superficially resemble involute variants of *Koninckites khoorensis* but clearly differ by their tabulate venter.

As explained by Diener (1909), von Krafft intended to create the species *Meekoceras noetlingi* for specimens belonging to the present species, and indeed used this denomination in several papers. He should have made the formal description in this monograph, but could not complete it. When Diener revised his notes, he considered it as identical to *Proptychites markhami*. Hence, the species *Meekoceras noetlingi* was never formally described and remains a *nomen nudum*. The type specimens of *Proptychites markhami* have a suture line typical of Proptychitidae with indentations on the sides of the lobes, whereas the specimens described in von Krafft & Diener (1909) are strictly identical to ours. The specimens cited by Noetling (1901, 1904), curated at the University of Tübingen (Germany), are identical to the specimens described here. *Vavilovites* sp. indet., *in* Ware *et al.* (2011) corresponds to the inner whorls of thick variants. In Brühwiler *et al.* (2012), one specimen coming from a float block in Mud clearly corresponds to this species. It was then considered to be derived from the 'Parahedenstroemia beds'. It is also encrusted by numerous bivalves. Associated facies is also similar to that of our specimens from Mud, hence it must derive from the same bed as the specimens here under consideration. The few specimens from Pakistan described as *Vavilovites* cf. *V. sverdrupi* by Ware & Bucher (2018b) also correspond to inner whorls of the present species.

Occurrence. – Guling Village, sample Gul18 (*n* = 2); Lalung Cliff, sample Lal58 (*n* = 16), Mud Bottom Section, samples Mud29 (*n* = 8), Mud60 (*n* = 6), Mud63 (*n* = 5) and Mud202 (*n* = 3); Mud Top Section, sample Mud19′ (*n* = 7).

Genus *Koninckites* Waagen, 1895

Type species. – *Koninckites vetustus* Waagen, 1895.

Koninckites vetustus Waagen, 1895;

Plate 8, figures 6–8; Figure 25

1895   *Koninckites vetustus* n. gen., n. sp. Waagen, pp. 261, 262, pl. 27, figs 4 (lectotype), ?5.

v 2018b   *Koninckites vetustus* Waagen, 1895; Ware & Bucher, pp. 67–69, figs 35–37; pl. 14, figs 1–36, pl. 15, figs 1–20, pl. 16, figs 1–15, pl. 17, figs 1–4 (*cum syn.*).

Material. – 26 specimens.

Measurements. – See Figure 25.

Occurrence. – Lalung Cliff, samples Lal63 (5) and Lath10 (1); Mud Bottom section, sample Mud17

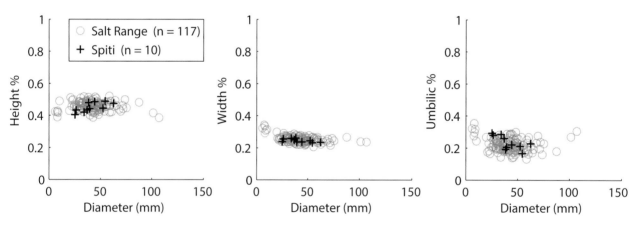

Fig. 25. *Koninckites vetustus* Waagen, 1895. Scatter diagrams of H/D, W/D, and U/D (abbreviations as in Fig. 17).

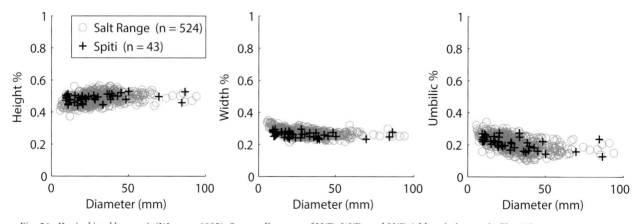

Fig. 26. *Koninckites khoorensis* (Waagen, 1895). Scatter diagrams of H/D, W/D, and U/D (abbreviations as in Fig. 17).

(n = 2); Mud Top sections, samples Mud31 (n = 17) and Mud62 (n = 1).

Koninckites khoorensis (Waagen, 1895)

Plate 8, figures 9–20; Plate 9, figures 1–13; Figure 26

> 1895 *Proptychites khoorensis* n. gen., n. sp. Waagen, pp. 176–178, pl. 20, fig. 4 (holotype).
>
> v 2018b *Koninckites khoorensis* (Waagen, 1895); Ware & Bucher, pp. 69–74, figs 35, 38–40, pl. 17, figs 5–17, pl. 18, figs 1–31, pl. 19, figs 1–12, pl. 20, figs 1–13 (*cum syn.*).

Material. – 219 specimens.

Measurements. – See Figure 26.

Occurrence. – Guling River, sample Gul30 (n = 25); Guling Village, sample Gul19 (n = 11); Lalung Cliff, samples Lal59 (n = 16), Lal60 (n = 22), Lal61 (n = 21), Lal62 (n = 3), Lath8 (n = 6) and Lath9 (n = 22); Mud Top section, samples Mud8 (n = 1) and Mud19 (n = 59); Mud Bottom Section, samples Mud30 (n = 7), Mud59 (n = 13) and Mud201 (n = 5); Tilling, sample Till1 (n = 8).

Genus *Radioceras* Waterhouse, 1996

Type species. – *Meekoceras radiosum* Waagen, 1895.

Radioceras truncatum (Spath, 1934)

Plate 9, figures 14, 15

> 1934 *Koninckites truncatus* n. sp. Spath, pp. 152, 153, figs 43c, 44 (holotype) [cop. *Koninckites davidsonianus* in Waagen, 1895].
>
> v 2018b *Radioceras truncatum* (Spath, 1934); Ware & Bucher, figs 35, 41, pp. 74–76, pl. 21, figs 1–6 (*cum syn.*).

Material. – 6 specimens.

Measurements. – See Ware & Bucher (2018b).

Occurrence. – Mud Bottom section, sample Mud31 (n = 4) and Mud Top section, sample Mud17 (n = 2).

Genus *Pashtunites* Ware & Bucher, 2018b

Type species. – *Koninckites kraffti* Spath, 1934.

Pashtunites kraffti (Spath, 1934)

Plate 9, figures 16–18

> 1934 *Koninckites kraffti* n. sp. Spath, pp. 155, 156, fig. 43c (holotype).
>
> v 2018b *Pashtunites kraffti* (Spath, 1934) n. gen. Ware & Bucher, pp. 76–77, figs 35, 42, pl. 21, figs 7–13 (*cum syn.*).

Material. – Two specimens.

Measurements. – No measurable specimen available.

Occurrence. – Lalung Cliff, sample Lal63 (n = 2).

Family Proptychitidae Waagen, 1895

Genus *Proptychites* Waagen, 1895

Type species. – *Ceratites lawrencianus* de Koninck, 1863.

Proptychites lawrencianus (de Koninck, 1863) *sensu* Waagen, 1895;

Plate 10, figures 7, 8

> ? 1863 *Ceratites lawrencianus* n. sp. de Koninck, p. 14, pl. 6, fig. 3 (holotype).
>
> v 2018b *Proptychites lawrencianus* (de Koninck, 1863); Ware & Bucher, pp. 83–85, figs 45, 47, pl. 22, figs 14–15, pl. 23, figs 1–11 (*cum syn.*).

Material. – One specimen.

Measurements. – No measurable specimen available.

Occurrence. – A float block above Lalung Ridge 2, sample Lal27.

Proptychites oldhamianus Waagen, 1895

Plate 10, figures 1–6

1895   *Proptychites oldhamianus* n. gen., n. sp. Waagen, pp. 166, 167, pl. 19, fig. 3 (holotype).

v 2018b   *Proptychites oldhamianus* Waagen, 1895; Ware & Bucher, pp. 85, fig. 45, pl. 24, figs 5–14 (*cum syn.*).

Material. – 11 specimens.

Measurements. – See Table 1.

Occurrence. – Guling River, sample Gul23 (*n* = 2); Lalung Ridge 1, samples Lal2 (*n* = 2) and Lal4 (*n* = 1); Lalung Ridge2, sample Lal52 (*n* = 2); Mud Bottom section, samples Mud51 (*n* = 1) and Mud52 (*n* = 4). Some poorly preserved specimens from sample Mud16 from Mud Top section, Gul24 from Guling River and Gul14 from Guling Village are tentatively assigned to this species.

Proptychites wargalensis Ware & Bucher, 2018b

Plate 10, figures 9, 10

v 2018b   *Proptychites wargalensis* n. sp. Ware & Bucher, pp. 85–86, fig. 45, pl. 24, figs 1–4.

Material. – 4 specimens.

Measurements. – No measurable specimen available.

Occurrence. – Guling Village, sample GU6 (*n* = 1); Mud Bottom section, sample Mud26 (*n* = 1); Mud Top section, sample Mud14 (*n* = 2). Some poorly preserved specimens from sample Gul14 from Guling Village, Lal4 from Lalung Ridge 1 and Lal52 from Lalung Ridge 2 are tentatively assigned to this species.

Proptychites ammonoides Waagen, 1895

Plate 10, figures 14–20; Plate 11, figures 1–3; Figure 27

1895   *Proptychites ammonoides* n. gen., n. sp. Waagen, pp. 171–173, pl. 17, fig. 1 (lectotype), pl. 19, fig. 2.

v 2018b   *Proptychites ammonoides* Waagen, 1895; Ware & Bucher, pp. 86–88, figs 45, 48, pl. 24, figs 15–17, pl. 25, figs 1–14, pl. 26, figs 1, 2 (*cum syn.*).

Material. – 39 specimens.

Measurements. – See Figure 27.

Occurrence. – Guling Village, samples Gul12 (*n* = 2), Gul15 (*n* = 6) and Gul16 (*n* = 2); Lalung Cliff, samples Lal54 (*n* = 1), Lal55 (*n* = 3), Lal56 (*n* = 1) and LATH6 (*n* = 3); Lalung Ridge1, sample Lal1 (*n* = 3); Lalung Ridge2, samples Lal7 (*n* = 4), Lal12 (*n* = 1) and Lal13 (*n* = 1); Mud Bottom section, samples Mud27 (*n* = 6), Mud58a (*n* = 4), Mud58b (*n* = 1) and Mud58.D3 (*n* = 1). Some poorly preserved specimens from sample Til3 from Tilling, Gul27 and Gul28 from Guling River are tentatively assigned to this species.

Proptychites cf. *P. pagei* Ware, Jenks, Hautmann & Bucher, 2011

Plate 11, figures 4–6

v ? 2011   *Proptychites pagei* n. sp. Ware, Jenks, Hautmann & Bucher, pp. 175, 176, fig. 18.

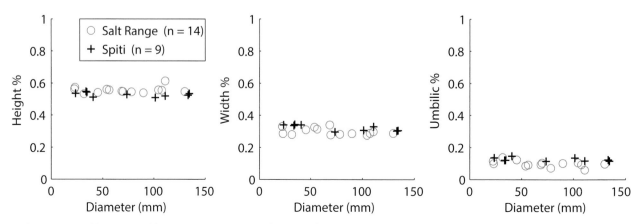

Fig. 27. Proptychites ammonoides Waagen, 1895. Scatter diagrams of H/D, W/D, and U/D (abbreviations as in Fig. 17).

v 2018b *Proptychites* cf. *P. pagei* Ware, Jenks, Hautmann & Bucher, 2011; Ware & Bucher, pp. 88, fig. 45, pl. 26, figs 3–8.

Material. – One single specimen.

Measurements. – No measurable specimen available.

Occurrence. – Tilling, sample Til2. One poorly preserved specimen from sample Lal57 from Lalung Cliff is tentatively assigned to this species.

Genus *Bukkenites* Tozer, 1994

Type species. – *Proptychites strigatus* Tozer, 1961.

Bukkenites sakesarensis Ware & Bucher, 2018b

Plate 11, figures 11–16

v 2018b *Bukkenites sakesarensis* n. sp. Ware & Bucher, pp. 81–83, figs 45, 46, pl. 21, figs 36–39, pl. 22, figs 1–3 (holotype), 4–13.

Material. – 4 specimens.

Measurements. – No measurable specimen available.

Occurrence. – Guling Village, sample GU5/1 ($n = 4$). Some poorly preserved specimens from sample Lal6 from Lalung Ridge 1 are tentatively assigned to this species.

Genus *Dunedinites* Tozer, 1963

Type species. – *Dunedinites pinguis* Tozer, 1963.

Dunedinites cf. *D. magnumbilicatus* (Kiparisova, 1961)

Plate 11, figures 7–10

? 1961 *Prosphingites magnumbilicatus* n. sp. Kiparisova, p. 114, text-fig. 78, pl. 25, fig. 4 (holotype).

? 2009 *Dunedinites magnumbilicatus* (Kiparisova, 1961); Shigeta & Zakharov, p. 104, figs 91, 92.1–92.4.

Material. – One specimen.

Description. – Cadiconic shell with a broadly rounded venter and a high vertical umbilical wall individualized by a narrowly rounded shoulder. No ornamentation visible. Suture line with deep and straight lobes and elongated saddles. Ventral lobe narrow. The three lateral saddles and two lateral lobes are nearly equally deep and narrow, the lobes bearing a few rather deep indentations at their base. Auxiliary series with one well individualized auxiliary lobe.

Measurements. – See Table 1.

Discussion. – This single specimen is too incomplete and poorly preserved for assignment at the species level. Its general shape is identical to the specimen of *Dunedinites magnumbilicatus* figured by Shigeta & Zakharov (2009), but it differs by its suture line with a clearly differentiated auxiliary lobe. It is also very close to the genus *Anotoceras* Hyatt, 1900, from which it differs by its broadly rounded venter and blunt umbilical shoulder. It should be noted here that the genus *Anotoceras* is of uncertain affinities. It has only been figured by Diener (1897), based only on a couple of specimens, the age of which is uncertain, as they come from the so-called 'Otoceras beds', which in Diener's sense correspond both to the lower limestone member and the following Dienerian black shales. Its type species, *Anotoceras nala* (Diener, 1897), differs from our specimen only by its slightly tectiform venter and its sharp umbilical shoulder.

These two features lead most previous authors to assign it to Otoceratidae in spite of its very simple suture line. Waterhouse (1994) created the family Anotoceratidae for this genus. These Griesbachian and Dienerian cadiconic ammonoids are very rare. Inclusive of our specimen, only nine specimens are known and figured worldwide. Additional material with detailed stratigraphy would be necessary to decipher their affinity and establish any synonymy between *Anotoceras* and *Dunedinites*.

Occurrence. – Mud Bottom section, sample Mud52.

Family Mullericeratidae Ware, Jenks, Hautmann & Bucher, 2011

Genus *Mullericeras* Ware, Jenks, Hautmann & Bucher, 2011

Type species. – *Aspidites spitiensis* von Krafft, 1909.

Mullericeras spitiense (von Krafft, 1909)

Plate 11, figures 17–23

1909   *Aspidites spitiensis* n. sp. von Krafft, p. 54, pl. 4, figs 4 (lectotype), 5, pl. 16, figs 3–8.

v 2018b   *Mullericeras spitiense* von Krafft, 1909; Ware & Bucher, pp. 89–90, figs 49, 50, pl. 27, figs 1–9 (cum syn.).

Material. – 12 specimens.

Measurements. – See Table 1.

Discussion. – An up-to-date detailed description of this species can be found in Ware *et al.* (2011).

Occurrence. – Guling Village, sample Gul17 ($n = 1$); Lalung Cliff, sample Lal57 ($n = 1$); Mud Bottom section, samples Mud28 ($n = 1$), Mud29 ($n = 2$), Mud60 ($n = 2$), Mud63 ($n = 2$) and Mud202 ($n = 2$); Mud Top section, sample Mud19′ ($n = 1$).

Mullericeras shigetai Ware & Bucher, 2018b

Plate 12, figures 1–5

v 2018b   *Mullericeras shigetai* n. sp. Ware & Bucher, pp. 90–92, figs 49, 51, pl. 27, figs 10–13 (holotype), 14–28 (cum syn.).

Material. – 12 specimens.

Measurements. – See Table 1.

Occurrence. – Lalung Cliff, sample LATH6 ($n = 1$); Lalung Ridge2, sample Lal7 ($n = 8$); Mud

Bottom section, sample Mud58b ($n = 3$). Poorly preserved specimens from sample Gul16 from Guling Village are tentatively assigned to this species.

Mullericeras niazii Ware & Bucher, 2018b

Plate 12, figures 6–11; Figure 28

v 2018b   *Mullericeras niazii* n. sp. Ware & Bucher pp. 92–93, figs 49, 52, pl. 28, figs 10–18.

Material. – 14 specimens.

Measurements. – See Figure 28.

Discussion. – The description provided by Ware & Bucher (2018b) was based on incomplete, rather small specimens (i.e. without adult body chamber). Here, some larger complete specimens were found, revealing that the maximal whorl width, which was already observed to shift from the inner third of the flanks in the juveniles to the middle of the flanks in sub-adults, shifts further on the external third of the flanks in large adult specimens. Our material otherwise agrees perfectly with the original description of this species.

Occurrence. – Guling River, sample Gul29 ($n = 1$); Guling Village, sample Gul17 ($n = 1$); Lalung Cliff, sample Lal57 ($n = 1$); Mud Bottom section, samples Mud28 ($n = 6$), Mud61 ($n = 3$) and Mud203 ($n = 2$).

Genus *Ussuridiscus* Shigeta & Zakharov, 2009

Type species. – *Meekoceras (Kingites) varaha* Diener, 1895.

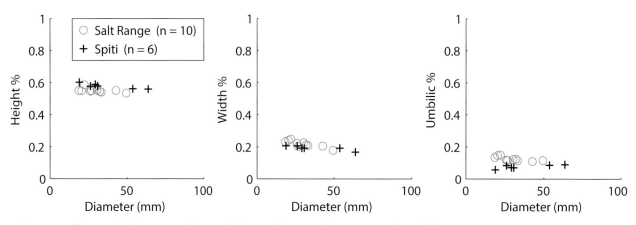

Fig. 28. Mullericeras niazii Ware & Bucher, 2018b. Scatter diagrams of H/D, W/D, and U/D (abbreviations as in Fig. 17).

Ussuridiscus varaha (Diener, 1895)

Plate 12, figures 12, 13

1895 *Meekoceras (Kingites) varaha* n. sp.
  Diener, p. 52, pl. 1, fig. 2 (holotype).

v 2018b *Ussuridiscus varaha* (Diener, 1895); Ware
  & Bucher, pp. 94–95, figs 49, 53, pl. 28,
  figs 19–31 (*cum syn.*).

Material. – Three specimens.

Measurements. – No measurable specimen available.

Occurrence. – Guling Village, sample GU5/1 ($n = 1$); Mud Bottom section, sample Mud25 ($n = 1$); Mud Top section, sample Mud13 ($n = 1$).

Ussuridiscus ensanus (von Krafft, 1909)

Plate 12, figures 14–20; Figure 29

1909 *Aspidites ensanus* n. sp. von Krafft, pp.
  56, 57, pl. 5, figs 3, 4, 5 (lectotype), 6, 7,
  pl. 6, fig. 1, pl. 14, fig. 6.

v 2018b *Ussuridiscus ensanus* (von Krafft, 1909);
  Ware & Bucher, pp. 95–96, figs 49, 54, pl.
  3, fig. 1 pl. 28, figs 32–41, pl. 29, figs 1–
  14 (cum syn.).

Material. – 39 specimens.

Measurements. – See Figure 29.

Occurrence. – Guling River, sample Gul23 ($n = 1$); Guling Village, sample Gul13 ($n = 5$); Lalung Cliff, samples Lal67.1 ($n = 1$) and Lal67.2 ($n = 3$); Lalung Ridge 1, samples Lal2 ($n = 4$), Lal3 ($n = 2$) and Lal4 ($n = 4$); Lalung Ridge 2, samples Lal8 ($n = 2$) and Lal52 ($n = 1$); Mud Bottom section, samples Mud26 ($n = 2$), Mud51 ($n = 3$) and Mud52 ($n = 6$); Mud Top section, samples Mud3 ($n = 3$), Mud4 ($n = 1$) and Mud5 ($n = 1$). Some poorly preserved specimens from sample Gul22 from Guling River, Gul14 from Guling Village, Lal10 from Lalung Ridge 2 and Lal5 from Lalung Ridge 1 are tentatively assigned to this species.

Ussuridiscus ornatus Ware & Bucher, 2018b

Plate 12, figures 23, 24

v 2018b *Ussuridiscus ornatus* n. sp. Ware &
  Bucher, pp. 97–98, fig. 49, pl. 29, figs 15–
  17, 18–21 (holotype) (*cum syn.*).

Material. – 4 specimens.

Measurements. – No measurable specimen available.

Occurrence. – Guling River, sample Gul24 ($n = 1$) and Mud Top section, sample Mud14 ($n = 3$).

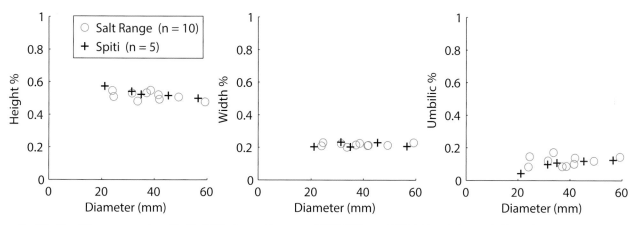

Fig. 29. Ussuridiscus ensanus (von Krafft, 1909). Scatter diagrams of H/D, W/D, and U/D (abbreviations as in Fig. 17).

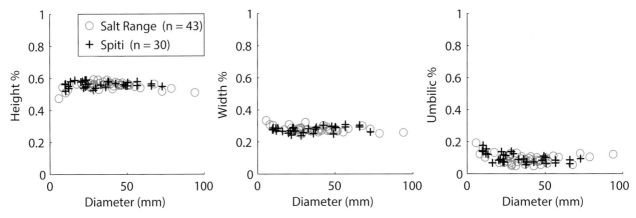

Fig. 30. Kingites davidsonianus (de Koninck, 1863). Scatter diagrams of H/D, W/D, and U/D (abbreviations as in Fig. 17).

Ussuridiscus ventriosus **Ware & Bucher, 2018b**

Plate 12, figures 21, 22

v 2018b *Ussuridiscus ventriosus* n. sp. Ware &
Bucher pp. 96–97, fig. 49, pl. 29, figs 28–
31 (holotype).

Material. – One specimen. Some poorly preserved specimens from sample Gul24 from Guling River are tentatively assigned to this species.

Measurements. – No measurable specimen available.

Occurrence. – Mud Bottom section, sample Mud53.

Family *incertae sedis*

Genus *Kingites* **Waagen, 1895**

Type species. – *Kingites lens* Waagen, 1895.

Kingites davidsonianus **(de Koninck, 1863)**

Plate 13, figures 1–8; Figure 30

1863 *Ceratites davidsonianus* n. sp. de Koninck,
p. 13, pl. 6, fig. 2 (holotype).

v 2018b *Kingites davidsonianus* (de Koninck,
1863); Ware & Bucher, pp. 99–100, fig. 55,
pl. 30, figs 1–18, pl. 31, figs 1–4 (*cum syn.*).

Material. – 64 specimens.

Measurements. – See Figure 30.

Occurrence. – Guling River, sample Gul30 ($n = 3$); Lalung Cliff, samples LATH9 ($n = 1$) and Lal60 ($n = 5$); Mud Bottom section, samples Mud30 ($n = 7$), Mud59 ($n = 5$) and Mud201 ($n = 2$); Mud Top section, sample Mud19 ($n = 37$); Tilling, sample Till1 ($n = 4$).

Kingites korni **Brühwiler, Ware, Bucher, Krystyn & Goudemand, 2010**

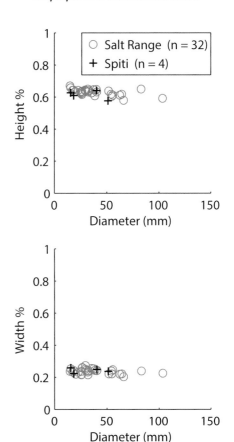

Fig. 31. Pseudosageceras simplelobatum Ware & Bucher, 2018b. Scatter diagrams of H/D and U/D (abbreviations as in Fig. 17).

Plate 13, figures 9–13

v 2010   *Kingites korni* n. sp. Brühwiler, Ware, Bucher, Krystyn & Goudemand, p. 734, fig. 15 (holotype).

v 2018b   *Kingites korni* Brühwiler, Ware, Bucher, Krystyn & Goudemand, 2010; Ware & Bucher, pp. 100–101, fig. 56, pl. 31, figs 5–9.

Material. – Three specimens.

Measurements. – See Table 1.

Occurrence. – Lalung Cliff, sample Lal63 ($n = 1$) and Mud Bottom section, sample Mud31 ($n = 2$).

Superfamily Sagecerataceae Hyatt, 1884
Family Hedenstroemiidae Hyatt, 1884

Genus *Pseudosageceras* Diener, 1895

Type species. – *Pseudosageceras* sp. indet. Diener, 1895.

Pseudosageceras simplelobatum Ware & Bucher, 2018b

Plate 13, figs 14–17; Figure 31

v 2018b   *Pseudosageceras simplelobatum* n. sp. Ware & Bucher, pp. 103–105, figs 49, 58, 59, pl. 32, figs 5–16, pl. 33, figs 1–3 (holotype), 4–10 (*cum syn.*).

Material. – 10 specimens.

Measurements. – See Figure 31.

Occurrence. – Guling River, sample Gul30 ($n = 2$); Lalung Cliff, sample Lal61 ($n = 3$); Mud Bottom section, samples Mud59 ($n = 1$) and Mud201 ($n = 1$); Mud Top section, samples Mud17 ($n = 2$) and Mud19 ($n = 1$).

Acknowledgements. – O. N. Bhargava (Haryana, India) and T. Galfetti (Oxford) are acknowledged for their help in the field. James M. Neenan (Zürich) improved the English text. Claude Monnet (Lille) is thanked for providing his statistical analyses software. Nicolas Goudemand and Séverine Urdy (Lyon) are thanked for their help with Matlab. Mike Orchard from the Geological Survey of Canada (Vancouver) is thanked for providing access to the collections and archives of E.T. Tozer. Technical support for preparation and photography was provided by Markus Hebeisen and Rosemarie Roth (Zürich). Last but not least, Beli Parkash (Manali, India) is thanked for his assistance in the field. This work is a contribution to the Swiss National Science Foundation project 200020-135446 to HB. Additional field work was funded by the Swiss National Science Foundation project 200020-135075 to Winand Brinkmann. Earlier fieldwork of L.K. was sponsored by the Austrian Academy of Sciences, National Committee for IGCP, within project 467 (Triassic Time).

References

Agassiz, L. 1847: Lettres sur quelques points d'organisation des animaux rayonnés. *Comptes Rendus de l'Académie des Sciences 25*, 677–682.

Arthaber, G.V. 1911: Die Trias von Albanien. *Beiträge zur Paläontologie und Geologie Österreich-Ungarns und des Orients 24*, 169–276.

Bengtson, P. 1988: Open nomenclature. *Palaeontology 31*, 223–227.

Bhargava, O.N., Krystyn, L., Balini, M., Lein, R. & Nicora, A. 2004: Revised Litho- and sequence stratigraphy of the Spiti Triassic. *Albertiana 30*, 21–39.

Brayard, A., Bucher, H., Escarguel, G., Fluteau, F., Bourquin, S. & Galfetti, T. 2006: The Early Triassic ammonoid recovery: paleoclimatic significance of diversity gradients. *Palaeogeography, Palaeoclimatology, Palaeoecology 239*, 374–395.

Brühwiler, T., Brayard, A., Bucher, H. & Guodun, K. 2008: Griesbachian and Dienerian (Early Triassic) Ammonoid Faunas from Northwestern Guangxi and Southern Guizhou (South China). *Palaeontology 51*, 1151–1180.

Brühwiler, T., Ware, D., Bucher, H., Krystyn, L. & Goudemand, N. 2010: New Early Triassic ammonoid faunas from the Dienerian/Smithian boundary beds at the Induan/Olenekian GSSP candidate at Mud (Spiti, Northern India). *Journal of Asian Earth Sciences 39*, 724–739.

Brühwiler, T., Bucher, H. & Krystyn, L. 2012: Middle and late Smithian (Early Triassic) ammonoids from Spiti, India. *Special Papers in Palaeontology 88*, 115–174.

Cuvier, G.L.C.F.D. An 6 1797: *Tableau élémentaire de l'histoire naturelle des animaux 14*, 710 pp. Baudouin, Paris.

Dagys, A.S. & Ermakova, S. 1996: Induan (Triassic) ammonoids from North-Eastern Asia. *Revue de Paléobiologie 15*, 401–447.

Diener, C. 1895: Triadische Cephalopodenfaunen der Ostsibirischen Küstenprovinz. *Mémoires du Comité Géologique St. Pétersbourg 14*, 1–59.

Diener, C. 1897: Part I: The Cephalopoda of the Lower Trias. *Palaeontologia Indica, Series 15. Himalayan Fossils 2*, 1–181.

Griesbach, C.L. 1901: *General Report on the Work Carried on by the Geological Survey of India for the Period from the 1st April 1900 to the 31st March 1901*, pp. 36. Office of the Superintendent, Government Printing, Calcutta, India.

Hautmann, M., Ware, D. & Bucher, H. 2017: Geologically oldest oysters were epizoans on Early Triassic ammonoids. *Journal of Molluscan Studies 2017*, 1–8. https://doi.org/10.1093/mollus/eyx018.

Hyatt, A. 1884: Genera of fossil cephalopods. *Proceedings of the Boston Society of Natural History 22*, 253–338.

Hyatt, A. 1900: Cephalopoda. 502–592. In Zittel, K.A.V. (ed.): *Textbook of Palaeontology, Vol. 1.* 1st English edition, 839 pp. Eastman, C. R., London.

Kiparisova, L.D. 1961: Paleontological fundamentals for the stratigraphy of Triassic deposits of Primorye region. 1. Cephalopod Mollusca. *Trudy Vsyesoyuzhogo Nauchno-isslyedovatyel'skogo Geologichyeskogo Instityta (VSEGEI). Novaya seriya 48*, 1–278 [In Russian].

de Koninck, L.G. 1863: Description of some fossils from India, discovered by Dr. A. Fleming, of Edinburgh. *The Quarterly Journal of the Geological Society of London 19*, 1–19.

von Krafft, A. 1909: see Krafft, A. von & Diener, C. 1909.

von Krafft, A. & Diener, C. 1909: Lower Triassic cephalopoda from Spiti, Malla Johar, and Byans. *Palaeontologia Indica 6*, 1–186.

Krystyn, L., Balini, M. & Nicora, A. 2004: Lower and Middle Triassic stage and substage boundaries in Spiti. *Albertiana 30*, 40–53.

Krystyn, L., Bhargava, O.N. & Richoz, S. 2007a: A candidate GSSP for the base of the Olenekian Stage: Mud at Pin Valley; district Lahul & Spiti, Himachal Pradesh (Western Himalaya), India. *Albertiana 35*, 5–29.

Krystyn, L., Richoz, S. & Bhargava, O.N. 2007b: The Induan-Olenekian Boundary (IOB) in Mud – an update of the candidate GSSP section M04. *Albertiana 36*, 33–45.

Matthews, S.C. 1973: Notes on open nomenclature and synonymy lists. *Palaeontology 16*, 713–719.

Noetling, F. 1901: Beiträge zur Geologie der Salt Range, insbesondere der permischen und Triassischen Ablagerungen. *Neues Jahrbuch für Mineralogie, Geologie und Paläontologie, Beilage-band 14*, 369–471.

Noetling, F. 1904: Ueber das Alter der Otoceras-Schichten von Rimkin Paiar (Painkhanda) im Himalaya. *Neues Jahrbuch für Mineralogie, Geologie und Paläontologie, Beilage-band 18*, 528–555.

Romano, C., Ware, D., Brühwiler, T., Bucher, H. & Brinkmann, W. 2016: Marine Early Triassic Osteichthyes from Spiti, Indian Himalayas. *Swiss Journal of Palaeontology 135*, 275–294.

Shigeta, S. & Zakharov, Y.D. 2009: Cephalopods. 44–140. In Shigeta, Y., Zakharov, Y.D., Maeda, H. & Popov, A.M. (eds) *The Lower Triassic System in the Abrek Bay Area, South Primorye, Russia*, 218 pp. National Museum of Nature and Science Monographs 38, Tokyo.

Spath, L.F. 1930: The Eotriassic Invertebrate Fauna of East Greenland. *Meddelelser om Grønland 83*, 1–90.

Spath, L.F. 1934: *Catalogue of the Fossil Cephalopoda in the British Museum (Natural History), part IV: The Ammonoidea of the Trias*, pp. 521. The Trustees of the British Museum, London.

Tozer, E.T. 1961: Triassic Stratigraphy and faunas, Queen Elizabeth Islands, Arctic Archipelago. *Memoir of the Geological Survey of Canada 316*, 1–116.

Tozer, E.T. 1963: Lower Triassic ammonoids from Tuchodi Lakes and Halfway River areas, northeastern British Columbia. *Bulletin of the Geological Survey of Canada 96*, 1–28.

Tozer, E.T. 1971: Triassic time and ammonoids. *Canadian Journal of Earth Sciences 8*, 989–1031. Errata and addenda, 1611 pp.

Tozer, E.T. 1994: Canadian Triassic Ammonoid Faunas. *Bulletin of the Geological Survey of Canada 467*, 1–663.

Vavilov, M.N. 1976: see Vavilov, M.N. & Zakharov, Y.D. 1976.

Vavilov, M.N. & Zakharov, Y.D. 1976: A revision of the Early Triassic genus *Pachyproptychites*. In *Morphology and systematics of Soviet Far East fossil invertebrates*, 60–67, Proceedings of the Institute of Biology and Pedology, Far-East Science Centre, Academy of Sciences of the USSR 42 (145). [In Russian]

Waagen, W. 1895: Salt ranges fossils. vol. 2: Fossils from the Ceratites formation - Part I – Pisces, Ammonoidea. *Palaeontologia Indica 13*, 1–323.

Ware, D. & Bucher, H. 2018a: Foreword: Dienerian (Early Triassic) ammonoids and the Early Triassic biotic recovery: a review. *Fossils & Strata 63*, 3–9.

Ware, D. & Bucher, H. 2018b: Systematic Palaeontology. In Ware, D., Bucher, H., Brühwiler, T., Schneebeli-Hermann, E., †Hochuli, P.A., Roohi, G., Rehman, K. & Yaseen, A. 2018: *Griesbachian and Dienerian (Early Triassic) ammonoids from the Salt Range*, Pakistan. *Fossils & Strata 63*, 13–175.

Ware, D., Jenks, J.F., Hautmann, M. & Bucher, H. 2011: Dienerian (Early Triassic) ammonoids from the Candelaria Hills (Nevada, USA) and their significance for palaeobiogeography and palaeoceanography. *Swiss Journal of Geoscience 104*, 161–181.

Ware, D., Bucher, H., Brayard, A., Schneebeli-Hermann, E. & Brühwiler, T. 2015: High-resolution biochronology and diversity dynamics of the Early Triassic ammonoid recovery: the Dienerian faunas of the Northern Indian Margin. *Palaeogeography, Palaeoclimatology, Palaeoecology 440*, 363–373.

Ware, D., Bucher, H., Brühwiler, T., Schneebeli-Hermann, E., †Hochuli, P.A., Roohi, G., Rehman, K. & Yaseen, A. 2018: Griesbachian and Dienerian (Early Triassic) ammonoids from the Salt Range, Pakistan. *Fossils & Strata 63*, 13–175.

Waterhouse, J.B. 1994: The Early and Middle Triassic ammonoid succession of the Himalayas in western and central Nepal. Part 1. Stratigraphy, classification and Early Scythian ammonoid systematics. *Palaeontographica A232*, 1–83.

Waterhouse, J.B. 1996: The Early and Middle Triassic ammonoid succession of the Himalayas in western and central Nepal. Part 2. Systematic studies of the Early Middle Scythian. *Palaeontographica A241*, 27–100.

Plate 1 All figures natural size unless otherwise indicated; asterisks indicate the position of the last septum

1–7: ***Ghazalaites roohii* Ware & Bucher, 2018b**

1, 2: Lateral and ventral views. PIMUZ30829.
       Loc. Mud14, middle part of the Lower Limestone Member, Mud Top Section, *Gyronites plicosus* Regional Zone; early Dienerian.

3, 4: Lateral and ventral views. PIMUZ30830.
       Loc GU5/1, upper part of the Lower Limestone Member, Guling Village, *Gyronites dubius* Regional Zone; early Dienerian.

5–7:   (5, 6) Lateral and ventral views. (7) Suture line at H = 14.8 mm, × 1.5 (mirrored image). PIMUZ30831.
       Loc. Mud52, top of the Lower Limestone Member, Mud Bottom Section, *Gyronites frequens* Regional Zone; early Dienerian.

8–18: ***Gyronites frequens* Waagen, 1895**

8, 9: Lateral and apertural views. PIMUZ30832.
       Loc. Mud52, top of the Lower Limestone Member, Mud Bottom Section, *Gyronites frequens* Regional Zone; early Dienerian.

10–12: Lateral, apertural and ventral views. PIMUZ30833.
       Loc. Mud52, top of the Lower Limestone Member, Mud Bottom Section, *Gyronites frequens* Regional Zone; early Dienerian.

13–15: Lateral, apertural and ventral views. PIMUZ30834.
       Loc. Lal1, top of the Lower Limestone Member, Lalung Ridge 1, condensed bed with *Gyronites frequens* horizon (early Dienerian) and *Ambites atavus* horizon (middle Dienerian).

16–18: Lateral, apertural and ventral views. PIMUZ30835.
       Loc. Mud25, upper part of the Lower Limestone Member, Mud Bottom Section, *Gyronites dubius* Regional Zone; early Dienerian.

Plate 2   All figures natural size unless otherwise indicated; asterisks indicate the position of the last septum.

1–9:   *Gyronites plicosus* Waagen, 1895
1–3:   Lateral, apertural and ventral views. PIMUZ30839.
        Loc. Gul23, upper part of the Lower Limestone Member, Guling River, *Gyronites plicosus* Regional
        Zone; early Dienerian.
4, 5:   Lateral and ventral views. PIMUZ30838.
        Loc. Mud51, upper part of the Lower Limestone Member, Mud Bottom Section, *Gyronites plicosus*
        Regional Zone; early Dienerian.
6–9:   (6–8) Lateral, apertural and ventral views. (9) Suture line at H = 8.9 mm, × 2 (mirrored image).
        PIMUZ30840.
        Loc. Mud15, upper part of the Lower Limestone Member, Mud Top Section, *Gyronites plicosus*
        Regional Zone; early Dienerian.

10–12: *Gyronites sitala* (Diener, 1897)
        Lateral, apertural and ventral views. Complete specimen. PIMUZ30841.
        Loc. Mud3, middle part of the Lower Limestone Member, Mud Top Section, *Gyronites plicosus*
        Regional Zone; early Dienerian.

13–21: *Gyronites levilatus* n. sp.
13, 14: Lateral and ventral views. PIMUZ30842. Paratype.
        Loc. Mud5, upper part of the Lower Limestone Member, Mud Top Section, *Gyronites plicosus* Regional
        Zone; early Dienerian.
15–18: (15–17) Lateral, apertural and ventral views. (18) Suture line at H = 11.6 mm, × 2 (mirrored image).
        PIMUZ30843. Paratype.
        Loc. Mud5, upper part of the Lower Limestone Member, Mud Top Section, *Gyronites plicosus* Regional
        Zone; early Dienerian.
19–21: Lateral, apertural and ventral views. PIMUZ30844. Holotype.
        Loc. Mud3, upper part of the Lower Limestone Member, Mud Top Section, *Gyronites plicosus* Regional
        Zone; early Dienerian.

22–28: *Gyronites bullatus* n. sp.
22–25: (22–24) Lateral, apertural and ventral views. (25) Suture line at H = 4.1 mm, × 4 (mirrored image).
        PIMUZ30846. Holotype.
        Loc. Mud52, top of the Lower Limestone Member, Mud Bottom Section, *Gyronites frequens* Regional
        Zone; early Dienerian.
26–28: Lateral, apertural and ventral views. × 2. PIMUZ30847. Paratype.
        Loc. Mud52, top of the Lower Limestone Member, Mud Bottom Section, *Gyronites frequens* Regional
        Zone; early Dienerian

29–30: *Gyronites schwanderi* Ware & Bucher, 2018b
        Lateral and ventral views. PIMUZ30848.
        Loc. Gul13, top of the Lower Limestone Member, Guling Village, condensed bed with *Gyronites
        frequens* horizon (early Dienerian) and *Ambites atavus* horizon (middle Dienerian).

31–34: *Ambites tenuis* Ware & Bucher, 2018b
31, 32: Lateral and ventral views. PIMUZ30849.
        Loc. Gul13, top of the Lower Limestone Member, Guling Village, condensed bed with *Gyronites
        frequens* horizon (early Dienerian) and *Ambites atavus* horizon (middle Dienerian).
33, 34: Lateral and ventral views. PIMUZ30850.
        Loc. Lal1, top of the Lower Limestone Member, Lalung Ridge 1, condensed bed with *Gyronites frequens*
        horizon (early Dienerian) and *Ambites atavus* horizon (middle Dienerian).

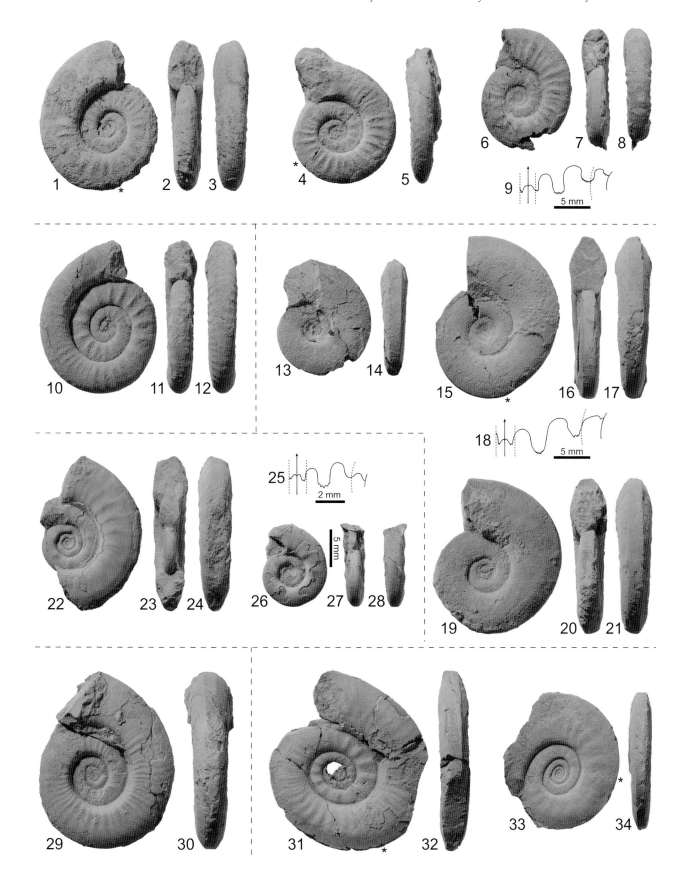

Plate 3 All figures natural size unless otherwise indicated; asterisks indicate the position of the last septum.

1–4: ***Ambites discus* Waagen, 1895.**
  (1–3) Lateral, apertural and ventral views. (4) Suture line at H = 14.8 mm, × 1.5.
  PIMUZ30851.
  Loc. Mud58b, base of the Limestone and Shale Member, Mud Bottom Section, *Ambites discus* Regional
  Zone; middle Dienerian.
5–7: Lateral, apertural and ventral views. PIMUZ30852.
  Loc. Mud58b, base of the Limestone and Shale Member, Mud Bottom Section, *Ambites discus* Regional
  Zone; middle Dienerian.
8, 9: Lateral and ventral views. PIMUZ30853.
  Loc. Lal55, base of the Limestone and Shale Member, Lalung Cliff, *Ambites discus* Regional Zone;
  middle Dienerian.

10–15: ***Ambites atavus* (Waagen, 1895)**
10, 11: Lateral and ventral views. PIMUZ30855.
  Loc. Lal7, base of the Limestone and Shale Member, Lalung Ridge 2, *Ambites atavus* Regional Zone;
  middle Dienerian.
12, 13: Lateral and ventral views. PIMUZ30856.
  Loc. Lal1, top of the Lower Limestone Member, Lalung Ridge 1, condensed bed with *Gyronites frequens*
  horizon (early Dienerian) and *Ambites atavus* horizon (middle Dienerian).
14, 15: Lateral and ventral views. PIMUZ30857.
  Loc. Gul12, base of the Limestone and Shale Member, Guling Village, *Ambites atavus* Regional Zone;
  middle Dienerian.

16–18: ***Ambites bojeseni* Ware & Bucher, 2018b**
  Lateral. apertural and ventral views. Complete specimen. PIMUZ30858.
  Loc. Mud27, base of the Limestone and Shale Member, Mud Bottom Section, *Ambites bojeseni*
  Regional Zone; middle Dienerian.

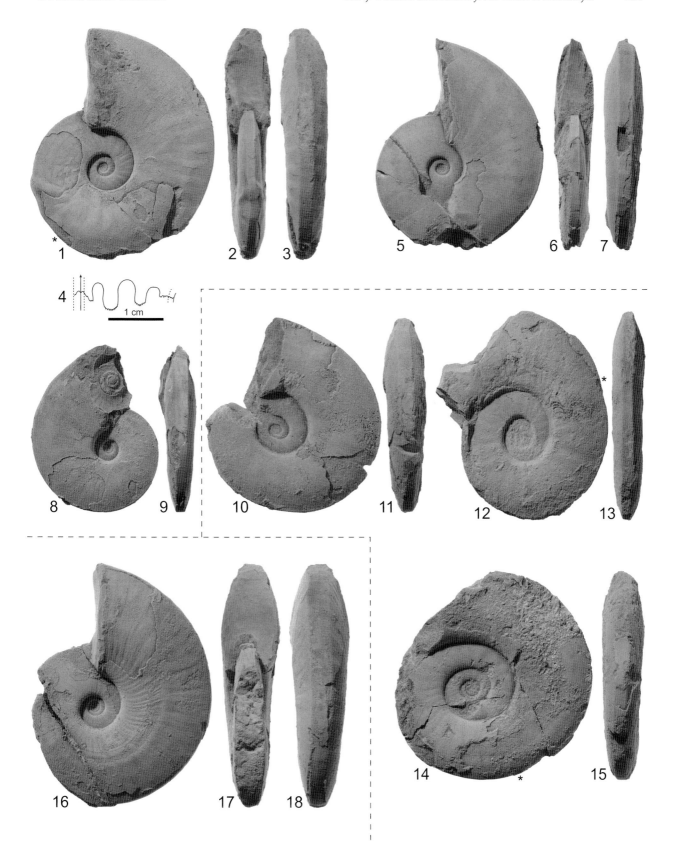

Plate 4 All figures natural size; asterisks indicate the position of the last septum.

1–15:   *Ambites bojeseni* **Ware & Bucher, 2018b**
1, 2:   Lateral and ventral views. PIMUZ30859.
         Loc. Gul15, base of the Limestone and Shale Member, Guling Village, *Ambites bojeseni* Regional Zone; middle Dienerian.
3–5:   Lateral, apertural and ventral views. Specimen with a bivalve encrusting the umbilicus on both sides. PIMUZ30860.
         Loc. Mud27, base of the Limestone and Shale Member, Mud Bottom Section, *Ambites bojeseni* Regional Zone; middle Dienerian.
6, 7:   Lateral and ventral views. PIMUZ30861.
         Loc. Mud27, base of the Limestone and Shale Member, Mud Bottom Section, *Ambites bojeseni* Regional Zone; middle Dienerian.
8–11:   (8–10) Lateral, apertural and ventral views. (11) Suture line at H = 24.6 mm. Complete specimen. PIMUZ30862.
         Loc. Mud27, base of the Limestone and Shale Member, Mud Bottom Section, *Ambites bojeseni* Regional Zone; middle Dienerian.
12–14:   (12, 13) Lateral and ventral views. (14) Suture line at H = 14.2 mm. PIMUZ30863.
         Loc. Mud27, base of the Limestone and Shale Member, Mud Bottom Section, *Ambites bojeseni* Regional Zone; middle Dienerian.
15:      Suture line at H = 23.5 mm (see also Pl. 5, fig. 1). PIMUZ30864.
         Loc. Gul15, base of the Limestone and Shale Member, Guling Village, *Ambites bojeseni* Regional Zone; middle Dienerian.

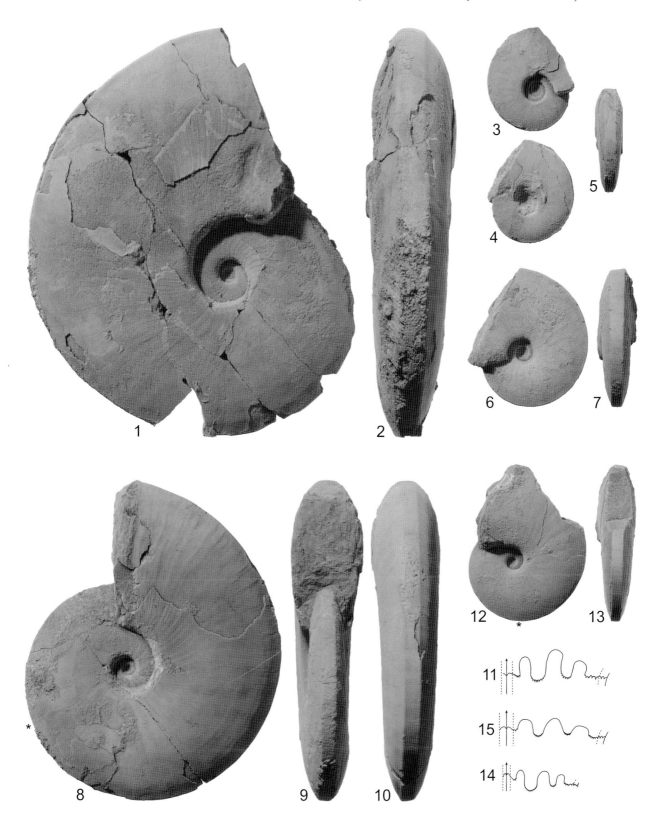

Plate 5 All figures natural size unless otherwise indicated; asterisks indicate the position of the last septum.

1–4:   *Ambites bojeseni* **Ware & Bucher, 2018b**
Lateral (1, 2), apertural (3) and ventral (4) views. Complete specimen with a bivalve encrusting the umbilicus on both sides. (See also Pl. 4, fig. 6). PIMUZ30864.
Loc. Gul15, base of the Limestone and Shale Member, Guling Village, *Ambites bojeseni* Regional Zone; middle Dienerian.

5–9:   *Ambites radiatus* **(Brühwiler *et al.*, 2008).**
5, 6:   Lateral and ventral views. PIMUZ30865.
Loc. Gul15, base of the Limestone and Shale Member, Guling Village, *Ambites bojeseni* Regional Zone; middle Dienerian.
7–9:   Lateral, apertural and ventral views. PIMUZ30866.
Loc. Gul15, base of the Limestone and Shale Member, Guling Village, *Ambites bojeseni* Regional Zone; middle Dienerian.

10–23: *Ambites nyingmai* **n. sp.**
10–13:   (10–12) Lateral, apertural and ventral views. (13) Suture line at H = 11.8 mm, × 2. PIMUZ30867. Holotype.
Loc. Mud58b, base of the Limestone and Shale Member, Mud Bottom Section, *Ambites discus* Regional Zone (middle Dienerian).
14, 15:   Lateral and ventral views. PIMUZ30868. Paratype.
Loc. Mud58b, base of the Limestone and Shale Member, Mud Bottom Section, *Ambites discus* Regional Zone; middle Dienerian.
16, 17:   Lateral and ventral views. PIMUZ30869. Paratype.
Loc. Mud58b, base of the Limestone and Shale Member, Mud Bottom Section, *Ambites discus* Regional Zone; middle Dienerian.
18–20:   Lateral, apertural and ventral views. PIMUZ30870. Paratype.
Loc. Mud58b, base of the Limestone and Shale Member, Mud Bottom Section, *Ambites discus* Regional Zone; middle Dienerian.
21–23:   Lateral, apertural and ventral views. PIMUZ30871. Paratype.
Loc. Mud58b, base of the Limestone and Shale Member, Mud Bottom Section, *Ambites discus* Regional Zone; middle Dienerian.

24, 25: *Ambites lilangensis* **(von Krafft, 1909)**
Lateral and ventral views. PIMUZ30874.
Loc. Mud205, base of the Limestone and Shale Member, Mud Bottom Section, *Ambites lilangensis* Regional Zone; middle Dienerian.

26–29: *Ambites* **cf. *A. frechi* (Tozer, 1994)**
(26–28) Lateral, apertural and ventral views. (29) Suture line at H = 10.7 mm, × 2 (mirrored image). PIMUZ30872.
Loc. Mud58b, base of the Limestone and Shale Member, Mud Bottom Section, *Ambites discus* Regional Zone; middle Dienerian.

30, 31: *Ambites subradiatus* **Ware & Bucher, 2018b**
Lateral and ventral views. PIMUZ30873.
Loc. Lal55, base of the Limestone and Shale Member, Lalung Cliff, *Ambites discus* Regional Zone; middle Dienerian.

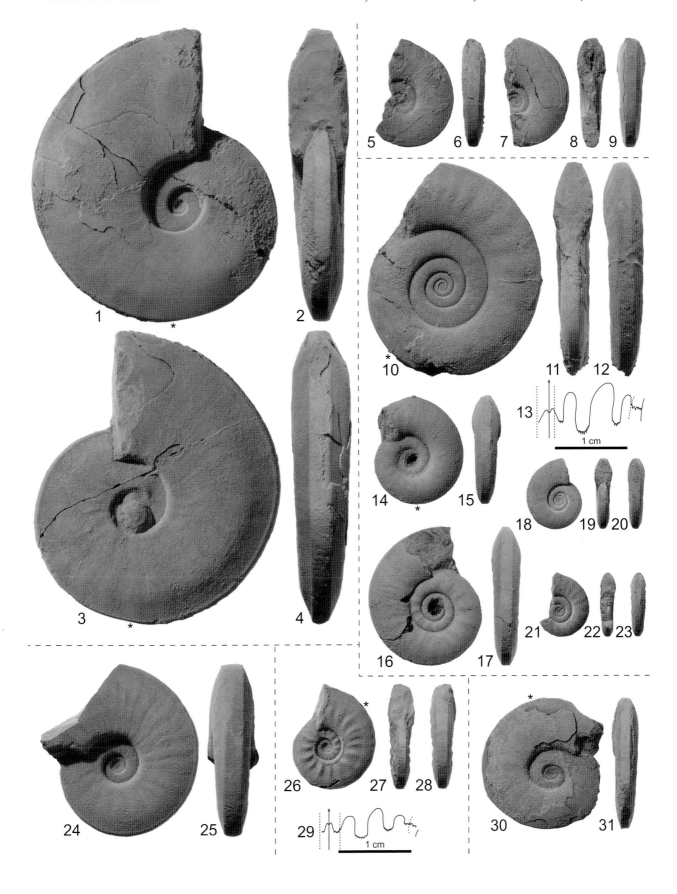

Plate 6   All figures natural size unless otherwise indicated; asterisks indicate the position of the last septum

1–24:   *Ambites lilangensis* (von Krafft, 1909)

1–4:   Complete specimen with encrusting bivalves on both sides of the flanks, which were prepared away on the left side. Lateral (1, 2), apertural (3) and ventral (4) views. PIMUZ30875.
Loc. Mud28, base of the Limestone and Shale Member, Mud Bottom Section, *Ambites lilangensis* Regional Zone; middle Dienerian.

5–7:   Complete specimen. Lateral, apertural and ventral views. PIMUZ30876.
Loc. Mud20, base of the Limestone and Shale Member, Mud Top Section, *Ambites lilangensis* Regional Zone; middle Dienerian.

8–10:   Complete specimen. Lateral, apertural and ventral views. PIMUZ30877.
Loc. Mud61, base of the Limestone and Shale Member, Mud Bottom Section, *Ambites lilangensis* Regional Zone; middle Dienerian.

11–13:   Complete specimen. Lateral, apertural and ventral views. PIMUZ30878.
Loc. Mud203, base of the Limestone and Shale Member, Mud Bottom Section, *Ambites lilangensis* Regional Zone; middle Dienerian.

14–16:   Lateral, apertural and ventral views. PIMUZ30879.
Loc. Mud28, base of the Limestone and Shale Member, Mud Bottom Section, *Ambites lilangensis* Regional Zone; middle Dienerian.

17–20:   Complete specimen with a bivalve encrusting the umbilicus. (17–19) Lateral, apertural and ventral views. (20) Suture line at H = 23.4 mm. PIMUZ30880.
Loc. Mud61, base of the Limestone and Shale Member, Mud Bottom Section, *Ambites lilangensis* Regional Zone; middle Dienerian.

21:      Suture line at H = 14.6 mm. × 1.5. PIMUZ30881.
Loc. Mud61, base of the Limestone and Shale Member, Mud Bottom Section, *Ambites lilangensis* Regional Zone; middle Dienerian.

22–24:   Lateral, apertural and ventral views. PIMUZ30882.
Loc. Mud28, base of the Limestone and Shale Member, Mud Bottom Section, *Ambites lilangensis* regional Zone; middle Dienerian.

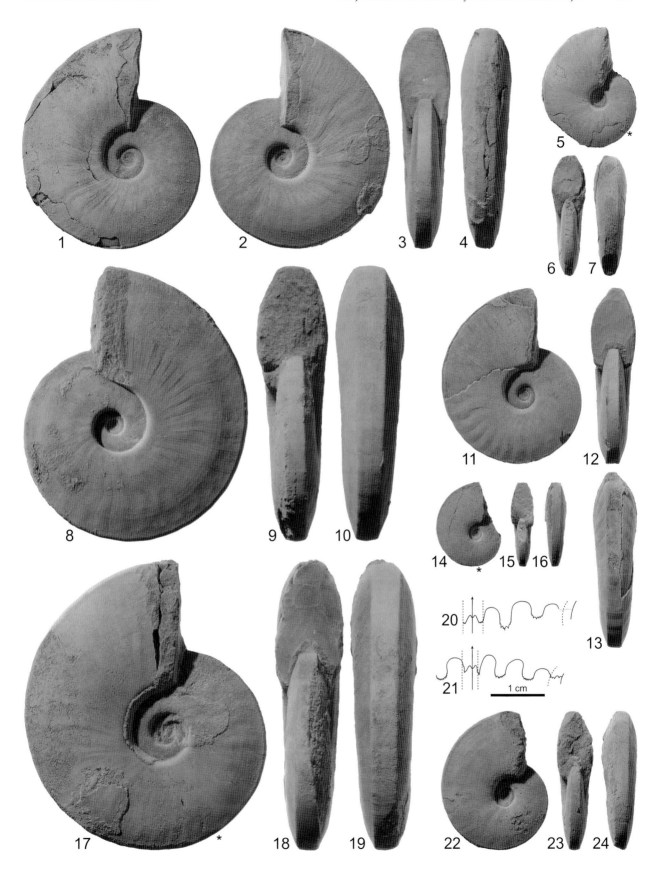

Plate 7   All figures natural size unless otherwise indicated; asterisks indicate the position of the last septum.

1–24:   *Vavilovites meridialis* n. sp.

1–3:   Nearly complete specimen heavily encrusted by numerous bivalves on both sides, except the last third of the whorl. Lateral (1, 2) and ventral (3) views. × 0.5. PIMUZ30883. Holotype.
Loc. Mud63, base of the Limestone and Shale Member, Mud Bottom Section, *Vavilovites meridialis* Regional Zone; late Dienerian.

4, 5:   Lateral and ventral views.PIMUZ30884. Paratype.
Loc. Mud19′, base of the Limestone and Shale Member, Mud Top Section, *Vavilovites meridialis* Regional Zone; late Dienerian.

6–8:   Specimen with a bivalve encrusting the umbilicus on both sides. Lateral, apertural and ventral views. PIMUZ30885. Paratype.
Loc. Mud63, base of the Limestone and Shale Member, Mud Bottom Section, *Vavilovites meridialis* Regional Zone; late Dienerian.

9, 10:   Lateral and ventral views. PIMUZ30886. Paratype.
Loc. Mud60, base of the Limestone and Shale Member, Mud Bottom Section, *Vavilovites meridialis* Regional Zone; late Dienerian.

11–13:   Lateral, apertural and ventral views. PIMUZ30887. Paratype.
Loc. Mud60, base of the Limestone and Shale Member, Mud Bottom Section, *Vavilovites meridialis* Regional Zone; late Dienerian.

14–16:   Nearly complete specimen with a bivalve encrusting the umbilicus on both sides. Lateral, apertural and ventral views. PIMUZ30888. Paratype.
Loc. Mud60, base of the Limestone and Shale Member, Mud Bottom Section, *Vavilovites meridialis* Regional Zone; late Dienerian.

17–19:   Nearly complete specimen with a bivalve encrusting the umbilicus on both sides. Lateral, apertural and ventral views. PIMUZ30889. Paratype.
Loc. Mud63, base of the Limestone and Shale Member, Mud Bottom Section, *Vavilovites meridialis* Regional Zone; late Dienerian.

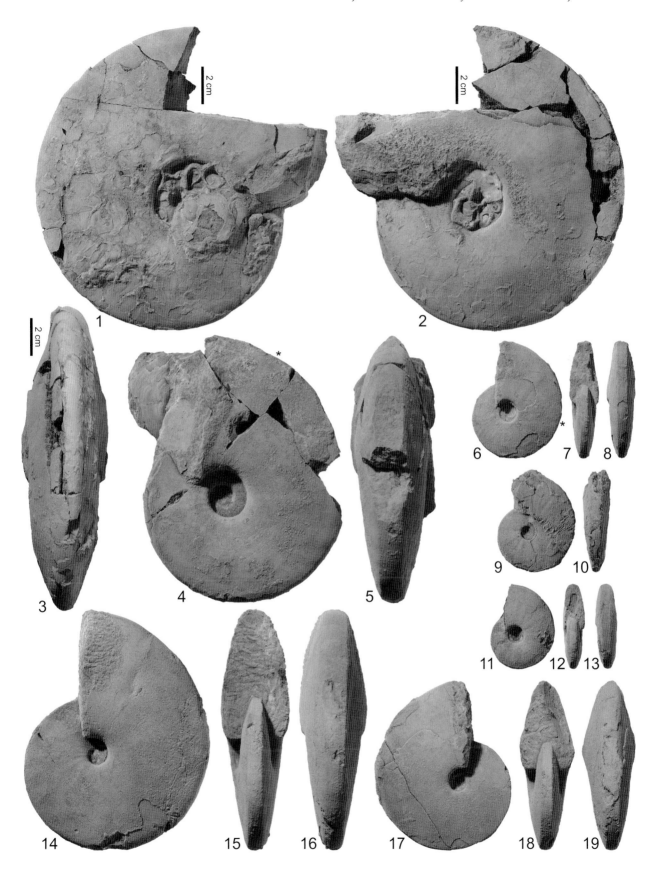

Plate 8 All figures natural size unless otherwise indicated; asterisks indicate the position of the last septum.

1– 5: ***Vavilovites meridialis* n. sp.**
  1 Lateral view. Nearly complete specimen with numerous encrusting bivalves prepared away on the left side to show the induced growth irregularity of the umbilicus. × 0.5. PIMUZ30890. Paratype.
  Loc. Mud202, base of the Limestone and Shale Member, Mud Bottom Section, *Vavilovites meridialis* Regional Zone; late Dienerian.

2–5: (2–4) Lateral, apertural and ventral views. (5) Suture line at H = 25.1 mm. PIMUZ30891. Paratype.
  Loc. Gul18, base of the Limestone and Shale Member, Guling Village, *Vavilovites meridialis* Regional Zone; late Dienerian.

6–8: ***Koninckites vetustus* Waagen, 1895**
  Lateral, apertural and ventral views. PIMUZ30892.
  Loc. Lal63, base of the Limestone and Shale Member, Lalung Cliff, *Koninckites vetustus* Regional Zone; late Dienerian.

9–11: ***Koninckites khoorensis* (Waagen, 1895)**
  Lateral, apertural and ventral views. Complete specimen with a bivalve encrusting the umbilicus on both sides. PIMUZ30893.
  Loc. Mud59, base of the Limestone and Shale Member, Mud Bottom Section, *Koninckites khoorensis* Regional Zone; late Dienerian.

12–14: Lateral, apertural and ventral views. *Koninckites khoorensis* (Waagen, 1895). PIMUZ30894.
  Loc. Mud19, base of the Limestone and Shale Member, Mud Top Section, *Koninckites khoorensis* Regional Zone; late Dienerian.

15–17: Lateral, apertural and ventral views. Nearly complete specimen with a bivalve encrusting the umbilicus on both sides. PIMUZ30895.
  Loc. Mud201, base of the Limestone and Shale Member, Mud Bottom Section, *Koninckites khoorensis* Regional Zone; late Dienerian.

18–20: Lateral, apertural and ventral views. Complete specimen with a bivalve encrusting the umbilicus on both sides. PIMUZ30896.
  Loc. Mud19, base of the Limestone and Shale Member, Mud Top Section, *Koninckites khoorensis* Regional Zone; late Dienerian.

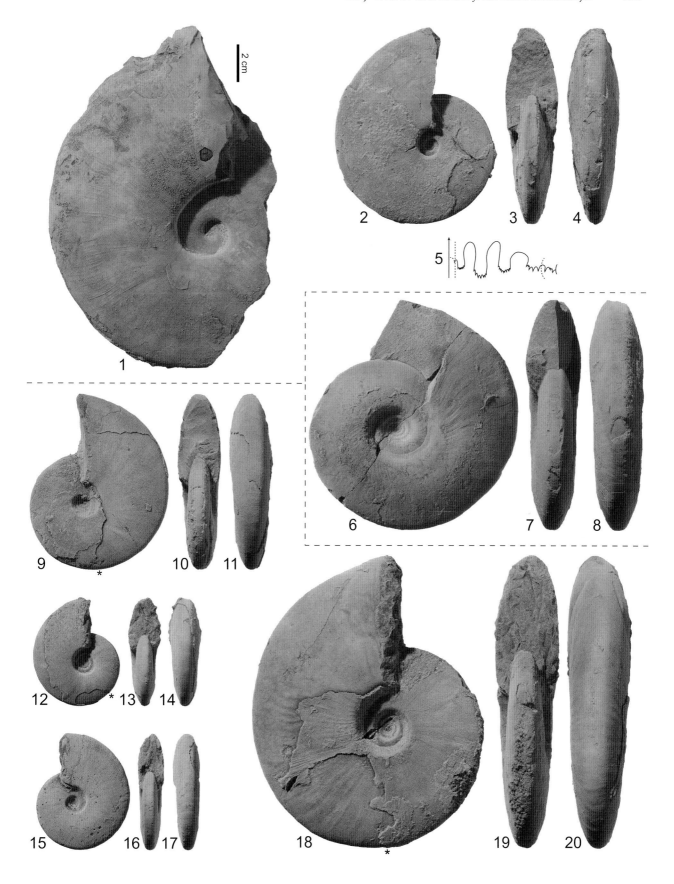

Plate 9 All figures natural size unless otherwise indicated; asterisks indicate the position of the last septum.

1–13: *Koninckites khoorensis* (Waagen, 1895)
    1–4:   (1–3) Lateral, apertural and ventral views. (4) Suture line at H = 25.7 mm (mirrored image). PIMUZ30897.
    Loc. Mud59, base of the Limestone and Shale Member, Mud Bottom Section, *Koninckites khoorensis* Regional Zone; late Dienerian.
    5–7:   Lateral, apertural and ventral views. PIMUZ30898.
    Loc. Mud19, base of the Limestone and Shale Member, Mud Top Section, *Koninckites khoorensis* Regional Zone; late Dienerian.
    8–10:   Lateral, apertural and ventral views. × 1.5. PIMUZ30899.
    Loc. Mud19, base of the Limestone and Shale Member, Mud Top Section, *Koninckites khoorensis* Regional Zone; late Dienerian. Nearly complete specimen.
    11–13:   Lateral, apertural and ventral views. × 1.5. PIMUZ30900.
    Loc. Mud19, base of the Limestone and Shale Member, Mud Top Section, *Koninckites khoorensis* Regional Zone; late Dienerian.

14–15: *Radioceras truncatum* (Spath, 1934)
    Lateral and ventral views. PIMUZ30901.
    Loc. Mud31, base of the Limestone and Shale Member, Mud Bottom Section, *Koninckites vetustus* Regional Zone; late Dienerian.

16–18: *Pashtunites krafti* (Spath, 1934)
    Lateral, apertural and ventral views. PIMUZ30902.
    Loc. Lal63, base of the Limestone and Shale Member, Lalung Cliff, *Koninckites vetustus* Regional Zone; late Dienerian.

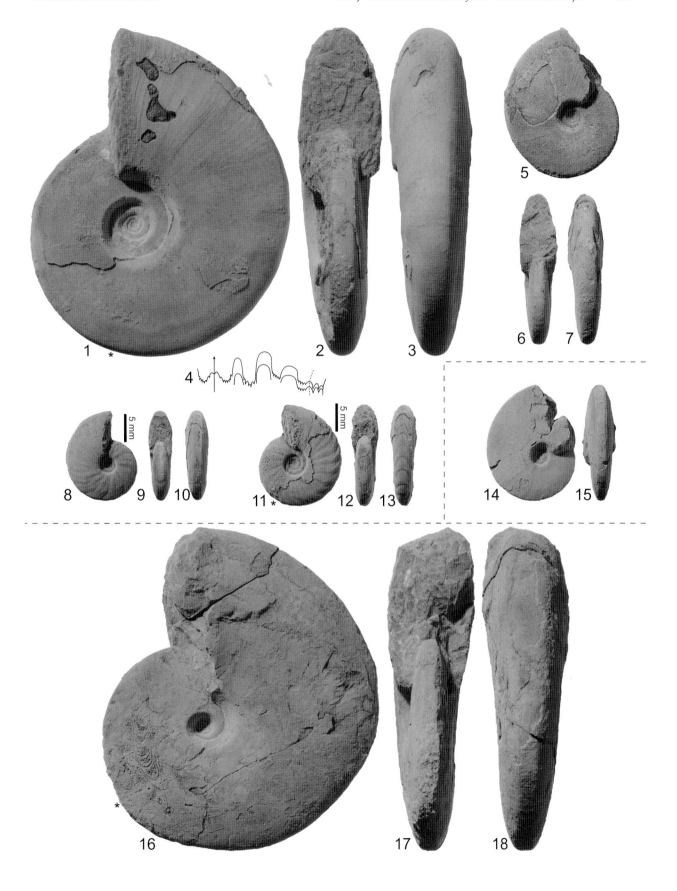

Plate 10   All figures natural size unless otherwise indicated; asterisks indicate the position of the last septum

1–6:   *Proptychites oldhamianus* **Waagen, 1895**
1–3:   (1–2) Lateral and ventral views. (3) Suture line at H = 22.3 mm. PIMUZ30903.
Loc. Mud52, top of the Lower Limestone Member, Mud Bottom Section, *Gyronites frequens* Regional Zone; early Dienerian.
4–6:   Lateral, apertural and ventral views. *Proptychites oldhamianus* Waagen, 1895. PIMUZ30904.
Loc. Mud52, top of the Lower Limestone Member, Mud Bottom Section, *Gyronites frequens* Regional Zone; early Dienerian.

7, 8:   *Proptychites lawrencianus* **(de Koninck, 1863)** *sensu* **Waagen, 1895**
Lateral and ventral views, × 0.5. PIMUZ30905.
Loc. Lal27, base of the Limestone and Shale Member, Lalung Ridge 2, from a floated block, precise stratigraphic position unknown (?middle Dienerian).

9, 10:   *Proptychites wargalensis* **Ware** & **Bucher, 2018b.**
Lateral and ventral views. PIMUZ30906.
Loc. GU6, top of the Lower Limestone Member, Guling Village, *Gyronites plicosus* Regional Zone; early Dienerian.

11, 12: *Proptychites ammonoides* **Waagen, 1895**
Lateral and ventral views. PIMUZ30907.
Loc. Mud27, base of the Limestone and Shale Member, Mud Bottom Section, *Ambites bojeseni* Regional Zone; middle Dienerian.
13:   Suture line at H = 30.2 mm. PIMUZ30908.
Loc. Mud58D3, base of the Limestone and Shale Member, Mud Bottom Section, *Ambites bojeseni* Regional Zone; middle Dienerian.
14–16: Apertural, ventral and lateral views. PIMUZ30909.
Loc. Mud58a, base of the Limestone and Shale Member, Mud Bottom Section, *Ambites bojeseni* Regional Zone; middle Dienerian.
17–20: Lateral (17, 18), apertural (19) and ventral (20) views. Specimen with bivalves encrusting the umbilicus removed to show the induced coiling irregularity. × 0.5. PIMUZ30910.
Loc. Mud58a, base of the Limestone and Shale Member, Mud Bottom Section, *Ambites bojeseni* Regional Zone; middle Dienerian.

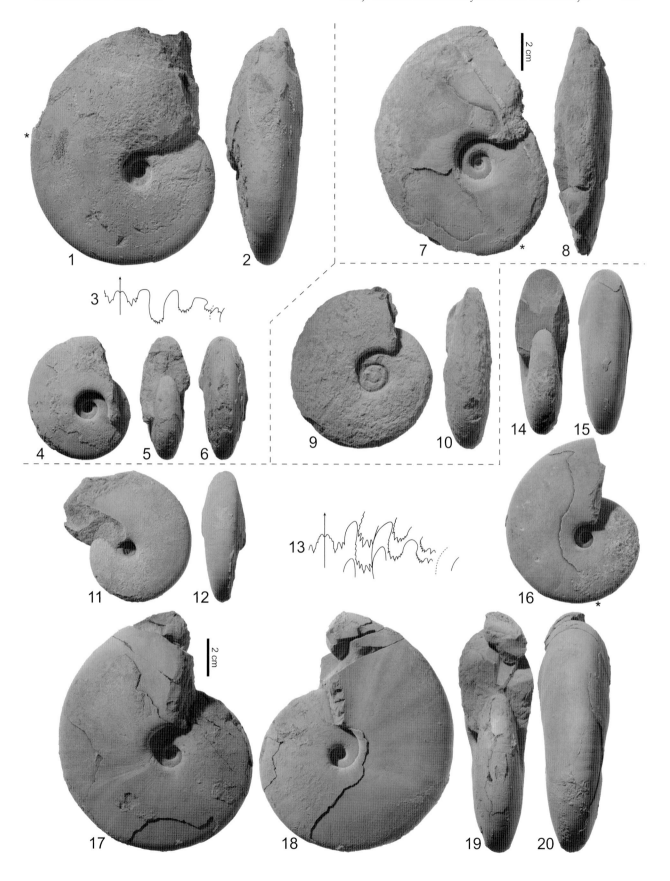

Plate 11   All figures natural size unless otherwise indicated; asterisks indicate the position of the last septum

1–3:   ***Proptychites ammonoides* Waagen, 1895**
Lateral, apertural and ventral views. Complete specimen with bivalves encrusting the umbilicus and the flanks except on the last sixth of the whorl, without inducing any coiling irregularity. × 0.5. PIMUZ30911.
Loc. Mud58a, base of the Limestone and Shale Member, Mud Bottom Section, *Ambites bojeseni* Regional Zone; middle Dienerian.

4–6:   ***Proptychites* cf. *P. pagei* Ware, Jenks, Hautmann & Bucher, 2011**
Lateral, apertural and ventral views. × 0.7. PIMUZ30912.
Loc. Til2, base of the Limestone and Shale Member, Tilling, *Ambites lilangensis* Regional Zone; middle Dienerian.

7–10:   ***Dunedinites* cf. *D. magnumbilicatus* (Kiparisova, 1961)**
(7–9)   Lateral, apertural and ventral views. × 2. (10) Suture line at H = 6.2 mm, × 2, PIMUZ30913.
Loc. Mud52, top of the Lower Limestone Member, Mud Bottom Section, *Gyronites frequens* Regional Zone; early Dienerian.

11, 12: *Bukkenites sakesarensis* Ware & Bucher, 2018b
11, 12: Lateral and ventral views. PIMUZ30914.
Loc. GU5/1, middle part of the Lower Limestone Member, Guling Village, *Gyronites dubius* Regional Zone; early Dienerian.
13–16   (13–15) Lateral, apertural and ventral views. (16) Suture line at H = 5.9 mm, × 3. PIMUZ30915.
Loc. GU5/1, middle part of the Lower Limestone Member, Guling Village, *Gyronites dubius* Regional Zone; early Dienerian.

17–19: *Mullericeras spitiense* (von Krafft, 1909).
17–19: Lateral, apertural and ventral views. Complete specimen. PIMUZ30916.
Loc. Mud29, base of the Limestone and Shale Member, Mud Bottom Section, *Vavilovites meridialis* Regional Zone; late Dienerian.
20–22: (20, 21) Lateral and ventral views. (22) Suture line at H = 15.9 mm, × 1.5. PIMUZ30917.
Loc. Mud28, base of the Limestone and Shale Member, Mud Bottom Section, *Ambites lilangensis* Regional Zone; middle Dienerian.
23:    Suture line at H = 17.9 mm, × 1.5 (mirrored image). PIMUZ30918.
Loc. Mud60, base of the Limestone and Shale Member, Mud Bottom Section, *Vavilovites meridialis* Regional Zone; late Dienerian.

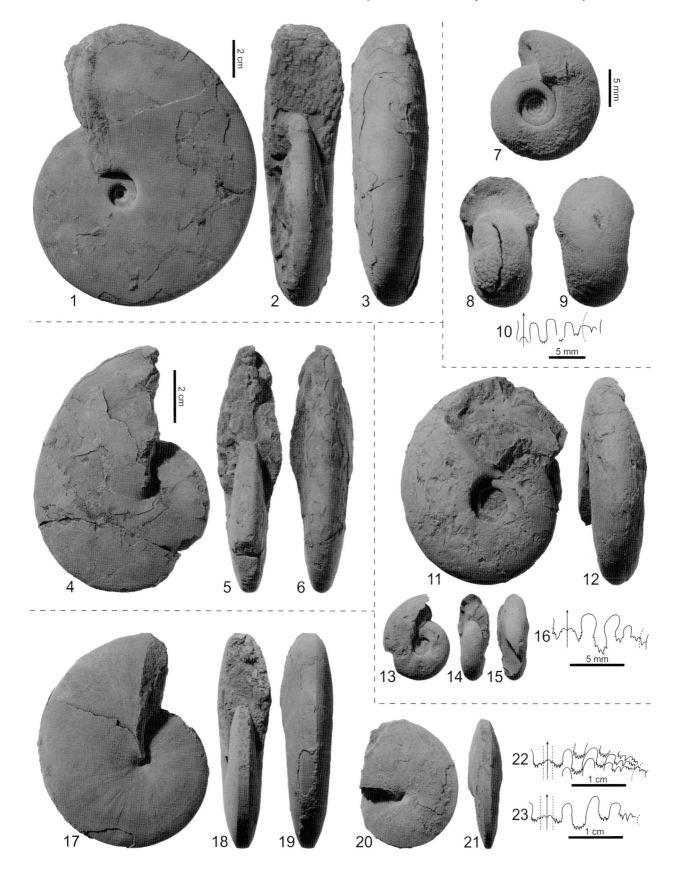

Plate 12   All figures natural size unless otherwise indicated; asterisks indicate the position of the last septum.

1–5:   *Mullericeras shigetai* **Ware & Bucher, 2018b**

1, 2:   Lateral and ventral views. PIMUZ30921.
    Loc. Lal7, base of the Limestone and Shale Member, Lalung Ridge 2, *Ambites atavus* Regional Zone; middle Dienerian.

3–5:   Lateral, apertural and ventral views. PIMUZ30922.
    Loc. Lal7, base of the Limestone and Shale Member, Lalung Ridge 2, *Ambites atavus* Regional Zone; middle Dienerian.

6–11:   *Mullericeras niazii* **Ware & Bucher, 2018b.**

6–8:   Lateral, apertural and ventral views. Nearly complete specimen. PIMUZ30923.
    Loc. Mud28, base of the Limestone and Shale Member, Mud Bottom Section, *Ambites lilangensis* Regional Zone; middle Dienerian.

9–11:   Lateral, apertural and ventral views. PIMUZ30924.
    Loc. Mud28, base of the Limestone and Shale Member, Mud Bottom Section, *Ambites lilangensis* Regional Zone; middle Dienerian.

12, 13: *Ussuridiscus varaha* **(Diener, 1895)**
    Lateral and apertural views. PIMUZ30925.
    Loc. GU5/1, middle part of the Lower Limestone Member, Guling Village, *Gyronites dubius* Regional Zone; early Dienerian.

14–20: *Ussuridiscus ensanus* **(von Krafft, 1909)**

14, 15:   Lateral and ventral views. PIMUZ30926.
    Loc. Gul13, top of the Lower Limestone Member, Guling Village, condensed bed with *Gyronites frequens* horizon (early Dienerian) and *Ambites atavus* horizon (middle Dienerian).

16–18:   (16–17) Lateral and ventral views. (18) Suture line at H = 16.7 mm, × 1.5. PIMUZ30927.
    Loc. Mud52, top of the Lower Limestone Member, Mud Bottom Section, *Gyronites frequens* Regional Zone; early Dienerian.

19, 20:   Lateral and ventral views. PIMUZ30928.
    Loc. Mud52, top of the Lower Limestone Member, Mud Bottom Section, *Gyronites frequens* Regional Zone; early Dienerian.

21, 22: *Ussuridiscus ventriosus* **Ware & Bucher, 2018b**
    Lateral and ventral views. PIMUZ30929.
    Loc. Mud53, upper part of the Lower Limestone Member, Mud Bottom Section, *Gyronites plicosus* Regional Zone; early Dienerian.

23, 24: *Ussuridiscus ornatus* **Ware & Bucher, 2018b**
    Lateral and ventral views. PIMUZ30930.
    Loc. Gul24, middle part of the Lower Limestone Member, Guling Village, *Gyronites plicosus* Regional Zone; early Dienerian.

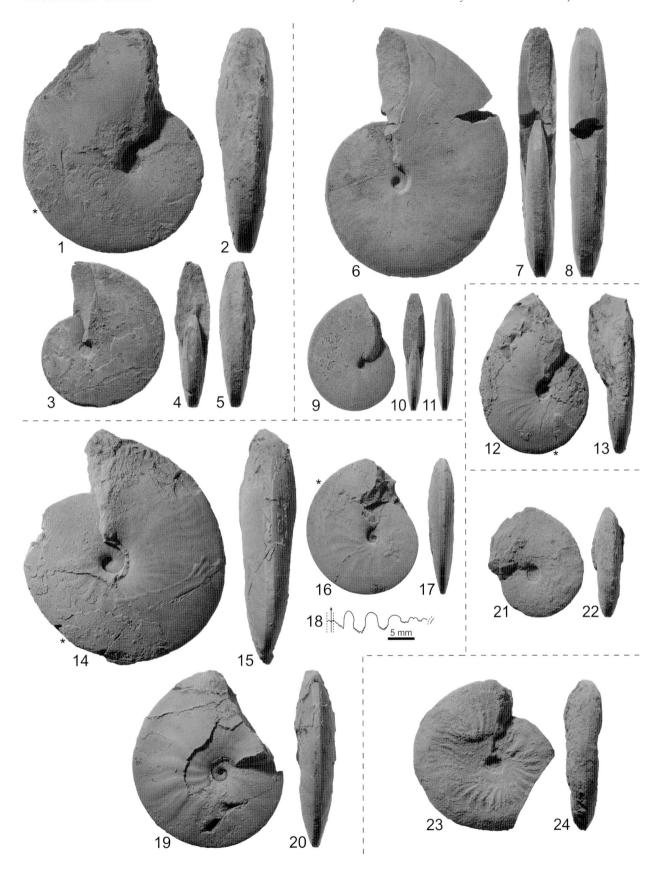

Plate 13   All figures natural size unless otherwise indicated; asterisks indicate the position of the last septum

1–8: ***Kingites davidsonianus* (de Koninck, 1863).**
1–4: (1–3) Lateral, apertural and ventral views. (4) Suture line at H = 22.7 mm, × 1.5. Complete specimen. PIMUZ30931.
Loc. Mud59, base of the Limestone and Shale Member, Mud Bottom Section, *Koninckites khoorensis* Regional Zone; late Dienerian.
5, 6: Lateral and ventral views. Complete specimen. PIMUZ30932.
Loc. Mud19, base of the Limestone and Shale Member, Mud Top Section, *Koninckites khoorensis* Regional Zone; late Dienerian.
7, 8: Lateral and ventral views. PIMUZ30933.
Loc. Mud19, base of the Limestone and Shale Member, Mud Top Section, *Koninckites khoorensis* Regional Zone; late Dienerian.

9–13: ***Kingites korni* Brühwiler *et al.*, 2010**
(9–12) Lateral (left and right), apertural and ventral views. (13) Suture line at H = 21 mm, × 2 (mirrored image). Complete specimen. Holotype (reproduced after Brühwiler *et al.* 2010). PIMUZ27863.
Loc. Mud31, base of the Limestone and Shale Member, Mud Bottom Section, *Koninckites vetustus* Regional Zone; late Dienerian.

14–17: ***Pseudosageceras simplelobatum* Ware & Bucher, 2018b**
(14–16) Lateral, apertural and ventral views. 17) Suture line at H = 23.7 mm, × 1.5 (mirrored image). PIMUZ30934.
Loc. Mud19, base of the Limestone and Shale Member, Mud Top Section, *Koninckites khoorensis* Regional Zone; late Dienerian.

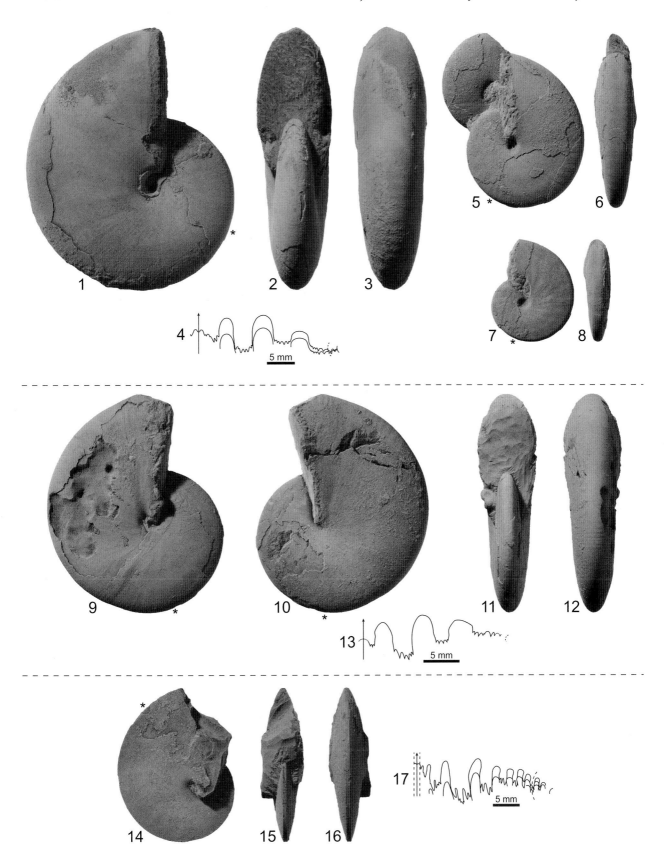